Springer Theses

Recognizing Outstanding Ph.D. Research

Aims and Scope

The series "Springer Theses" brings together a selection of the very best Ph.D. theses from around the world and across the physical sciences. Nominated and endorsed by two recognized specialists, each published volume has been selected for its scientific excellence and the high impact of its contents for the pertinent field of research. For greater accessibility to non-specialists, the published versions include an extended introduction, as well as a foreword by the student's supervisor explaining the special relevance of the work for the field. As a whole, the series will provide a valuable resource both for newcomers to the research fields described, and for other scientists seeking detailed background information on special questions. Finally, it provides an accredited documentation of the valuable contributions made by today's younger generation of scientists.

Theses are accepted into the series by invited nomination only and must fulfill all of the following criteria

- They must be written in good English.
- The topic should fall within the confines of Chemistry, Physics, Earth Sciences, Engineering and related interdisciplinary fields such as Materials, Nanoscience, Chemical Engineering, Complex Systems and Biophysics.
- The work reported in the thesis must represent a significant scientific advance.
- If the thesis includes previously published material, permission to reproduce this must be gained from the respective copyright holder.
- They must have been examined and passed during the 12 months prior to nomination.
- Each thesis should include a foreword by the supervisor outlining the significance of its content.
- The theses should have a clearly defined structure including an introduction accessible to scientists not expert in that particular field.

More information about this series at http://www.springer.com/series/8790

Miao Yu

Metallogenic Mechanism of the Galinge Polymetallic Iron Skarn Deposit, Qiman Tagh Mountains, Qinghai Province

Doctoral Thesis accepted by
Peking University, Beijing, China

Springer

Author
Dr. Miao Yu
Key Laboratory of Metallogenic Prediction of Nonferrous Metals and Geological Environment Monitoring, Ministry of Education, School of Geosciences and Info-Physics
Central South University
Changsha, Hunan, China

and

MLR Key Laboratory of Metallogeny and Mineral Assessment
Institute of Mineral Resources, CAGS
Beijing, China

and

China School of Earth and Space Sciences
Peking University
Beijing, China

Supervisors
Prof. Jingwen Mao
MLR Key Laboratory of Metallogeny and Mineral Assessment
Institute of Mineral Resources, CAGS
Beijing, China

Prof. Yongfeng Zhu
China School of Earth and Space Sciences
Peking University
Beijing, China

Prof. Chengyou Feng
MLR Key Laboratory of Metallogeny and Mineral Assessment
Institute of Mineral Resources, CAGS
Beijing, China

ISSN 2190-5053 ISSN 2190-5061 (electronic)
Springer Theses
ISBN 978-981-10-7906-1 ISBN 978-981-10-7907-8 (eBook)
https://doi.org/10.1007/978-981-10-7907-8

Library of Congress Control Number: 2018965886

This Springer imprint is published by the registered company Springer Nature Singapore Pte Ltd.
The registered company address is: 152 Beach Road, #21-01/04 Gateway East, Singapore 189721, Singapore

Supervisor's Foreword

Dr. Miao Yu started his master's degree in 2010 and conducted fieldwork in the Qiman Tagh Mountains, Qinghai Province, under the guidance of Researcher Chengyou Feng. Through the characterization of several typical deposits, their team proposed a model for ore formation and contributed significantly to exploration in the field. Dr. Yu was specifically responsible for the analysis of the Galinge deposit, which he continued to study with us (supervisors Fellow Jingwen Mao and Professor Yongfeng Zhu) for his doctoral degree starting in 2013. Because of his long experience working in the field, laboratory, and on modeling, Dr. Yu is exceptionally well qualified to answer the question of how and why the Galinge deposit formed.

The Galinge polymetallic iron skarn deposit, which has been mined since 2016, is the only large-scale iron skarn deposit with resources greater than one hundred million tons. It was found in 1970 by the provincial Qinghai Geophysical Prospecting Team while investigating aeromagnetic anomalies in the Qiman Tagh Mountains, Qinghai Province. After exploration on the basis of the 1:50000–1:5000 ground magnetic anomaly, electrical resistivity, gravity anomaly, and careful drill core analysis, the known reserves of the deposit had reached medium-sized scale. Based on the latest general and detailed exploration and prospecting from 2005 to 2012, the Galinge is now known to be a large-scale deposit, with inferred resources of 185 million tons iron ore and 390.6 thousand tons lead–zinc ore. As a representative iron polymetallic skarn deposit, research on the Galinge deposit provides essential information on the genesis and geological evolution of these valuable mineral resources.

This thesis describes the results of Dr. Miao Yu's doctoral study in the School of Earth and Space Sciences at Peking University. This pioneering research has uncovered the metallogenic mechanism of the Galinge deposit and established a thermodynamic model for the formation of iron polymetallic skarn deposits. With this fundamental understanding, Dr. Yu is further able to explain why all skarn deposits are distributed in the South Qiman Tagh Terrane, why only Triassic intrusions are responsible for the ore formation, and why the metal enrichment in the Galinge deposit is such large scale. The answers to these scientific problems are of critical significance in the exploration for similar ore bodies in the Galinge deposit and the entire Qiman Tagh area.

The expedition was led by Researcher Chengyou Feng who is Dr. Yu's sub-supervisor. The fieldwork encompassed sampling, mapping, drill core logging, and cross-sectional profiling. Through the careful guidance of Researchers Yiming Zhao and Daxin Li on microscopic identification and data analysis, Dr. Yu thoroughly mastered the whole process from fieldwork to laboratory analysis. Dr. Yu's comprehensive studies of the Galinge deposit have been published in four papers (SCI) in international journals including Earth-Science Reviews, Lithos, The Canadian Mineralogist, and Mineralogy and Petrology.

Beijing, China
June 2018

Prof. Jingwen Mao
Institute of Mineral Resources, CAGS
Peking University

Preface

Skarn deposits have been described in the scientific literature since at least the nineteenth century. Many scholars have contributed to revealing the metallogenic mechanism of skarn deposits, but limited accessibility of samples and complex geological history have prevented a general understanding of their formation. In our study, we refine and extend previous theories in many aspects including skarn mineralogy, igneous petrogenesis thermodynamic modeling, hydrothermal evolution, and tectonic setting. Such a systematic analysis provides a predictable method for evaluating prospects.

The large Galinge iron skarn deposit is located in the middle of the Qiman Tagh metallogenic belt where many skarn deposits with variable mineralization are present, including the Sijiaoyang, Tawenchahan, Yemaquan, Kendekeke, Hutouya, Kaerqueka, and Weibao deposits. Because of their close spatial associations and variable mineralization, it is necessary to find the differences and similarities among them and construct a general metallogenic model serving for exploration.

This study was jointly financially supported by the National Natural Science Foundation of China (Grant 41172076), the Program of High-level Geological Talents (201309) and Youth Geological Talents (201112) of the China Geological Survey, Geological Survey Program (Grant 1212011085528) from the China Geological Survey, and the Scientific Research Fund (Grant 201411025) of the Non-Commercial Unit from Ministry of Land and Resources, China.

This thesis reports a portion of the author's work on the Galinge iron polymetallic skarn deposit and other skarn deposits in Qiman Tagh belt. The work was completed under the general guidance of Prof. Dr. Chengyou Feng, Daxin Li, and Yiming Zhao, who are especially thanked for the direction of this study. Special thanks are due to Drs. Yanbo Cheng, Jia Sun, Wei Jian, Hui Wang, and Jiannan Liu, for their constructive critical comments, and Drs. Jianhou Zhou, Hongying Qu, Shihua Zhong, and Mingyu Zhang, for their expert assistance in sampling and experiments.

Thanks are due to the Editor-in-Chief Prof. Nelson Eby from Lithos, Prof. Carlo Doglioni from Earth-Science Reviews, Prof. Lhiric Agoyaoy and Johann Raith from Mineralogy and Petrology, and Associate Editor Yuanming Pan from The

Canadian Mineralogist for editorial handling. Cenozoic Geoscience Editing is acknowledged for their language polishing and scientific editing services. Thanks are due to Profs. D. C. Zhu, F. C. Meng, M. A. Parada, Georg Hoinkes, Ivan Baksheev, Bruce Yardley, Marta Franchini, Jeffrey Dick, and C. Li for their constructive and thoughtful comments.

The manuscript was completely reviewed by Prof. Zhaochong Zhang from China University of Geoscience (Beijing) and Researchers Guiqing Xie, Huishou Ye, Shunda Yuan, and Houmin Li from Institute of Mineral Resources, China Academy of Geological Sciences. The editor from Springer Nature helped greatly in the publishing process. They are thanked here for their support and encouragement.

The analysis of metallogenic mechanisms synthesizes many subjects including mineralogy, petrology, geochronology, geochemistry, thermodynamics, inclusion thermometry, geotectonics, and isotopes. Many viewpoints are not mature, some questions do not have certain answers, and some new references have not been caught in time for this manuscript. Therefore, in terms of the argument and author's ability, there are inevitably a few faults and deficiencies in the manuscript. Readers' corrections are most welcome!

Beijing, China Dr. Miao Yu

Parts of this thesis have been published in the following documents:

(1) **Yu, M.**, Feng, C.Y.*, Santosh M., Zhu, Y.F., Mao, J.W., Zhao, Y.M., Li, D.X., 2017. The Qiman Tagh Orogen as a window to the crustal evolution in northern Tibetan Plateau. Earth Science Reviews, 167: 103–123.
(2) **Yu, M.**, Feng, C.Y.*, Zhao, Y.M., Li, D.X., 2015. Genesis of post-collisional calc-alkaline and alkaline granitoids in Qiman Tagh, East Kunlun, China. Lithos, 239: 45–59.
(3) **Yu, M.**, Feng, C.Y.*, Zhu, Y.F., Mao, J.W., Zhao, Y.M., Li, D.X., 2016. Multistage amphiboles from the Galinge iron skarn deposit in Qiman Tagh, western China: evidence of igneous rocks replacement. Mineralogy and Petrology, 111: 81–97.
(4) **Yu, M.**, Feng, C.Y.*, Zhu, Y.F., Mao, J.W., Zhao, Y.M., Li, D.X., 2017. Multistage skarn-related tourmaline from the Galinge deposit, Qiman Tagh, western China: a fluid evolution perspective. The Canadian Mineralogist, 55: 3–19.
(* indicates the corresponding author)

Contents

Chapter 1
The Qiman Tagh Orogen as a Window to the Crustal Evolution in Northern Qinghai-Tibet Plateau

The Qiman Tagh Orogen, forming the northern part of the Qinghai-Tibet plateau, provides important insights for understanding the tectonic evolution and uplift history of the northeastern part of the world's largest plateau and its relation with the Proto and Paleo-Tethys Ocean. Lots of studies have been carried out in this region during the past three decades focusing on the complex large scale deformation patterns resulting from sequential accretion and collision, from north to south, of several microcontinents, flysch complexes, and island arcs onto the southern margin of Eurasia since the Early Paleozoic [1, 9, 30, 96, 103, 105, 120, 124, 138, 168, 169, 180, 202]. The Eastern Kunlun Mountains, located in the central Qinghai-Tibet Plateau, is a composite orogenic belt tracking the effects of several Wilson cycles, preserving the crustal evolution history of the Qinghai-Tibet plateau [80, 90, 97, 106, 204]. Most of the existing studies, however, are focused on the pre-Cenozoic geological evolution of the eastern, central and southern sectors of Eastern Kunlun [50, 56, 79, 92, 152, 158], with less attention given to an overview of the northwestern region.

The Qiman Tagh area is located at the northwestern margin of the Eastern Kunlun Mountains (Figs. 1.1 and 1.2), and preserves the records of the local tectonic evolution spanning from the Proterozoic to the Mesozoic. The complex geologic features of this region, including fragments of island arcs, oceanic islands, seamounts, accretionary complexes, old metamorphic basement, long-lived magmatism (Neoproterozoic to Early Mesozoic), as well as regional metamorphism as recorded in the Qiman Tagh Orogen, provide important keys to evaluate the crustal evolution history of the Eastern Kunlun as well as the Qinghai-Tibet plateau. Large orogenic belts are in general characterized by multistage and multiphase crustal evolution, including continent–continent or arc–continent collision with accompanying crustal thickening, and post-tectonic extension and thinning of the previously thickened crust as exemplified by the four-phase model proposed for the Sveconorwegian orogen [5].

Recently, several studies have addressed the tectonic evolution of Qiman Tagh area [26, 36, 76, 91, 143, 144], although there is no consensus and with the lack of a systematic compilation of the geology and crustal evolution history from Proterozoic

M. Yu, *Metallogenic Mechanism of the Galinge Polymetallic Iron Skarn Deposit, Qiman Tagh Mountains, Qinghai Province*, Springer Theses,
https://doi.org/10.1007/978-981-10-7907-8_1

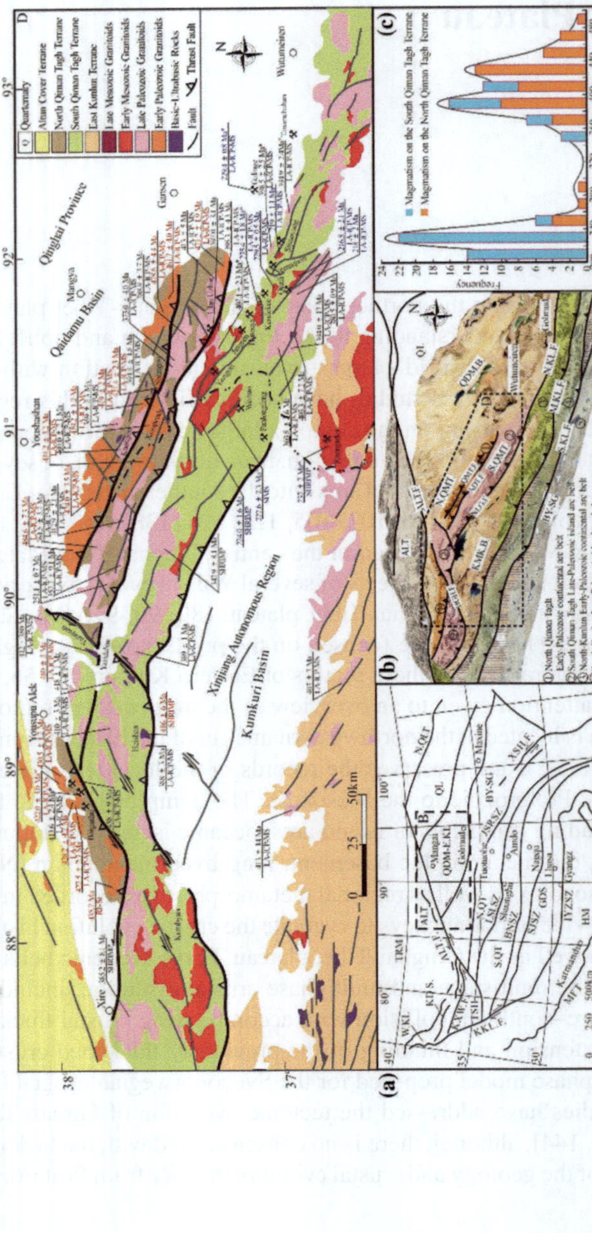

Fig. 1.1 **a** Tectonic outline of the Tibet Plateau showing the locations of the main suture zones; **b** satellite image of the East Kunlun showing major tectonic units; **c** U–Pb intrusion ages from the North Qiman Tagh belt and South Qiman Tagh belt. **d** Tectonic-magmatic sketch map of the Qiman Tagh showing the main tectonic subdivisions, the South Qiman Tagh magma belt and the North Qiman Tagh magma belt. They are bordered by the Adatan thrust fault in the east and Baiganhu thrust fault in the west. Age data with "*" are from our research team. Age data in different colors aims to easily distinguish Early Paleozoic, Late Paleozoic and Early Mesozoic data

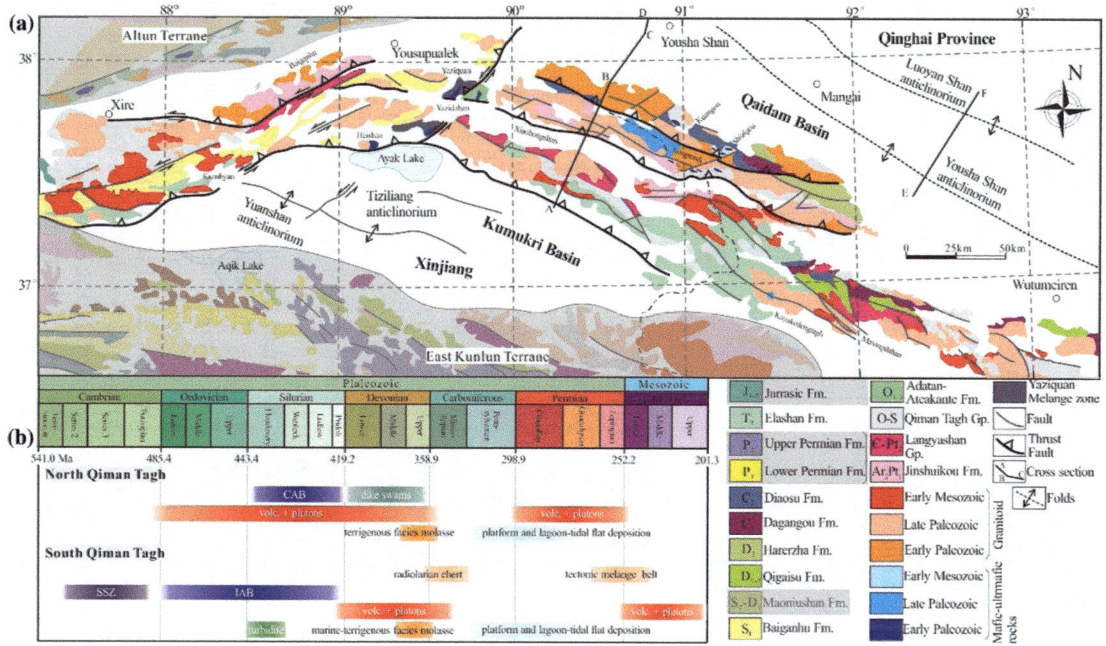

Fig. 1.2 **a** Simplified geological map showing the distinct geological records from Late Archean to Early Mesozoic within the Qiman Tagh Terrane, and the locations of two cross sections of A–D and E–F in the Fig. 1.4; **b** time scale showing the age of sedimentation, timing of magmatism and metamorphism in the North Qiman Tagh and South Qiman Tagh Terranes. See text for references. Characteristic geochemical signature of magmatic events is mentioned: CAB: continental arc basalt; IAB island arc basalt; SSZ: supra-subduction zone. Time scale is from Gradstein et al. [44]

to Mesozoic. Several critical issues thus remain unresolved, such as: (i) was there a single suture zone in the Qiman Tagh Orogen? (ii) was there a Paleozoic intra-oceanic island arc chain in the Paleo-Tethyan Qiman Tagh Ocean? (iii) when and how did the ocean close? (iv) what is the timing and what are the mechanisms of the oroclinal bending of the western Qiman Tagh Orogen (Fig. 1.1). It is also not clear whether the present curvature of the orogen is an original geometry or instead resulted from a secondary oroclinal bending of thrust faults that were originally linear [35, 180]. With these points in mind, the objective of this paper is to provide a coherent picture of the pre-Cenozoic geologic framework of the Qiman Tagh Orogen, and present a new interpretation of the spatial and temporal relationships among the various terranes within this orogen. We also attempt to reconstruct the paleogeographic evolution of the Tethyan domain on the basis of the ophiolite, flysch/molasse sedimentation patterns, and syn-collisional magmatism, which suggest the location of the Paleo-Tethyan suture in the southern margin of the Asian continent.

1.1 Lithostratigraphy

The Qiman Tagh Terrane can be divided into the South Qiman Tagh Terrane (SQT) and North Qiman Tagh Terrane (NQT) by the Adatan Fault that runs in the middle of the Qiman Tagh Terrane (Fig. 1.1) [142]. Wang et al. [149] suggested instead that the Qiman Tagh Terrane can be separated into the East and West Qiman Tagh Terranes by the NE trending Baiganhu Fault. The Qiman Tagh Terrane in between displays a prominent oroclinal bending (Fig. 1.1). The West Qiman Tagh Terrane contains evidence for coexisting Silurian S- and A-type granites, whereas the East Qiman Tagh contains only Late Paleozoic granitoids. Here we prefer to separate the North Qiman Tagh Terrane from the South Qiman Tagh Terrane to outline the broader Qiman Tagh Orogen, as they are defined by distinct stratigraphic units and magmatic series. The North Qiman Tagh Terrane is separated from the South Qiman Tagh Terrane by the Adatan Fault in the east and Baiganhu Fault in the west, and it was disrupted by the deformation and movement of the Yaziquan tectonic mélange belt (Fig. 1.1). The South Qiman Tagh Terrane is bounded by the Nalingele Fault in the east and Kumbyan Fault in the west (Fig. 1.1). This subdivision is parallel to that of the East Kun Terrane, which adjoins the eastern part of the Qiman Tagh Terrane, between the Jinsha Suture and the North Kunlun Thrust [9]. The East Kunlun Terrane has also been divided into the North Kunlun Terrane and South Kunlun Terrane (Fig. 1.1), separated by the Middle Kunlun Suture. Many of the stratigraphic features of the East Kunlun Terrane are comparable with those of the Qiman Tagh Terrane.

The Precambrian Jinshuikou Formation, recognized as the old metamorphic basement of the Qiman Tagh Orogen, is locally exposed in the North Qiman Tagh Mountains, but is exposed only poorly in the South Qiman Tagh Mountains (Fig. 1.2) [55]. The composition and source of the rocks in this basement are similar to those in the East Kunlun Mountains [92], and are predominantly represented by the Upper Schist Formation and Lower Gneisses Formation, encompassing amphibolite- to

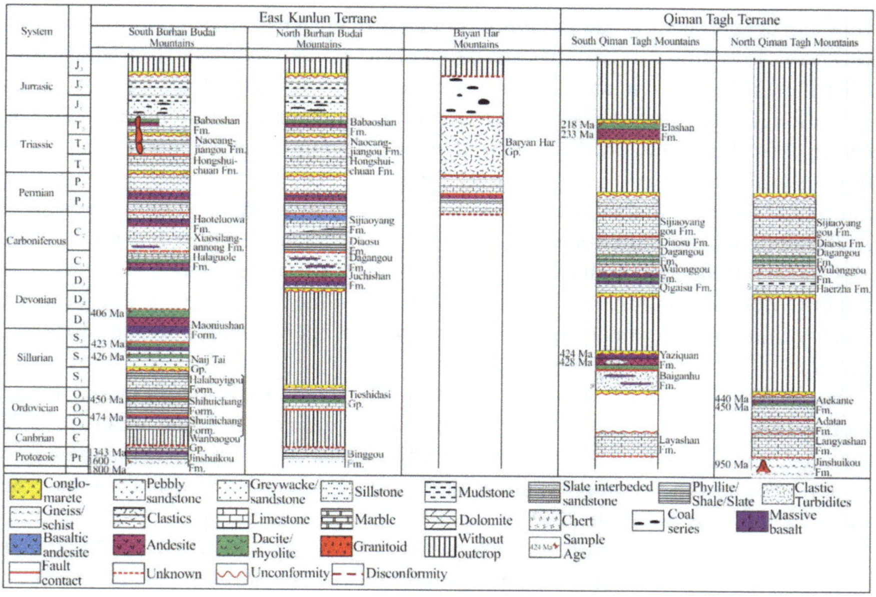

Fig. 1.3 Generalized stratigraphic columns for the various tectonic-stratigraphical units from the East Kunlun Mountains and Qiman Tagh Mountains. Thicknesses of different units are not to scale. Timescale is from Gradstein et al. [44]. See text for details. The strata from the Burhan Budai Mountains in the East Kunlun terrane is based on [20, 61, 87, 172]

granulite-facies metamorphosed granites and minor limestones that formed in the Paleoproterozoic and Mesoproterozoic (Fig. 1.3) [51, 187]. Liu and Ye [84] have identified gray granodioritic, tonalitic and trondhjemitic gneisses (Archean TTG) in the Jinshuikou Formation. These are overlain by a continuous, thick succession of Cambrian-Ordovician marine sediments interbedded with intermediate-mafic volcanics [104, 181]. The inherited zircon grains in the gneiss are dominated by Paleoproterozoic ages in the range of 1600–1800 Ma [187]. In the East Kunlun Mountains, the Jinshuikou basement covers large areas in the North Kunlun Mountains, with few exposures in the South Kunlun Mountains (He et al. [50, 51, 84]). Meng et al. [92, 93] reported concordant zircon U–Pb ages from the gneissic granites emplaced in the Jinshuikou Formation in the Qiman Tagh belt at 938 ± 5 Ma (SHRIMP method) and 938 ± 2 Ma (LA-ICP-MS method).

The Middle Proterozoic-Cambrian Langyashan Group is mainly composed of massive stromatolitic carbonates, and a few interbedded iron-bearing quartzites. They were strongly folded and thrusted during the Paleo-Tethys Ocean subduction in the Early Paleozoic.

The Ordovician-Silurian strata in the Qinghai province are termed Tanjianshan Group, but in the Xinjiang province these are known as Qiman Tagh Group. Here we use the term Qiman Tagh Group, which is mainly composed of clastic rocks at the base, carbonate rocks in the middle and volcanic rocks in the uppermost parts.

Although the chronostratigraphy is long debated, evidence from paleontology and zircon chronology suggests that these rocks can be classified into four formations [147, 174]: the Early Ordovician Adatan Formation, Middle-Late Ordovician Atekante Formation, Early Silurian Baiganhu Formation, and Middle Silurian Yaziquan Formation (Fig. 1.3). The Ordovician strata exposed in the East Kunlun Terrane were originally divided into four formations (Fig. 1.3) [107, 181], which from bottom to top, are: (i) the Halabayigou Formation (O_1, mainly composed of sandy slates); (ii) the Shuinichang Formation (lower Upper Ordovician, composed of phyllite and greywacke interbedded with marble and limestone); (iii) the Shihuichang Formation (upper Ordovician, composed of marble and crystalline limestone); and (iv) the turbidite formation (O), consisting of sandy turbidite intercalated with limestone and shale.

The Adatan Formation, mainly composed of terrigenous clastic rocks and acidic-intermediate volcanic rocks, is distributed in the Yaziquan-Atekante area (Fig. 1.2). The Atekante Formation is a carbonate sedimentary formation that involves gray crystalline limestone, dark gray dolomitic limestone interbedded with carbonaceous slate and siliceous dolomite [147].

The Baiganhu Formation, distributed widely along the Yaziquan-Baiganhu-Kumbyan area in the south of the Baiganhu fault, in the western part of the South Qiman Tagh terrane (Fig. 1.2), is mainly composed of sand, mud, arc-derived volcano clastics and volcanic ashes, which shows a turbidite sedimentary sequence [66, 77]. Li et al. [65] discovered an Early-Silurian graptolite fauna from the lower parts of the Baiganhu formation. The sedimentary features indicate that it formed in a back-arc basin environment, and was sourced from an island arc in the south as inferred on the basis of the fan-shaped paleocurrent direction [66]. In addition, the grain size of the turbidite sedimentary sequence becomes finer from bottom to top, the sandstones become lesser in volume, and the slate and/or shale become more abundant, indicating progressively deeper domains of the basin, and reflects the dispersion of the turbidite fan to the sea.

Qi [114] reported zircon SHRIMP U–Pb ages from the basalt and andesite from the upper volcanic formation (Yaziquan Formation) between 427.7 ± 4.4 Ma and 424.4 ± 5.3 Ma in the Yaziquan area, respectively. The Yaziquan Formation occurs only in the Yaziquan tectonic mélange zone, and is mainly composed of intermediate to basic igneous rocks, including meta-basalts, basaltic andesites, andesites, diorites and gabbros. They are considered as subduction-related igneous rock assemblages [174]. The Ordovician Adatan Formation, occurs in the eastern part of Yaziquan area and is mainly composed of sandstone, schist and volcanic rocks.

The Qiman Tagh Mountains contain a series of Upper-Devonian molasse deposits that uncomfortably overlie the Ordovician-Silurian Qiman Tagh Group, and are covered by biogenetic carbonates [81, 157]. The sequence is dominated by terrigenous facies and abyssal to shallow sea facies clastic rocks, carbonatite and intermediate-acidic volcanic clastic assemblages. Shi et al. [127] noted that the petrogenetic characteristics of the molasse formation in the NQT and SQT are different. The molasse formation in the NQT, named Haerzha Formation, carries abundant plant and brachiopod fauna, which reflects an environment alternating between sea and

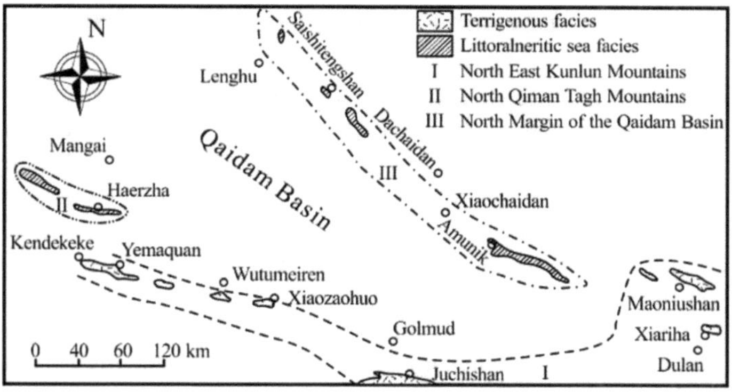

Fig. 1.4 Sketch geological map showing the distribution of the Devonian Molasse Formation in the mountains around the Qaidam Basin. Modified based on Shi et al. [127]

river facies (Fig. 1.4) [81, 157]. However, the Qigaisu Formation that occurs in the SQT is composed of terrigenous deposits (Fig. 1.4) [127]. Pan et al. [106, 107] consider this formation to be similar to the Upper-Devonian Qizilafu Formation in the West Kunlun Mountains, representing a post-orogenic molasse formation. However, Chen et al. [18] suggested that the rocks are petrogenetically similar to those of the Maoniushan Formation in the East Kunlun Mountains, which characterized by sedimentation of terrigenous facies and littoral-neritic sea facies including clastic rocks, carbonatites and intermediate-acidic volcanic assemblages. Lu et al. [87] suggested that the Maoniushan Formation was formed in Middle Silurian-Early Devonian (ca. 426–403 Ma). Some studies suggests that it was formed in a collisional orogenic environment, which would indicate that the Early Paleozoic ocean might have closed in the Middle Silurian-Early Devonian [87, 111, 112, 193, 199]. However, Chen et al. [18] correlated the molasse formation to a stretching rift system, indicative of the beginning of basin evolution in the late Paleozoic.

The Carboniferous deposits are mainly composed of the Dagangou Fm., Diaosu Fm. and Sijiaoyanggou Fm., which unconformably cover the Devonian or older strata. The sedimentary environment can be divided into three facies: a shore facies, a carbonate platform facies, and a neritic marine basin facies [99]. Two marine ingressions in the Early Carboniferous and a long-lasting transgression in Late Carboniferous were also recognized. With the continuous expansion of transgression, the sedimentary environment gradually evolved from offshore to epeiric-sea, and mainly developed in a paralic sedimentary environment [99]. Until the formation of the upper Carboniferous-lower Permian Sijiaoyanggou Fm., the depositional system was active with large sediment thickness, lateral distribution and stability [78].

1.2 Structural Framework of the Qiman Tagh Orogen

The Qiman Tagh Mountains preserve major regional structures, which, from north to south, include the north-dipping Qiman Tagh imbricate thrust zone, the north-dipping Adatan-Baiganhu thrust system, and the north-dipping Nalingelehe-Kumbyan thrust system (Fig. 1.5). These terrane-bounding important thrust faults shaped the orogen and controlled the structural architecture of the orogen after the initial accretion and final collision. The North Qiman Tagh Mountains are dominated by the north-dipping Adatan-Baiganhu and Qiman Tagh thrusts (Fig. 1.2) [179]. The Adatan thrust fault strikes northwest but sharply swings to the northeast Baiganhu thrust fault at its western termination in Xinjiang Province, locally placing Precambrian and Paleozoic rocks directly over Eocene strata. The Baiganhu Fault system experienced early stage thrusting to form a series of nappes and folds. Later it was effected by ductile shearing with left-lateral strike-slip features, following the rotation and curvature of early folds [68, 71]. The Baiganhu shear zone is mainly composed of mylonite, phyllonite, tectonic schist, fault breccia and fault gouge, and is up to 4–5 km thick [67, 68]. The Adatan-Baiganhu thrust fault system was separated by the Yaziquan-Yazidaban tectonic mélange belt, and the map-view pattern of the fault system displays a Z-shaped oroclinal bending structure from the Yaziquan-Yazidaban tectonic mélange belt to the Xire area (Fig. 1.2). In addition, the Qiman Tagh thrust runs across the middle of the North Qiman Tagh Mountains, placing Proterozoic gneisses and Cambrian limestones over Ordovician-Carboniferous strata (Fig. 1.2). To the east, the Qiman Tagh thrust pinches into the Qaidam basin.

In the norther flank of the Qiman Tagh Mountains, several folds and south-dipping thrust were developed including the Gas Hure syncline, Yousha Shan-Luoyan Shan anticlinorium system, and Arlar Thrust (Figs. 1.2 and 1.5). All of these fold systems developed in the Cenozoic. The Yousha Shan anticlinorium is a south-verging fold, with a steep to overturned southern limb (Fig. 1.5). Contraction structures in this fold complex have an eastward decreasing magnitude of shortening. The fold is still active and has been propagating eastward in the Quaternary (Fig. 1.5). The Gas Hure syncline is the southern continuation of the active Yousha Shan anticlinorium (Fig. 1.5). The south-dipping Arlar Thrust is a minor blind structure, well-imaged seismically below the Gas Hure syncline, which formed in the early Miocene and Pliocene (Fig. 1.5).

The Yaziquan-Yazidaban tectonic mélange belt experienced multiphase, multi-system, and multilayer tectonic deformations during the Carboniferous-Permian. It results from early stage ductile shear deformation and later brittle overprinting [185]. Various structural features indicate a left lateral strike-slip deformation, combined with thrusting (Fig. 1.1).

The South Qiman Tagh Terrane is dominated by Nalingele thrust fault in the east and Kumbyan thrust fault in the west as boundaries (Fig. 1.1) [67]. Both thrust faults placed Proterozoic, Paleozoic and Triassic rocks directly over Quaternary sediments [179]. The Nalingele thrust fault is mostly covered by Quaternary sediments (Fig. 1.5). The Kumbyan strike-slip and thrust fault is located in the southwestern

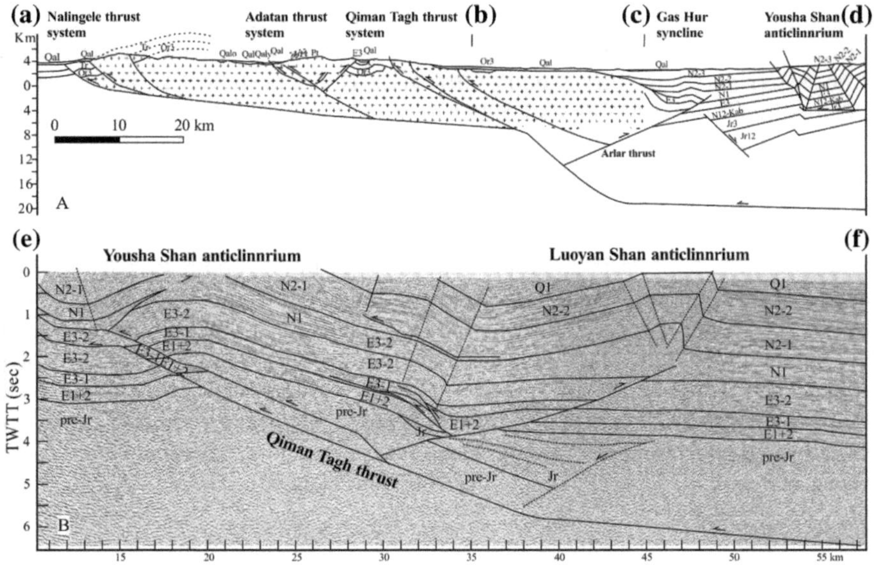

Fig. 1.5 **a** Geologic cross section across the Qiman Tagh–Yousha Shan thrust belt (modified after Yin et al. [179]). Geology across northern flank of the Qiman Tagh Mountains is constrained by drill-hole and seismic data. See text for details. **b** Seismic section across the eastern tip of the north-dipping Qiman Tagh thrust zone and the eastern segments of the Yousha Shan anticlinorium with some interpretation (modified after Yin et al. [179]) Abbreviations: Qaly-active fluvial channel deposits; Qal-Quaternary alluvial deposits; Qalo-older Quaternary alluvial deposits; N2−1 and N2−2-older and younger Miocene sedimentary units; N1-Late Oligocene strata; E3-Eocene–Oligocene strata; E3−1-Middle to Late Eocene strata; E3−2-Early Oligocene strata; E12-Kab-Cretaceous and lower Eocene strata; E1+2-Paleocene–Early Eocene strata; Jr—Jurassic strata; Jr1-2 and Jr3-Jurassic sedimentary units; pre-Jr—pre-Jurassic rocks; Tr-Triassic sedimentary and volcanic rocks; C-Carboniferous sedimentary unit; Or3-Upper Ordovician metavolcanic and metagraywacke sequences; Pt-Proterozoic gneisses; gr-granites

part of the Yaziquan-Yazidaban tectonic mélange belt (Fig. 1.1) and trends NE-NEE for 46 km length and inclines as NW with 50°–70° dip angle. It shows of multiple reactivation with an early stage dominated by thrusting in Paleozoic, and late stage ductile overprints, especially in Mesozoic and Cenozoic [49].

1.3 Metamorphism

High-grade metamorphic rocks have not been reported from the interior of the Qiman Tagh region. However, in the Suhaitu area between the eastern part of the Qiman Tag Terrane and the East Kunlun Terrane, along the middle Kunlun fault, a high-pressure metamorphic belt has been discovered [116]. The high pressure rocks are generally composed of eclogite and garnet amphibolite [116]. The eclogite-facies metamorphic conditions were estimated at 2 Gpa and 600–700 °C, whereas amphibolite facies retrogression occurred at 0.7 Gpa and 550 °C [116]. In addition, in the Wenquan area, in the eastern part of the East Kunlun Terrane, the eclogites show peak-metamorphic P-T conditions of >1.6 GPa and 590–650 °C [93]. The peak metamorphic ages of eclogite-facies rocks in Suhaitu and Wenquan are 411.1 ± 1.9 Ma (LA-ICP-MS zircon U–Pb dating) and 428 ± 2 Ma (SHRIMP zircon U–Pb dating), respectively. The eclogites in the Suhaitu area formed later than that in the Wenquan area, and the P-T conditions of formation in the Suhaitu are higher than those in the Wenquan area. In addition, a suite of basic granulites intercalated with the ophiolitic mélange belt have been reported in the Qingshuiquan area within the Middle East Kunlun fault [15, 17, 70]. Li et al. [70] reported a SHRIMP U–Pb zircon age of 507.7 ± 8.3 Ma for the granulite facies metamorphism, and suggested that the Qingshuiquan granulites formed under temperatures of 760–880 °C and pressures of 0.08–1.2 Gpa. These granulites, together with a series of high-pressure metamorphic rocks represented by eclogites in the Central East Kunlun Fault zone, represent subduction and collision events, and provide new constraints for the Central Kunlun suture zone in Early Paleozoic [16, 93, 116].

1.4 Magmatism

1.4.1 Proterozoic Magmatism

Neoproterozoic granitic gneisses, widely occurring in the East Kunlun, Qilian and Altun terranes, might be related to the assembly of different blocks, and the response of the formation of the global supercontinent Rodinia [88, 92, 187]. In the Qiman Tagh terrane, Neoprotozoic granitic gneisses (ca. 1000–820 Ma) are largely distributed in both the North and South Qiman Tagh Mountains [92, 134, 186]. In addition, these rocks are involved in separate metamorphic and cooling events during the Silurian-Devonian [92]. Their geochemical features are similar to those of the Jinshuikou metamorphic basement, which formed during Neoarchean to Mesoproterozoic (Fig. 1.6) [20, 89]. Meng et al. [92, 93] inferred that the source of the granites might be the Baishahe Formation of the Proterozoic Jinshuikou Group. The low La/Yb and Sr/Y ratios of the Neoprotozoic granitic gneisses (Fig. 1.7), indicate a convergent margin setting rather than magma generation within a continental extensional setting. These geochronological and geochemical data have been correlated with the global Rodinia supercontinent [51].

Furthermore, available Sm–Nd and Lu–Hf isotopic data also indicate magma derivation from a crustal source for the Neoproterozoic granitoids (Figs. 1.8 and 1.10), possibly through remelting of the Jinshuikou metamorphic basement, representing typical continent-continent syn-collisional granites [38, 92]. Integrated with the regional geology, the Meso- to Neoproterozoic plate collisional event has been correlated with the assembly of the Rodinia supercontinent [51].

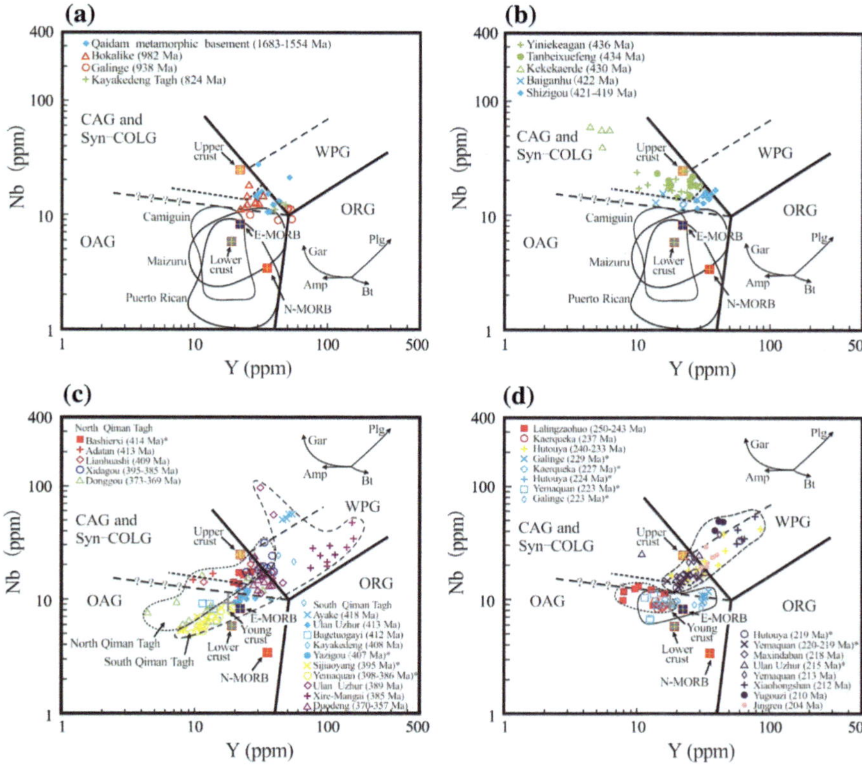

Fig. 1.6 Nb–Yb discrimination diagram [110] for the Qiman Tagh granitoids. **a** Meso-Neo Proterozoic granitoids; **b** early Paleozoic granitoids; **c** late Paleozoic granitoids; **d** triassic granitoids. Geochemical data with "*" are from this study. Ocean ridge granites (ORG), within-plate granites (WPG), volcanic arc granites (VAG), syn-collisional granites (Syn-COLG), oceanic arc granites (OAG), and continental arcs (CAG). The boundary between OAG and CAG is postulated based on the calc-alkaline lava composition from Puerto Rican, Maizuru and Camiguin oceanic island arc. The average upper crust component is from Rudnick and Fountain [121]; the average lower crust component is from Taylor and Mclennan [137]; the average N-MORB component is from Hofmann [52]. The average of E-MORB component is from Sun and McDonough [133]. The component plotting area of calc-alkaline lavas from the Camiguin ocean island arc is from Castillo et al. [8], the Maizuru Camiguin ocean island arc is from Suda [132], and the Puerto Rican ocean island arc is from Jolly et al. [62]

1.4.2 Early Paleozoic Magmatism

The Early Paleozoic granitoids, which are restricted to the North Qiman Tagh Mountains, are largely distributed in the northern part of the Baiganhu and Adatan thrust fault [40, 69, 76, 82, 86, 142, 149]. The rocks are preferentially aligned along NE and NWW trends (Fig. 1.1). Liu et al. [82] suggested that these rocks might have formed in a collisional environment. Zheng et al. [197] invoked post-subduction continental

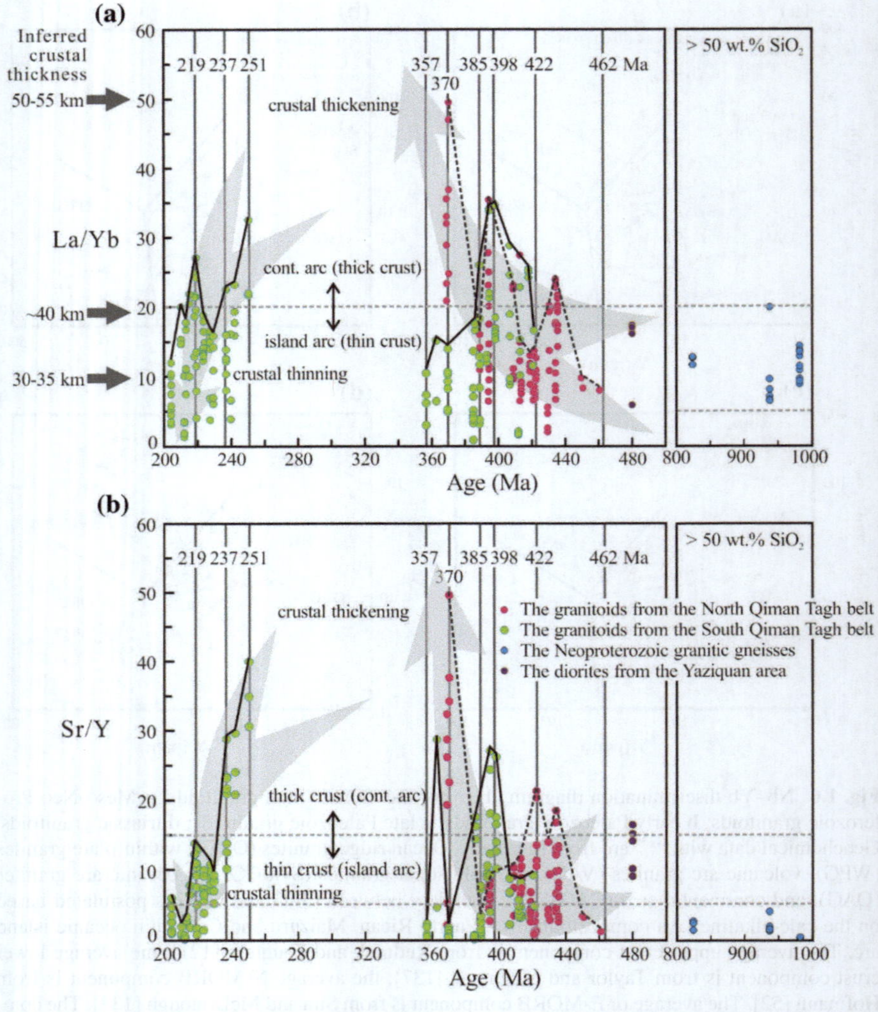

Fig. 1.7 Plot of **a** La/Yb versus Age and **b** Sr/Y versus Age of magmatic rocks (>50 wt% SiO_2) from Neoproterozoic to Early Mesozoic age. The green circle samples (c. 421–204 Ma) with an interval c. 357–251 Ma samples are from the South Qiman Tagh terrane; The red circle samples (c. 461–360 Ma) are from the North Qiman Tagh terrane; The blue circle represents Neoproterozoic granitic gneiss; The purple circle stands for the diorite samples in the Yaziquan-Yazidaban mélange belt. Correlating crustal thicknesses are from Haschke et al. [48]

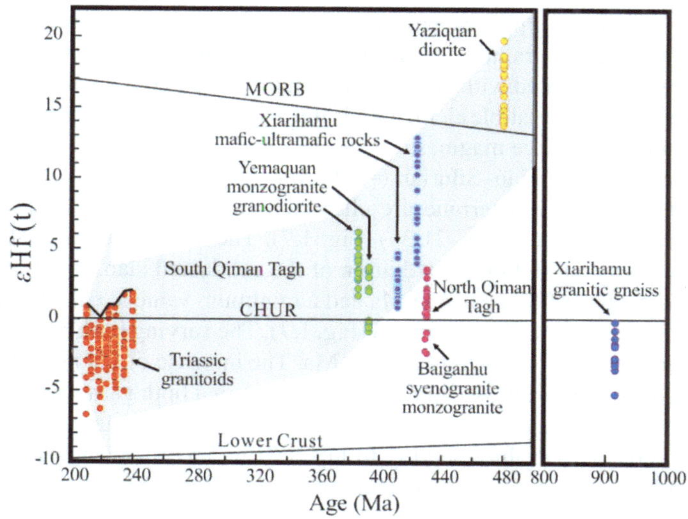

Fig. 1.8 Plots of εHf (t) versus U–Pb ages of the igneous rocks (ca. 200–1000 Ma) from the South and North Qiman Tagh Terrane. Xiarihamu is located in the central region of the Eastern Kunlun Orogen. The Xiarihamu region contains four mafic–ultramafic complexes that have intruded Neoproterozoic granitic gneiss and Paleoproterozoic Jinshuikou metamorphic basement. Ni–Cu sulfide ores are hosted in the Xiarihamu mafic-ultramafic rocks. The Xiarihamu sample data are from Li et al. [64] (411.6 ± 2.4 Ma), Peng et al. [111, 112] (424 ± 1 Ma) and Gan [38] (915.2 ± 2.3 Ma). The shadow area shows a significant decrease trend of εHf (t) values from Ordovician to Triassic granitoids in the South Qiman Tagh terrane

collision in the Late Silurian to explain the formation of granitic magmas sourced from remelting of the metamorphosed Proterozoic sediments. Li et al. [76] considered that the formation of these rocks was related with the northward subduction of the Tethyan Qiman Tagh Ocean. Debates surround the polarity of subduction of the ocean during the Early Paleozoic, with Li et al. [75] suggesting southward subduction in the Early Paleozoic, and consequently some micro-continental blocks being accreted to the northern margin of the Gondwana supercontinent. Wang et al. [143] reported bimodal volcanic rocks in the Kuangou-Xiaolangyanshan area, consisting of two groups of basalt + basaltic andesite (450 Ma) and dacite + lithoidite (440 Ma) association, which are geochemically similar to island arc tholeiites. Gao and Li [40] suggested an extensional environment during the Sillurian to explain the occurrence of metaluminous-peraluminous A-type granites. Zhou et al. [198] proposed that the Qiman Tagh Ocean was closed during the Late Ordovician based on their studies of Late Ordocinian adakitic granitic rocks.

In this study, a compilation of all the available geochemical data shows CAG (continental arc granitoid) and syn-collisional affinity (Fig. 1.6b). However, the Late Silurian (after ca. 422 Ma) samples are more similar to the Proterozoic magmatic rocks (Fig. 1.6b), suggesting that the Late Silurian granites probably resulted from remelting of the metamorphic basement in a syn-collisional environment. In contrast with the Late Silurian granites, the Early Silurian granites are more enriched Nb, which might be associated with oceanic lithosphere subduction. During subduction, the fluid enriched incompatible elements Nb and Sr would be transferred from oceanic lithosphere into the source magmas.

Furthermore, Ordovician–Silurian (−422 Ma) magmatic rocks (>50 wt% SiO_2) in the North Qiman Tagh terrane are characterized by a slight increase in Sr/Y values (from 1.3–3.51 to 10.48–21.35) (Fig. 1.7). The increase in Sr content might be associated with amphibole dehydration of the subducted slab. The La/Yb ratios show a peak (6.66–23.98) at ca. 435 Ma and a minimum value at ca. 422 Ma, with a continuous increase from 422 to 398 Ma (Fig. 1.7). The varying La/Yb ratios indicate that collision probably persisted to ca. 398 Ma. The increase of La/Yb following the progressive decrease in Yb concentrations, indicates that both garnet and amphibole were major residual phases in the source with minor or no plagioclase [118, 119]. The Baiganhu zircon εHf(t) values varying from −2.31 to 5.57 indicate that the granites were derived from the melting of ancient felsic crust (Fig. 1.8) [40].

1.4.3 Late Paleozoic Magmatism

1.4.3.1 The North Qiman Tagh Terrane (NQT)

Late Paleozoic granitoids are ubiquitous in the whole Qiman Tagh belt. In the NQT they are spatially accompanied by the Early Paleozoic granitoids, and are topographically similar to the Early Paleozoic granitoids trending in the NE and NWW directions (Fig. 1.1). However, most of the granitoids are concentrated to the north of the Adatan Fault [76, 114, 142, 155], and only a few occur to the north of the Baiganhu Fault [200]. Zhou et al. [200] suggested that the Baiganhu magmatic series shows a continuous evolution from Early Paleozoic S-type to Late Paleozoic A-type granites, which indicates a post-orogenic setting in Early Devonian. The NQT contains abundant mafic-ultramafic dike swarms of Middle-Late Devonian age, which implies that the North Qiman Tagh might have experienced an episode of intracontinental extension at that time [115, 142].

In Fig. 1.6c, the Early Devonian granitoid data plot in the CAG field and are similar to the Early Silurian rocks (~398 Ma), which indicates that they were probably derived from the melting of old crust in a syn-collisional setting, and melt was probably contaminated by upper crust material. However, the Middle Devonian samples plot in both CAG and WPG (within plate granitoid) fields (Fig. 1.6c), which implies a transitional stage from post-orogenic to within-plate conditions. In addition, the Late Devonian granitoids plotting in the OAG field are intensely depleted in Nb and Y (Fig. 1.6c), which might reflect melting of E-MORB type source rocks. Both La/Yb and Sr/Y ratios of magmatic rocks belonging to Early-Late Devonian in the NQT show minimum age of ca. 385 Ma and maximum ranges at ca. 370 Ma (Fig. 1.7), which indicates that the sources of the Middle Devonian rocks are different from that of the Late Devonian suite.

1.4.3.2 The South Qiman Tagh Terrane (SQT)

In the SQT, Late Paleozoic granitoids are similar to those in the North Qiman Tagh Terrane with regard to their topography, and trend in the NE and NWW directions. However, they are spatially accompanied or intruded by Early Mesozoic granitoids, with U–Pb ages ranging from 418 Ma to 348 Ma, clustering mainly in the Devonian (Fig. 1.1c) [10, 42, 45, 47, 91, 135, 143, 145, 192]. Hao et al. [46] and Zhao et al. [196] proposed that the Early Devonian intrusive rocks were formed in a post-collision environment, and suggested that the collision-related orogeny took place during the Silurian. However, only a few Silurian granitoids have been reported in the South Qiman Tagh belt. Gao et al. [42] and Yao et al. [178] suggested that the Early-Middle Devonian magmatic rocks might have formed by underplating and magma mixing in the Late Paleozoic collision-post collision stage. Tan et al. [135] found Early Devonian (413 Ma) peraluminous granites that are indicative of a syn-collisional environment. In addition, the Qiman Tagh Late Devonian intrusive rocks were suggested to have formed in an extensional setting through underplating of mantle-derived magma during the post-orogenic or late orogenic stage [45, 47, 145, 155].

In this study, the Early Devonian (older than 398 Ma) granitoid samples mainly plot in the OAG field, and a few samples plot in the WPG field (Fig. 1.6c), suggesting that oceanic lithosphere subduction and dehydrated melting played a major role, although locally, for the magma generation in the SQT. Nevertheless, the Early Devonian (older than 407 Ma) granitoids contains slightly higher Nb and Y concentrations than those of the late Early Devonian granitoids (Fig. 1.6c). The magma generated by subducted slab melting would rise through the mantle wedge. Part of these magma could be emplaced directly into the crust; another part would reside at lower crustal depth and differentiate to form new intermediate-acidic magma, and mafic relics would form young lower crust. Generally, the lower crust is strongly depleted in incompatible elements [28]. Nb and Y are highly incompatible, and are preferentially retained and enriched in the residual mantle during the formation of continental crust, but their abundances are anomalously high in both OIB and MORB [52]. However, the Middle-Late Devonian (after 389 Ma) samples plot mainly in the CAG and WPG fields (Fig. 1.6c). Moreover, a transition can be observed from the Early Devonian to the Middle Devonian granitoids from enriched mantle source to lower crust source (Fig. 1.6c).

Both La/Yb and Sr/Y ratios of magmatic rocks belonging to the Early-Late Devonian phase in the South Qiman Tagh terrane show peak age values at ca. 398 Ma, which indicate that the Early Devonian rocks were probably originated from enriched mantle source with high Sr and low Y content. Furthermore, the positive εHf(t) values also show different sources between Early Devonian and Middle Devonian granitoids (Fig. 1.8). The εHf(t) values of the Yemaquan Early Devonian granitoids range from

−0.8 to +3.7, which indicates lower crustal sources. However, the εHf(t) values of the Yemaquan Middle Devonian granitoids range from 1.7 to 6.2, indicating significant involvement of depleted mantle components in their petrogenesis [42].

1.4.4 Early Mesozoic Magmatism

Early Mesozoic granitoids, are restricted to the SQT, and are largely distributed along the NWW direction in the South of the Adatan thrust fault, and NE direction in the southern part of the Baiganhu thrust fault [11, 36, 42, 72, 74, 85, 114, 125, 176, 184, 191, 201]. Chen et al. [12] suggested that Middle Triassic was a transition stage from subduction to collision, with slab breakoff (or oceanic crust delamination) as a critical mechanism of the crust-mantle mixing associated with Middle Triassic magmatism. However, Gao et al. [41] considered the transition stage as prolonged one, ranging from 242 to 218 Ma. Feng et al. [36] inferred that the tectonic transformation from compression to extension was initiated at ca. 240 Ma, to Late Triassic, when the region evolved into post-collisional setting. Yu et al. [182] identified calc-alkaline and alkaline granitoids including granodiorites, monzogranites and syenogranites that were successively emplaced during Late Triassic, which reflects a post-collisional environment.

In Fig. 1.6d, the Early-Middle Triassic (older than 237 Ma) granitoid data straddle the boundaries between OAG and CAG fields. The Midle Triassic Hutouya granitoid plots in the CAG and WPG fields. The Late Triassic granitoids could be geochemically divided into two stages. The early stage (oleder than 219 Ma) plotting in the OAG field shows a continuous evolution from the Early Triassic rocks (Fig. 1.6d), suggesting a similar magma source with juvenile lower crust components; and the late stage (219–204 Ma) plots in the CAG and WPG fields, suggesting a crustal thinning process, with the magma evolving from juvenile to old lower crust. The La/Yb and Sr/Y ratios of magmatic rocks from the South Qiman Tagh terrane shows a slight decrease (La/Yb from 21.94–32.5 to 3.12–11.96 and Sr/Y from 30.61–39.36 to 0.88–6) though Early-Triassic to Late-Triassic (Fig. 1.7). The relatively high La/Yb and Sr/Y ratios of the Early-Middle Triassic rocks suggest a deep juvenile lower crust. Nevertheless, the low La/Yb and Sr/Y ratios of the Late Triassic rocks show a peak value at 219 Ma, suggesting a possible mixed source involving lithospheric mantle and old Lower crust.

In addition, the mostly strong negative zircon εHf(t) values of the Triassic granitoids range from −6.71 to 1.83 (Fig. 1.8), indicating that their parent magmas were most likely derived from the partial melting of juvenile crust [41, 72, 117, 164], but the contamination of mantle source can not be excluded. Moreover, the strongly negative εNd (t) values (Fig. 1.10), showing a continuous increasing trend from Early Triassic to Late Triassic granitoids, suggest a crustal source but increasing effect of enriched lithosphere mantle in Late Triassic [11, 36, 156, 177]. In addition, the Triassic granitoids show εNd (t) range similar to that of the contemporaneous volcanic rocks (Fig. 1.10), suggesting that they were probably derived from the same crustal source [83]. The dioritic enclaves are widely developed in the Triassic granitoids, and the felsic host rocks and mafic enclaves generally have the same age and εNd (t) range. Gao et al. [39] suggested that they were derived from the partial melting of lower crust and lithospheric mantle, respectively, and eventually underwent a magma mixing process. Moreover, the εNd (t) range of the Triassic granitoids resemble that of the Xiarihamu mafic–ultramafic rocks (Fig. 1.10), but the latter one were generally associated with underplated mantle-derived magma with contamination of crustal materials [111, 112, 195].

1.4.5 Ophiolites

The South Qiman Tagh Terrane contains a series of ultramafic-mafic rocks in the Heishan-Yaziquan-Yazidaban realm. Some researchers consider these rocks as fragments of ophiolites formed in mid-ocean ridge environment [13, 126, 173]. However, others suggest that they were formed in an island arc environment, with magma derivation through partial melting of the mantle wedge [26, 91, 159, 185].

The Heishan ophiolite (Figs. 1.1 and 1.2), first reported by Chen et al. [13], carries an incomplete ophiolitic succession including meta-peridotites, olivine pyroxenites, pyroxenites, and cumulate gabbros, cumulate diabases, massive basalts and pillow basalts, carbonates, and clastic rocks [13, 49, 91, 146]. Zircons grains from the cumulate gabbros in Heishan ophiolite mélange belt yielded ages in the range of 816–486 Ma [13, 91]. In Fig. 1.9, the geochemical data of the Heishan basic-ultrabasic rocks plot in the area of the average primary mantle (PM) components and are distinct from typical mid-ocean ridge basalt components that are concentrated in the area of N-MORB [32, 109]. They are also different from plume type ophiolite components that are mainly scattered in the area of the E-MORB [32]. Thus, these rocks are more likely to be primary island arc basalt components, and geochemical features are consistent with their evolution in a suprasubduction regime with no significant crustal contamination [123]. Thus, they may represent suprasubduction zone (SSZ) type ophiolites. The Heishan ophiolites have a high $^{143}Nd/^{144}Nd$ ratios range (0.51303–0.51311), and high εNd (t) values in the range of 7.5–8.9 (Fig. 1.10). The volcanic rocks with high average εNd value of 8.1 were probably derived from subducted oceanic crust, corresponding to "immature" island arcs [102]. The typical sources of MORB have a higher average εNd (t) value of +10.

The Yaziquan-Yazidaban ophiolite (Figs. 1.1 and 1.2) is compositionally dominated by moderately metamorphosed and strongly deformed basic-ultrabasic rocks, such as olivine-rich peridotite, pyroxene-peridotite, gabbro and diabase [173], and are interbedded with Lower Carboniferous deep-sea sediments. In addition, Li et al. [77] reported Early Carboniferous radiolarian fauna in dark gray chert at the Yaziquan area. The pyroxenite in the mélange belt yielded zircon SHRIMP U–Pb age of 384.4 ± 4.4 Ma [185]. However, Xiao [159] considered the unit as a tectonic mélange belt rather than a real ophiolite, and proposed that the rocks formed in a mantle wedge environment within an island arc setting.

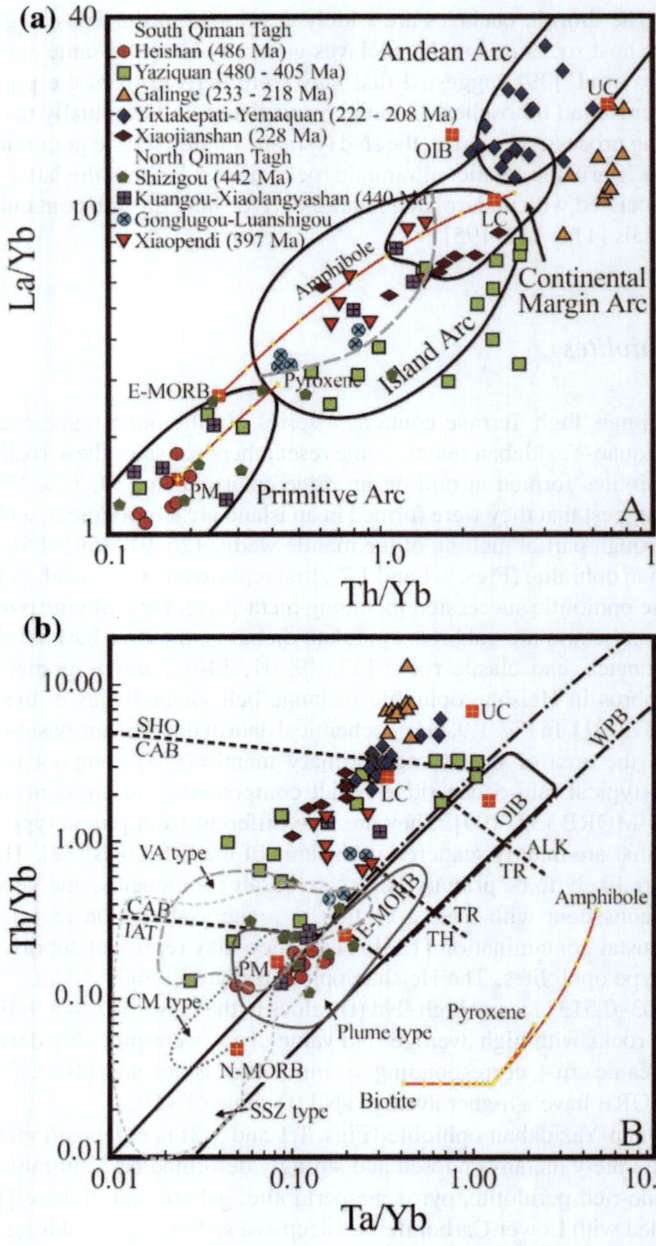

Fig. 1.9 **a** Discrimination diagram for volcanic rocks based on Th/Yb versus La/Yb [22]. **b** The Ta/Yb versus Th/Yb plot [108], showing the geochemical variations of different ophiolite types [32]. P type (plume type); MOR type (mid-ocean ridge type); CM type (Continental Margin type); VA type (volcanic arc type); SSZ type (Suprasubduction zone type); Compositional fields: TH, tholeiite; TR, transitional; ALK, alkaline basalt; IAT, island arc tholeiite; CAB, calc-alkaline basalt; SHO, shoshonitic

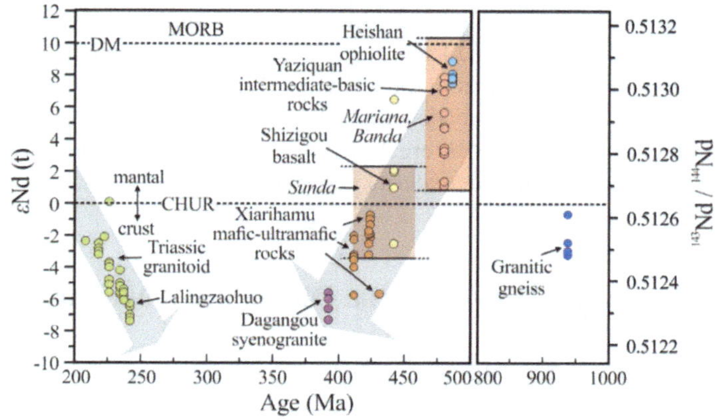

Fig. 1.10 $^{143}Nd/^{144}Nd$ variations versus age (Ma) for Ordovician-Triassic (ca. 500–200 Ma) volcanics and granitoids from the Qiman Tagh Mountains. DM represents the average εNd (0) value (εNd (0) = 10) for depleted mantle as determined from mid-ocean ridge basalts; CHUR ("chondritic uniform reservoir") stands for present-day reference values (εNd (0) = 0) of chondrite [60]. The range of εNd (t) values for the volcanic rocks from the Sunda continental margin arc are based on White and Patchett [151], and for Mariana and Banda island arcs are based on Ikeda et al. [58] and Vroon et al. [140]. The neodymium isotopes (εNd (t) = 0) can be used to separate mantle and crustal origin

The εNd (t) values of these rocks range from 7.9 to 1.1 (Fig. 1.10), which is markedly similar to the values from the Mariana and Banda intra-oceanic island arc magmatic rocks (Fig. 1.10). Most of the data on volcanic arcs built on oceanic crust fall within a small range of εNd (t) from about 6.5 to 10 [29]. In Fig. 1.9, most of the geochemical data plot in the island arc field, and a few values fall in the primary arc environment like SSZ type ophiolite.

The North Qiman Tagh belt also contains a series of mafic-ultramafic rocks in the Shizigou-Kuangou-Xiaopendi area (Figs. 1.1 and 1.2). The Shizigou ophiolite occurs as blocks in clastic rocks belonging to the Qiman Tagh Formation, and mainly contains serpentinite, gabbro, diabase, basalt and chert [115, 130, 142]. They experienced least three tectonic deformation and metamorphic events, and were finally exposed during the Early-Silurian. The basalt and gabbro from the Shizigou ophiolite were dated at 442 ± 16 Ma and 449 ± 34 Ma by Sm–Nd [25, 130]. Geological and geochemical evidence suggest that the Shizigou ophiolite is a fragment of a submarine back-arc basin complex formed above a short-lived subduction zone during the Late Ordovician [25, 130]. However, Wang et al. [143] suggested that the Late Ordovician ophiolite in this area formed through calc-alkaline volcanism that generated in a back-arc rift setting. Their $^{143}Nd/^{144}Nd$ ratios (0.512523–0.5127795) are within the range of the igneous rocks from the Sunda continental margin arc (Fig. 1.10) [151]. In this study, our geochemical data of the ophiolites in the Kuangou-Shizigou area are similar to VA type ophiolites that straddle the boundary between the IAT and calc-alkaline (CA) fields (Fig. 1.9). Generally, VA type ophiolite is considered to be subduction-initiation-related [32]. In addition, Wang et al. [143] discovered Nb-rich basalt in the Gonglugou-Luanshigou region, which is related with mantle partial melting above a subduction zone [27, 54, 113, 122]. Mafic dike swarms of Late-Devonian age are abundantly developed in the Xiaopendi area [115]. Their geochemical signature correlates with break-off of the subducted ocean slab, followed by upwelling of enriched mantle material along deep faults.

1.5 The Evolution History of the Qiman Tagh Mountains

The overall structural grain of the Qiman Tagh Mountains is approximately parallel to the direction of the Qinghai-Tibet Plateau. The orogen can be spatially and temporally divided into two distinct tectonic-magmatic belts: the North Qiman Tagh belt, which is mainly composed of Paleozoic granitoids and lacks Early Mesozoic magmatic rocks; and the South Qiman Tagh belt, which is dominated by Late Paleozoic-Early Mesozoic granitoids and is barren of Early Paleozoic intermediate-acidic intrusions. It is important to understand the relation between the North and South Qiman Tagh magmatic belt so as to be able to evaluate whether they formed as a single magmatic arc or represent different arcs.

The two tectonic domains are separated by the Adatan Thrust in the east and the Baiganhu Thrust Fault in the west. The North Qiman Tagh belt is disrupted by the formation of the Yaziquan mélange zone in the center (Fig. 1.1), and displays a prominent oroclinal bending. In the southwestern part of the Qaidam basin, several apparently oroclinal folds and faults are present in the west of the Qiman Tagh orogen, to the south of the Altun Tagh fault. Whether the oroclinal structure is an original configuration controlled by the presence of a thrust ramp, or resulted instead from the secondary bending of the original linear fabrics induced by thrusting and shearing is an important question.

The Qiman Tagh granitoids can be geochronologically divided into four main age groups, which are primarily defined on the basis of the different geochemical features previously described (Fig. 1.7): 480–422, 422–398, 398–357 and 251–204 Ma; and 4 subgroups: 398–385, 398–357, 251–237 and 237–204 Ma. These different stages are considered to reflect different tectonic events and environments and magma sources. We outline below a four-stage conceptual model for the development of these rocks.

1.5.1 *The North Qiman Tagh as an Active Continental Margin*

The Qiman Tagh Terrane was mainly composed of several independent linear components such as arcs and microcontinents with Proterozoic basement and cover rocks. The different arcs include the Cordillera-type continental arcs and Mariana-type intra-oceanic arcs. The North Qiman Tagh Terrane, developed initially as an active continental margin (Fig. 1.11a), was generated by long-lasting magmatic activity, subduction, deformation and terrane accretion processes. Recent studies have identified the Adatan suture zone that outcrops intermittently and extends from the northwestern Yazidaban area to southeastern Shizigou area across the Qiman Tagh Mountains (Fig. 1.1) [25, 130, 173]. Although this suture zone lacks MOR type ophiolites, several other features suggest the presence of a paleo subduction zone. Ages of magmatic rocks show that abundant Early Paleozoic granitoids formed predominantly to the north of the Adatan fault, with only a few in the southern part. This difference indicates that the Paleo-Tethys Ocean lithosphere likely subducted northward along the Adatan Fault rather than the Nalingelehe Fault in Early Paleozoic period. In addition, most of granitoids from the Ordovician to the Early Devonian developed in the North Qiman Tagh Terrane and show a strong affinity with CAG (continental arc granitoid), whereas the Early Devonian to the Middle Devonian granitoids developed in the South Qiman Tagh Terrane are characterized by OAG (oceanic island arc granitoid) affinity (Fig. 1.6). If both these groups formed in an active continental margin setting, their geochemical features would be similar. The Middle Ordovician-Early Devonian CAGs and Middle Devonian within-plate granites from the North Qiman Tagh Terrane confirm continuing subduction activity for at least 70 Ma, indicating at the same time that the subducted ocean might have been of considerable width. In addition, the Early Silurian volcanic-arc type ophiolites occurring in the Shizigou-Kuangou-Xiaopendi area (Figs. 1.1 and 1.2) further demonstrate that the North Qiman Tagh Terrane developed as an active continental margin [115, 130, 142]. The formation of VA ophiolites is generally controlled by (1) partial melting of the mantle above the subduction zone, and (2) the dehydration of and element flux from the subducted slab into the overlying mantle [32].

1.5.2 The South Qiman Tagh as an Oceanic Island Arc

The South Qiman Tagh Terrane is characterized by contrasting geological features in comparison to the North Qiman Tagh Terrane, and is thus treated as an independent terrane. Its tectonostratigraphic analysis indicates that it developed as a composite of several intra-oceanic island arcs from the Early Ordovician until the termination of the Variscan-age orogeny (Fig. 1.11b); the constituent elements are linearly scattered in the Qiman Tagh Ocean. During the Early Paleozoic, as the old oceanic lithosphere resisted subduction, the new oceanic lithosphere was obducted or overthrust to create the South Qiman Tagh volcanic island arc chains built on oceanic crust as it was too buoyant to sink ("forced subduction"), similar to the current Mariana Islands and Philippine Island in the Pacific and the Lesser Antilles in the Atlantic [37]. In addition, the Paleozoic evolution of the eastern Tien Shan orogen in China indicates that it contains some intra-oceanic arcs that formed an archipelago in the Paleo-Asian Ocean in the Paleozoic [161–163, 194].

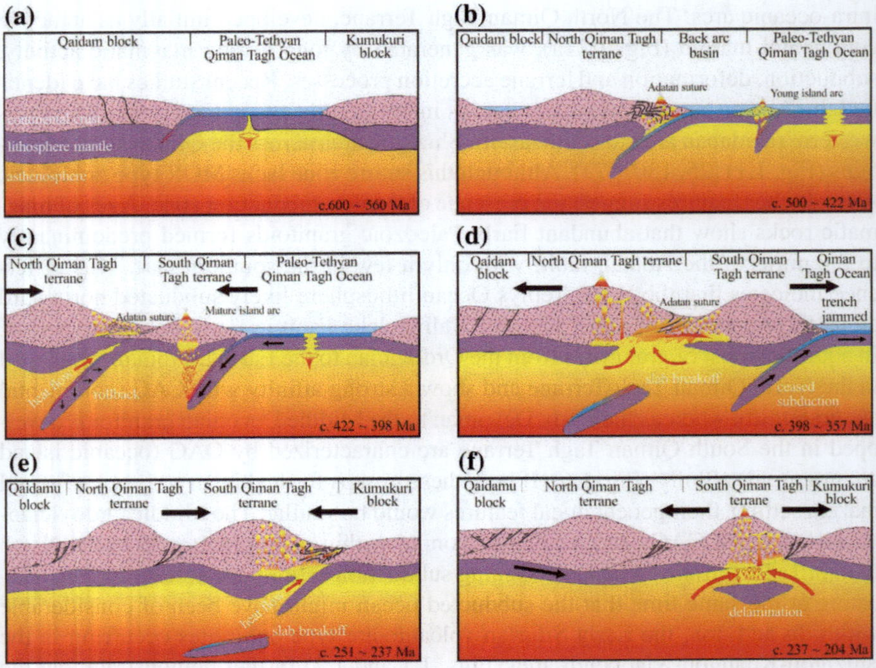

Fig. 1.11 Schematic cross sections to illustrate the major events during the evolution of Qiman Tagh Mountains divided into six time snapshots: **a** Qiman Tagh ocean prior to Qiman Tagh orogeny; **b** the formation of the North Qiman Tagh continental arc and the South Qiman Tagh young island arc accompanying the formation of Adatan backarc basin; **c** the development of the South Qiman Tagh mature island arc and the closure of the backarc basin; **d** following continent-arc collision along the Adatan suture zone, delamination of the subducted, dense oceanic lithosphere occurred in a post-collisional environment; At the same time, reconstruction of orogeny in the South Qiman Tagh Terrane; After that during 357–251 Ma, a trench jam resulted in weakening or cease of subduction, following a magmatism gap of ca. 100 Ma; **e** closure of the Qiman Tagh Ocean causing final collision; **f** intense post-collision magmatic activity

The ancient volcanic arcs related to intra-oceanic subduction zones are normally not preserved, because they are destroyed by subsequent orogenic events or covered by modern volcanic arc material. However, a complete series of mafic-ultramafic rocks developed as a series of ophiolite in the Heishan-Yaziquan-Yazidaban area along a c. NE-SW direction. The Yaziquan mélange zone, as a part of the South Qiman Tagh Terrane, its deformation and movement truncated the Adatan suture zone and disrupted the North Qiman Tagh Terrane (Fig. 1.2).

Although debates surround their tectonic setting environment either MOR (mid-ocean ridge) or SSZ (supra-subduction zone), our study correlates them to the SSZ type, following secondary small scale oceanic crust subduction in the Early Paleozoic and not of the mid-ocean ridge. Furthermore, Xiong et al. [165] reported the Late Ordovician mafic appinites derived from the mantle wedge in the South Kunlun

Terrane, which indicate Proto-Tethyan subduction along the South Kunlun suture zone.

The formation mechanism of an SSZ ophiolite is similar to those forming VA-type ophiolites, and therefore, they share many geochemical similarities (Fig. 1.9b) [32, 123]. An SSZ type ophiolite can form in forearc, backarc, and incipient arc settings. Forearc ophiolites show time-progressive compositional and geochemical evolution, from the oldest mid-ocean ridge-like (MORB-like) to the island arc tholeiite (IAT), and to the youngest calc-alkaline lavas [32, 59]. The time span from the Heishan SSZ to the Yaziquan island arc diorite is less than 10 Ma. Generally, SSZ ophiolites have a relatively shorter subduction history (<10 Ma) as opposed to a relatively longer subduction history (~20–30 Ma) of VA ophiolites [31]. However, from the Yaziquan island arc tholeiite (IAT) to the youngest calc-alkaline lavas, the time span is longer, around 70 Ma.

Young intra-oceanic island arcs, such as the Marianas, only produce basalts and basaltic andesites at the initial stage. However, through island arc evolution over long time periods, the magmas differentiate at depth and tend to become acidic in composition. This is possibly why the South Qiman Tagh Terrane has only substantial Early Ordovician intermediate-basic island arc igneous rocks (Yaziquan Formation), including gabbros, basalts, diorites and basaltic andesites [26], but only limited amounts of acidic rocks. Nevertheless, some researchers consider that these intermediate-basic igneous rock suites belong to ophiolites representing remnants of the Qiman Tagh ocean crust. Generally, subduction-related magmatism appears to have been the most important process for the formation of continental crust even from the early history of the Earth [123, 175]. During the evolution process to a mature island arc, the primary 10- to 20-km thick basaltic crust of the intra-ocean island arc is thickened to more than 20 km, as it chemically evolves towards a calc-alkaline composition, dominated by average andesitic composition of continental crust. Holbrook et al. [53] suggested that volcanic arcs constitute the building blocks of continental crust (see also Condie and Kröner [23]). For example, the modern Aleutian arc has the requisite thickness (30 km) to build continental crust. The island arc crust transformation into proper continental crust probably occurs during arc-continent collision or through the subsequent establishment of a continental arc.

The South Qiman Tagh Terrane is considered to be an exotic and buoyant mass, rather than the seaward extension of the continental lithosphere. Its stratigraphy indicates diverse origins with the North Qiman Tagh Terrane (Fig. 1.3). The North Qiman Tagh belt consists of ancient metamorphic basement blocks (Jinshuikou Gp.) that formed during the Precambrian. It is overlain by a continuous marine sedimentary sequence that was deposited from the Cambrian to the Ordovician and is mainly exposed in large areas along two sides of the Adatan thrust fault and Baiganhu area (Fig. 1.2), and dominates in the North Qiman Tagh terrane. However, these strata scarcely appear in the South Qiman Tagh terrane [184]. If the South Qiman Tagh Terrane was split from the old continental lithosphere, it should contain the old metamorphic basement. Therefore, the South Qiman Tagh Terrane might have formed above the oceanic crust rather than at the continental margin.

In the western part of the South Qiman Tagh Terrane, the Early-Silurian Baiganhu turbidite formation is considered to have formed in a back arc basin environment, with the material sourced from an island arc to the south as inferred from the paleocurrent direction [66]. Therefore, the Baiganhu turbidites were most probably formed in the back arc basin after the South Qiman Tagh island arc generation in the Ordovician. If the South Qiman Tagh Terrane was rifted from an old continental lithosphere, the Baiganhu turbidites were most reasonably formed in a foreland basin environment rather than a back arc basin setting.

1.5.3 Collision Between the South and North Qiman Tagh Terrane

The formation of the South Qiman Tagh oceanic island arc was spontaneously accompanied by the generation of a backarc basin. The persistent subduction that promoted the island arc evolution eventually resulted in the closure of the ocean. During closure of the backarc basin, the continental island arc collided with the approaching South Qiman Tagh passive margin (Fig. 1.11c). Therefore, the South Qiman Tagh Terrane is considered as an exotic terrane, with a major role for continental growth [4, 23, 101, 128]. This process is similar with present-day plate motion and associated processes in the northwest Pacific, with Ben-Avraham and Cox [4] predicting that the Shatsky Rise in the Pacific Ocean will collide with the Japanese arc in the future.

Single zircon U–Pb ages of 414–409 Ma on the granites from the North Qiman Tagh Terrane suggest their derivation from the anatexis of a Proterozoic crust within a collisional environment during the Early Devonian (Fig. 1.6c). Some consider that the collision between the South and North Qiman Tagh Terrane started in Late Silurian times (ca. 422 Ma in this study) [197], a time span which has now been extended up to ca. 398 Ma. Moreover, following collision (after 398 Ma), the subducted oceanic slab broke off resulting in asthenospheric upwelling. This resulted in heating of the basement and the formation of large volumes of WPG (within-plate granitoid) magmas in an extensional environment. Generally, oceanic lithosphere delamination is different from continental lithosphere delamination [6, 98, 190]. The former is active and fast, causing hot asthenosphere upwelling, and would thus result in a series of geological phenomena such as magma production, crustal uplift, tectonics relaxation and orogenic collapse. Orogenic collapse is characterized by substantial volumes of collisional granite [98]. In contrast, continental lithosphere delamination is a slow and passive process, and only causes thinning of the lower crust and collapse of plateaus and/or mountains.

Both in the South and North Qiman Tagh Terranes, Upper-Devonian molasse formations which unconformably overlie the Qiman Tagh Group strata are developed as a result of orogen collapse [81, 170]. However, the Haerzha molasse Formation in the North Qiman Tagh Terrane is chronologically close to the Qizilafu Formation in the West Kunlun Terrane, which is more likely to have formed in a post-orogenic environment after the closure of the back-arc ocean. The Qigaisu Formation in the South Qiman Tagh Terrane is similar to the Maoniushan Formation in the East Kunlun Terrane [127], which was formed in a collisional environment. Their depositional environment displays moderate differences, indicative of the closure of the ocean and the initiation of another tectonic cycle [14, 107]. In terms of the stratigraphic gap from the Middle-Silurian to Middle-Devonian in the Qiman Tagh terrane, the Qiman Tagh might have been in a rapid uplift stage.

Large volumes of deep-water flysch sediments (the Baiganhu Formation) were thrust over the continental shelf during the collision between the North and South Qiman Tagh Terrane. The colliders are deeply buried under kilometers of sediment cover and are believed to be oceanic arcs that fringed accreting terranes. Such orogenic style is commonly termed "soft-collision", which is in contrast with the collisional style in the eastern part of the Qiman Tagh Mountains where evidence for strong uplift of island arc-continent collision is present similar to the modern collision between the Australian continent and an the Banda Arc [7]. In addition, Qi [114] suggested that the East Kunlun Orogenic belt has also not experienced continent-continent collision stage, and only the amalgamation of island arc formed by discontinuous northward subduction. In addition, both the South and North Qiman Tagh Terranes show absence of Late Silurian sedimentary units, supporting the model of collision between South and North Qiman Tagh Terrane during the Late Silurian.

Furthermore, based on the peak metamorphism ages of eclogites found in the East Kunlun Mountains, which range between 428 and 411 Ma [93, 116], we infer the timing of collision in the East Kunlun area as Late Silurian and to Early Devonian.

1.5.4 Renewal of Orogeny in the South Qiman Tagh Terrane

Following arc accretion and renewal of subduction, the trench migrated from the oceanic side and a new marginal sea could be formed; a new volcanic arc was then built on the accreted terrane (Fig. 1.11d). Persistent subduction generated large volumes of OAGs and WPGs that intruded into the ocean floor and thickened the island arc sequences in the Early-Middle Devonian (414–389 Ma). In addition, the magmatic source of OAGs shows a transition from E-MORB to lower crust. Such time-related succession from low-K calc-alkaline, through high-K calc-alkaline, to alkaline granitoids is produced by the transition from subduction-related magmatism to within-plate magmatism at the end of orogenesis [3]. Almost at the same time, the trench is greatly diminished and jammed in depth, and consequently subduction will cease [4, 101]. This would result in a gap in magmatism across a vast region [4, 101]. We consider this as the possible scenario for the sudden termination of the granitoid magmatism (ca. 357–251 Ma) recorded in the South Qiman Tagh Terrane. Trench jamming might have preferably occurred on either an oceanic plateau or a microcontinent of a sizeable mass that was too buoyant to be subducted [100, 101].

1.5.5 Final Closure of the Paleo Tethyan Qiman Tagh Ocean

The final closure of the Paleo Tethyan Qiman Tagh Ocean resulted in the collision of two continents (Fig. 1.11e, f). The South Qiman Tagh Terrane became sandwiched between the colliding continental blocks, and was accreted as part of the Paleo-Asian continent. The two continental margins and the intermediate island arc systems became overthrust and underthrust, respectively, and folded and metamorphosed as they were dragged into depth. The southern Kumukuri microcontinent block split from the Qaidam continental plate in Late Sinian (Ediacaran)–Ordovician, and the Qiman Tagh ocean basin developed into a limited ocean basin [49]. During the Early Paleozoic orogenic event, the South Qiman Tagh ocean island arc chain which existed in the ocean between the Kumukuri block and North Qiman Tagh continental arc from Ordovician to Mid-Silurian, collided with the North Qiman Tagh continental arc in the Late Silurian to Mid-Devonian. After that, the Kumukuri microcontinent started amalgamation along the Nalingele Fault in Carboniferous–Permian, following strong strike-slip deformation and rotation of the Yaziquan-Yazidaban tectonic mélange belt, and the orocline of the Qiman Tagh orogen. The left-lateral strike-slip fault system in the Yaziquan-Yazidaban tectonic mélange belt was formed in this period [185]. Moreover, both the Baiganhu and Kumbyan thrust faults experienced secondary ductile shearing deformation, showing left-lateral strike slip features. Furthermore, the Early Devonian granitoid plutons shows intensive shearing deformation and mimics the shape of the orocline (Figs. 1.1 and 1.2). Therefore, we propose that the secondary deformation of the Baiganhu and Kumbyan thrust faults probably occurred also in the Carboniferous–Permian, and eventually caused the rotation and deformation of the Yaziquan-Yazidaban mélange belt and orocline of the Qiman Tagh Orogen in the Triassic as a consequence (Fig. 1.13). However, Dupont Nivet et al. [35] proposed that the arc shape is an original configuration produced by transport above an arcuate thrust ramp. Most of the hard collision was accommodated by thrusting, lateral extrusion and orocline, which is similar to the collision occurred in the East African Orogen [131], the Himalaya Orogen [167], and the western Altai Orogen [160]. Weil and Sussman [150] defined that orocline implies a secondary bending of an originally linear belt during a subsequent deformation event. In addition, the strong thrusting in the south of the Kumukuri basin, which was initiated in the Late Carboniferous [24], might have further influenced the Kumukuri microblock squeezing into the South Qiman Tagh Terrane.

The simple model presented here (Fig. 1.13) based on the available geological and geochemical data, explains the Qiman Tagh Orogen curvature as a rotation orocline that formed in response to the continental collision, and controlled the anticlockwise-rotation of the mélange belt. The interaction of regional strike-slip and thrust faults are strengthened by the model. The model of orocline implies that the originally linear Qiman Tagh Orogen was subsequently bent by the collision of the asymmetric, triangular, north-blunting Kumukuri block. Following accretion to the continent, the western part of the South Qiman Tagh Terrane was transported northward along the strike-slip fault such as the Kumbyan fault. The terranes would have probably reached

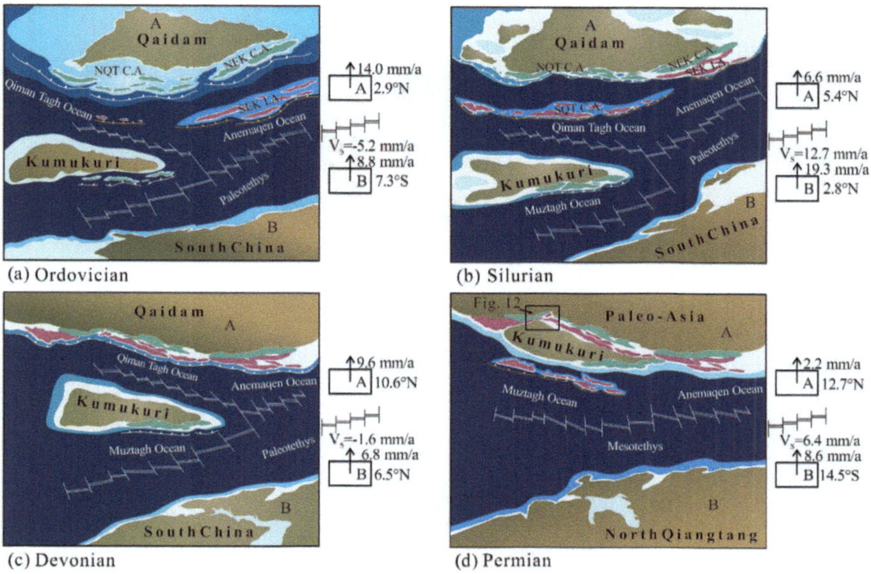

Fig. 1.12 Paleogeographic maps that showing the evolution of the Qiman Tagh Mountains during the Paleozoic: **a** Ordovician; the elements of the North Qiman Tagh Terrane are widely distributed along the Qaidam margin; Qaidam block located close to the equator and in the low Latitude region from the Sinian (Ediacaran) to the Carboniferous [136], and closed to the South China block in the south. The South China block is composed of the Yangtze and Cathaysia sub-blocks that amalgamated in the Proterozoic around 860 Ma; **b** Silurian; the North Qiman Tagh volcanic arcs accreted to the Qaidam. South Qiman Tagh island arcs were scattered in the Qiman Tagh Ocean, and separated from Qaidam by backarc basin; **c** Devonian; the North Qiman Tagh Terrane and South Qiman Tagh Terrane were assembled by the Late Devonian; The Qaidam Block had already accreted to the Alashan/Hexi Corridor terrane by the Late Devonian as a part of the Paleo-Asia [129, 161, 163]. **d** Permian; the Kumukuri Terrane northward drift into the Qiman Tagh Terranes closing the Qiman Tagh Ocean. The North Qiangtang-Qamdo-Simao block as a composite terrane had already separated from South China–Indochina by back-arc spreading since the Carboniferous [95], but Cheng et al. [21] suggested that the Qamdo and North Qiangtang block were independent of each other during the Late Carboniferous to the Early Permian, and collided in the Late Permian-Early Triassic. The paleogeographic data indicates that this block was situated at low to intermediate latitudes in the Southern Hemisphere in the Late Carboniferous-Late Permian [21]. NQT C.A. = North Qiman Tagh continent arc; SQT I.A. = South Qiman Tagh island arc; NEK C.A. = North East Kunlun continental arc; SEK C.A. = South East Kunlun island arc; Doglioni et al. [33] reviewed that the subduction velocity versus can be simply calculated by the hinge velocity V_H minus the lower plate velocity V_L ($V_S = V_H - V_L$). The motion of the hinge zone can tentatively be approximated with that of the subduction trench. The convergence rate between two plates is usually assumed to be equal to the subduction rate which is strongly dependent on the rate and migration direction of the subduction hinge [33, 34]. The subduction rate (1) increases when the hinge diverges, and (2) decreases when hinge converges. The paleomagnetic results are from Huang et al. [57], Wu et al. [153, 154], Li et al. [73], Xu et al. [166], Wang et al. [141]

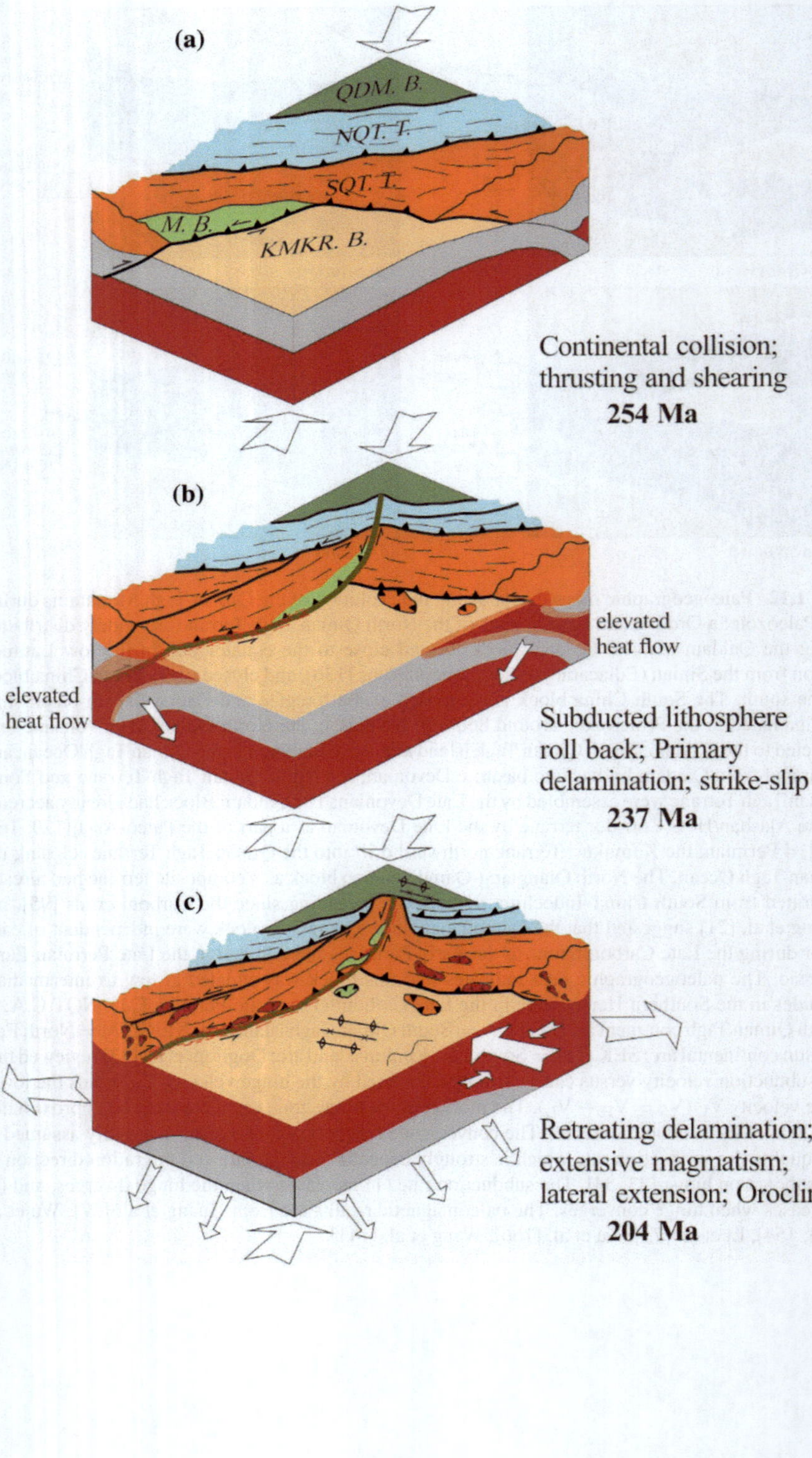
(a)
QDM. B.
NQT. T.
SQT. T.
M. B.
KMKR. B.
Continental collision; thrusting and shearing
254 Ma
(b)
elevated heat flow
elevated heat flow
Subducted lithosphere roll back; Primary delamination; strike-slip
237 Ma
(c)
Retreating delamination; extensive magmatism; lateral extension; Orocline
204 Ma

◀**Fig. 1.13** Models proposed to account for the observed orocline of the western Qiman Orogen: **a** the collision between the southern Kumukuri microblock and the South Qiman Tagh Terrane occurred in the Late Permian (before ca. 254 Ma), and coevally accompanied extensive thrusting and shearing activity; **b** following the accretion to the continent, the mélange belt (M.B.) were transported northward along the strike-slip fault beside its margin along such as the Kumbyan fault, and counterclockwise rotated. The belt would probably reach their present position in the Late Triassic. Therefore, the North Qiman Tagh Terrane were disrupted by the mélange belt. In the Early-Middle Triassic (ca. 254–237 Ma), the subducted oceanic lithosphere rolled back to initiate a delamination process, and a coeval lateral extensional shear zone were renewed along the northwestern margin of the Kumukuri block; **c** in the Late Triassic (ca. 237–204 Ma), voluminous granitoid magmatism were triggered by the elevated hear flow in response to the large delamination [139], and concentrated in the lateral extensional shearing zone

their present position in the Late Triassic. It carried the Yaziquan-Yazidaban mélange belt that rotated into its present configuration by the Late Triassic, in response to the formation of the orocline and truncation of the North Qiman Tagh Terrane. However, Dupont Nivet et al. [35] suggested that the crustal displacement is accommodated by strike-slip motion on the Altun Tagh fault rather than distributed shear within the northern Tibetan Plateau. However, the sinistral strike-slip displacement on the Altun Tagh fault is initiated in the Late Oligocene to the Earliest Miocene [19, 183]. These interpretations contradict with our concept that the bending happened much earlier in Triassic (Fig. 1.12). Moreover, both the orocline and transportation of crust would result in significant crustal shortening within the bend [43]. Whether these curvatures formed by bending of a quasi-linear belt, or as primary early Devonian arcuate features, is an unresolved question that warrants further paleomagnetic, geochronological and structural investigations.

Successive short-lived post-collisional episodes are induced by lithosphere stacking and slab breakoff, and mingling and mixing between mantle-derived magmas, with the crust-derived fluids play an important role [6]. Extensional within-plate episodes are promoted by lithosphere delamination, with increasing continental crust contamination [6]. Ueda et al. [139] proposed a model of elevated heat flow following lithosphere delamination in the East African orogen, to explain the observed widespread post-collisional granitoid magmatism (Fig. 1.13). During the Early-Middle Triassic (251–237 Ma), as the oceanic lithosphere was broken off, asthenospheric mantle materials were intruded into the boundary between lithosphere mantle and lower crust through a corridor (Fig. 1.11e). Consequently, partial melting of lower crust was initiated, producing a series of granitoids strongly depleted in Y, as garnet was retained in the restite within thickened lower crust. With increasing erosion by asthenosphere mantle, lithosphere mantle dropped into deep mantle (Fig. 1.11f). During this process, large amounts of calc-alkaline and alkaline granitoids were generated [182]. The transition from post-collisional to within-plate settings documents an important stage played by crustal anatexis in magma generation and a diminishing role of enriched OIB mantle sources [182].

The chert developed within or above the Early-Carboniferous Yaziquan ophiolite mélange belt was formed in a continental marginal island arc environment, which indicates that the closure of the Qiman Tagh paleo-ocean occurred after Early-Carboniferous [77]. The source materials are mainly derived from debris [25], rather than typical abyssal chert [185]. However, Lan et al. [63] reported early Carboniferous radiolarian cherts in the Muztag ophiolitic mélange zone in the west of the East Kunlun Mountains. Moreover, the early Permian radiolarian cherts, found in Anemaqen mélange zone in the Buqinshan area in the east of the East Kunlun Mountains, was defined as a product of marginal sea basin environment [148, 188, 189]. Therefore, the Paleo-Tethyan Ocean in the East Kunlun Mountains possibly had three branches, e.g., the Anemaqen Ocean, Muztagh Ocean and Qiman Tagh Ocean, and all of these were probably closed after Early-Permian.

The Carboniferous Dagangou Formation, Diaosu Formation and Sijiaoyanggou Formation, developed both in the South and North Qiman Tagh Terrane, are mainly composed of restricted platform carbonates and lagoon-tidal flat facies sandstone, siltstone and mudstone [99]. Their sedimentary environment evolved from offshore to epeiric sea and mainly formed in a paralic sedimentary environment [99], suggesting that the Qaidam block and its surrounding regions witnessed long-lasting subsidence [203]. This further proves that the North and South Qiman Tagh Terrane had been accreted together before the Carboniferous. After Early Permian, both the North and South Qiman Tagh Terranes lost their sedimentary records, indicating that they were uplifted after the final closure of the Paleo-Tethyan Qiman Tagh Ocean. As a consequence, we infer that the Kumukuri microcontinent collided with the South Qiman Tagh Terrane, and was incorporated into the Paleo-Asian supercontinent (Fig. 1.12).

In the South Qiman Tagh Terrane, the Upper Triassic Elashan Formation is exposed which is mainly composed of terrestrial volcanic rocks including andesite, dacite and rhyolite. The available geochemical characters show that the volcanic units formed during a post-collisional setting, and were produced by the partial melting of thickened crust, probably with minor mantle contamination. In addition, the contemporaneous Xiaojianshan gabbro dikes show a lithosphere mantle source [2]. Eventually, the Qiman Tagh orogenic process is terminated in the Late Triassic.

Paleomagnetic evidence shows that Qaidam, South China, North China and the Alashan–Hexi Corridor transition zone were all assembled together close to the equator in the Early Carboniferous, with a paleolatitude difference of ca. 10° [73, 141, 171]. By the Early Permian, the Paleo-Tethys became a wide ocean, connecting the Qiman Tagh Ocean, Anemaqen Ocean and Muztagh Ocean (Fig. 1.12) The South Qiangtang and Lhasa Blocks were still part of the Himalayan–Australian margin of Gondwana. However, by the late Early Permian, the South Qiangtang Block separated from the Gondwana supercontinent opening the Meso-Tethys behind it [94]. The Late Permian paleogeographic reconstruction indicates that the Songpan-Garze Sea was surrounded by Laurasia, the North China Block, the South China Block and the Qiangtang Terrane [57]. In addition, the Qaidam block has become a part of the Tarim block and both accreted to the Paleo-Asia craton since the Late Permian [95, 166].

1.6 Conclusion

(1) The Qiman Tagh Mountains can be divided into the North Qiman Tagh Terrane and South Qiman Tagh Terrane. They are separated by the Adatan thrust fault in the east and Baiganhu thrust fault in the west.
(2) The North Qiman Tagh Terrane, when active as continental margin, was most likely formed by the northward subduction of the Qiman Tagh Ocean lithosphere during the Early Paleozoic. During this process, an Early Paleozoic magmatic belt was formed that represents a continental island arc.
(3) The South Qiman Tagh Terrane, originally an oceanic island arc, might have been formed by intra-oceanic subduction in Early Paleozoic. It contains a series of mafic-ultramafic rocks including SSZ type ophiolites, island arc tholeiites and calc-alkaline basalts. With prolonged subduction, it evolved into a mature island arc with oceanic crust thickening, and developed a series of oceanic island arc granitoids in the Early-Middle Devonian (ca. 418–389 Ma).
(4) The collision between the South and North Qiman Tagh Terrane might have continued from the Late Silurian to the Early Devonian (ca. 422–398 Ma). The South Qiman Tagh Terrane is considered as an exotic terrane and its amalgamation is significant for the continental growth and trench seaward movement. A trench jam may have been responsible for the documented gap (ca. 357–251 Ma) in granitoid magmatism.
(5) Extensive Triassic magmatism with enhanced crustal contributions occurred within the South Qiman Tagh Terrane, which might be associated with the final closure of the Paleo-Tethyan Qiman Tagh Ocean. As a consequence, the Kumukuri microcontinent accreted into the South Qiman Tagh Terrane, following strong deformation and oroclinal bending.

References

1. Allegre CO, Courtillot V, Tapponnier P, Hirn A, Mattauer M, Coulon C, Achache J, Jaeger JJ, Scharer U, Marcoux J, Burg JP, Giradeau J, Armijo R, Gariepy C, Gopel C, Li TD, Xiao XC, Chang CF, Li GQ, Lin BY, Teng JW, Wang NW, Chen GM, Han TL, Wang XB, Deng WM, Sheng HB, Cao YG, Zhou J, Qiu HR, Bao PS, Wang SC, Wang BX, Zhou YX, Rong HX (1984) Structure and evolution of the Himalaya-Tibet orogenic belt. Nature 307:17–22
2. Ao C, Sun FY, Li BL, Wang G, Li L, Li SJ, Zhao JW (2015) U-Pb dating, geochemistry and tectonic implications of Xiaojianshan Gabbro in Qimantage Mountain, Eastern Kunlun Orogenic Belt. Geotectonica et Metallogenia 39:1176–1184
3. Barbarin B (1999) A review of the relationships between granitoid types, their origins and their geodynamic environments. Lithos 46:605–626
4. Ben-Avraham Z, Cox A (1981) Continental accretion: from oceanic plateaus to allochthonous terranes. Science 213:47–54
5. Bingen B, Nordgulen O, Viola G (2008) A four-phase model for the Sveconorwegian orogeny, SW Scandinavia. Norw J Geol 88:43–72
6. Bonin B (2004) Do coeval mafic and felsic magmas in post-collisional to within-plate regimes necessarily imply two contrasting, mantle and crustal, sources? A review. Lithos 78:1–24

7. Carter DJ, Audley-Charles MG, Barber AJ (1976) Stratigraphical analysis of island arc-continental margin collision in eastern Indonesia. J Geol Soc Lond 132:179–198
8. Castillo PR, Janney PE, Solidum RU (1999) Petrology and geochemistry of Camiguin Island, southern Philippines: insights to the source of adakites and other lavas in a complex arc setting. Contrib Mineral Petr 134:33–51
9. Chang CF, Chen NS, Coward MP, Deng WM, Dewey JF, Gansser A, Harris NB, Jin CW, Kidd WS, Leeder MR (1986) Preliminary conclusions of the Royal Society and Academia Sinica 1985 geotraverse of Tibet. Nature 323:501–507
10. Chen HW, Luo ZH, Mo XX, Zhang XT, Wang J, Wang BZ (2006) SHRIMP ages of Kayakedengtage complex in the East Kunlun Mountains and their geological implications. Acta Petrologica et Mineralogica 25:25–32
11. Chen J, Wang BZ, Li B, Zhang ZQ, Qiao BX, Jin TT (2015) Zircon U-Pb ages, geochemistry, and Sr-Nd-Pb isotopic compositions of Middle Triassic granodiorites from the Kaimuqi area, East Kunlun, Northwest China: implications for slab breakoff. Int Geol Rev 57:257–270
12. Chen J, Xie ZY, Li B, Li SP, Tan SX, Ren H, Zhang QM (2013) Geological and geochemical characteristics of the ore-bearing intrusions from the Lalingzaohuo Mo polymetallic deposit and its metallogenic significance. Geology and Exploration 49:813–824
13. Chen JL, Li DP, Li XL, Zhou XK, Wang XL, Wang XX, Dai XY, Du SX, Gao XP (2004) The discovery and the features of the Heishan ophiolite in the south margin of Qimantage Mountain, East Kunlun. Geology of Shanxi 22:35–45
14. Chen NS, Sun M, He L, Zhang KX, Wang GC (2002) Precise timing of the Early Paleozoic metamorphism and thrust deformation in the Eastern Kunlun Orogen. Chinese Sci Bull 47:1130–1133
15. Chen NS, Sun M, Wang QY, Zhang KX, Wan YS, Chen HH (2008) U-Pb dating of zircon from the central zone of the East Kunlun Orogen and its implications for tectonic evolution. Sci China, Ser D Earth Sci 51:929–938
16. Chen NS, Zhu J, Wang GC, Hou JG, Zhang KX, Zhu YH, Bai YS (1999) Metamorphic petrological features of high-grade metamorphic microlithons in Qingshuiquan region, eastern section of eastern Kunlun orogenic zone. Earth Sci-J China Univ Geosci 24:116–120
17. Chen NS, Zhu J, You ZD, Zhang KX (1998) A comparison about metamorphism among the oldest rock units from orogenic belts of Dabie, Eastern Qinling and Eastern Kunlun of the central mountain range. Earth Sci-J China Univ Geosci 23:449–455
18. Chen SJ, Li RS, Ji WH, Zhao ZM, Meng Y, Shi BD (2007) The deposition characteristics and tectono-paleogeographic environment of Kunlun orogenic belt in Late-Devonian. Geotectonica et Metallogenia 31:44–51
19. Chen Y, Gilder S, Halim N, Cogné J, Courtillot V (2002) New Mesozoic and Cenozoic data help constrain the age of motion on the Altyn Tagh fault and rotation of the Qaidam basin. Tectonics 21(5):1042
20. Chen YX, Liu CJ, Pei XZ, Li XB, Li RB, Li ZC, Chen GC, Pei L, Yang J (2013) Zircon U-Pb age, geochemical characteristics and tectonic significance of meta-volcanic rocks from Naij Tal Group, east section of East Kunlun. Earth Sci Front 20:240–254
21. Cheng X, Wu HN, Diao ZB, Wang HJ, Ma L, Zhang XD, Yang G, Hong JJ, Ji WH, Li RS (2013) Paleomagnetic data from the Late Carboniferous-Late Permian rocks in eastern Tibet and their implications for tectonic evolution of the northern Qiangtang-Qamdo block. Sci China Earth Sci 56:1209–1220
22. Condie KC (1989) Geochemical changes in basalts and andesites across the Archean-Proterozoic boundary: identification and significance. Lithos 23:1–18
23. Condie KC, Kröner A (2013) The building blocks of continental crust: Evidence for a major change in the tectonic setting of continental growth at the end of the Archean. Gondwana Res 23:394–402
24. Cui JW, Chen W, Li PW, Zhang XW, Li L (2003) Thrust propagation in the Aqqikkol lake area, the East Kunlun Mountains, northwestern China. Acta Geologica Sinica (English Edition) 77:468–478

25. Cui MH (2012) Petrogenesis of intermediate-basic igneous rocks and cherts from Yaziquan, Xinjiang Qimantag Mountain. Chinese Academy of Geological Science, Beijing, pp 1–81
26. Cui MH, Meng FC, Wu XK (2011) Early Ordovician island arc of Qimantag Mountain, eastern Kunlun: Evidences from geochemistry, Sm-Nd isotope and geochronology of intermediate-basic igneous rocks. Acta Petrol Sin 27:3365–3379
27. Defant MJ, Drummond MS (1993) Mount St. Helens: potential example of the partial melting of the subducted lithosphere in a volcanic arc. Geology 21:547–550
28. DePaolo DJ (1981) A neodymium and strontium isotopic study of the Mesozoic calc-alkaline granitic batholiths of the Sierra Nevada and Peninsular Ranges, California. J Geophys Res: Solid Earth 86:10470–10488
29. DePaolo DJ (1988) Neodymium isotope geochemistry: an introduction. Springer, Berlin, pp 1–183
30. Dewey JF, Shackleton RM, Chang CF, Sun YY (1988) The tectonic evolution of the Tibetan Plateau. Philos Trans R Soc London Ser A, Math Phys Sci 327:379–413
31. Dilek Y, Furnes H (2011) Ophiolite genesis and global tectonics: geochemical and tectonic fingerprinting of ancient oceanic lithosphere. Geol Soc Am Bull 123:387–411
32. Dilek Y, Furnes H (2014) Ophiolites and their origins. Elements 10:93–100
33. Doglioni C, Carminati E, Cuffaro M, Scrocca D (2007) Subduction kinematics and dynamic constraints. Earth-Sci Rev 83:125–175
34. Doglioni C, Panza GF (2015) Polarized plate tectonics. Adv Geophys 56:1–167
35. Dupont Nivet G, Robinson D, Butler RF, Yin A, Melosh HJ (2004) Concentration of crustal displacement along a weak Altyn Tagh fault: evidence from paleomagnetism of the northern Tibetan Plateau. Tectonics 23
36. Feng CY, Wang S, Li GC, Ma SC, Li DS (2012) Middle to Late Triassic granitoids in the Qimantage area, Qinghai Province, China: Chronology, geochemistry and metallogenic significances. Acta Petrol Sin 28:665–678
37. Frisch W, Meschede M, Blakey RC (2010) Plate tectonics: continental drift and mountain building. Springer Science & Business Media, Sturtz GmbH, Wurzburg, pp 1–207
38. Gan CH (2014) Petrology, geochemistry, U-Pb dating and Hf isotopic composition of zircons in igneous rocks from East Kunlun orogen, Qinghai. China University of Geosciences, Beijing, pp 1–75
39. Gao YB, Li K, Qian B, Li WY, Li DS, Su SS, Zhang CG, Zhang DM, Wang SM (2015) The genesis of granodiorites and dark enclaves from the Kaerqueka deposit in east Kunlun belt: evidence from zircon U-Pb dating, geochemistry and Sr-Nd-Hf isotopic compositions. Geol China 42:646–662
40. Gao YB, Li WY (2011) Petrogenesis of granites containing tungsten and tin ores in the Baiganhu deposit, Qimantage, NW China: constraints from petrology, chronology and geochemistry. Geochimica 40:324–336
41. Gao YB, Li WY, Ma XG (2012) Genesis, geochronology and Hf isotopic compositions of the magmatic rocks in Galinge iron deposit, eastern Kunlun. J Lanzhou Univ (Nat Sci) 48:36–47
42. Gao YB, Li WY, Qian B, Li K, Li DS, He SY, Zhang ZW, Zhang JW (2014) Geochronology, geochemistry and Hf isotopic compositions of the granitic rocks related with iron mineralization in Yemaquan deposit, East Kunlun, NW China. Acta Petrol Sin 30:1647–1665
43. Glen JMG (2004) A kinematic model for the southern Alaska orocline based on regional fault patterns. Spec Pap Geol Soc Am 383:161–172
44. Gradstein FM, Ogg JG, Schmitz MD, Ogg GM (2012) On the geologic time scale. Newsl Stratigr 45:171–188
45. Guo TZ, Liu R, Chen FB, Bai XD, Li HG (2011) LA-MC-ICP MS zircon U-Pb dating of Wulanwuzhuer porphyritic syenogranite in the Qimantag Mountain of Qinghai and its geological significance. Geol Bull China 30:1203–1211
46. Hao J, Liu XH, Sang HQ (2003) Geochemical characteristics and $^{40}Ar/^{39}Ar$ age of the Ayak adamellite and its tectonic significance in the East Kunlun, Xinjiang. Acta Petrol Sin 19:517–522

47. Hao NN, Yuan WM, Cao JH, Feng YL, Cheng XQ, Zhang AK (2014) The recognition of Late Devonian granodiorite and its geological implication in Kayakedengtage, East Kunlun Mountain, Qinghai-Tibet Plateau. Northwest Geol 47:107–115
48. Haschke M, Siebel W, Günther A, Scheuber E (2002) Repeated crustal thickening and recycling during the Andean orogeny in north Chile (21–26 S). J Geophys Res: Solid Earth 107
49. He BZ, Xu ZQ, Jiao CL, Cui JW, Wang SL, Wang GH, Li ZY, Qiu ZL (2009) Structural architecture and evolution of Kumkuli basin, north Tibet. J Earth Sci 20:464–476
50. He DF, Dong YP, Liu XM, Yang Z, Sun SS, Cheng B, Li W (2016) Tectono-thermal events in East Kunlun, Northern Tibetan Plateau: evidence from zircon U-Pb geochronology. Gondwana Res 30:179–190
51. He DF, Dong YP, Zhang FF, Yang Z, Sun SS, Cheng B, Zhou B, Liu XM (2016) The 1.0 Ga S-type granite in the East Kunlun Orogen, Northern Tibetan Plateau: implications for the Meso- to Neoproterozoic tectonic evolution. J Asian Earth Sci 130:46–59
52. Hofmann AW (1988) Chemical differentiation of the Earth: the relationship between mantle, continental crust, and oceanic crust. Earth Planet Sc Lett 90:297–314
53. Holbrook WS, Lizarralde D, McGeary S, Bangs N, Diebold J (1999) Structure and composition of the Aleutian island arc and implications for continental crustal growth. Geology 27:31–34
54. Hollings P (2002) Archean Nb-enriched basalts in the northern Superior Province. Lithos 64:1–14
55. Hu YD (2007) Geological characteristics and mineralizing prospective evaluation of Wulanwuzhuer copper deposit in the Eastern Kunlun orogenic belt. Qinghai province, Jilin University
56. Huang H, Niu Y, Nowell G, Zhao Z, Yu X, Zhu DC, Mo X, Ding S (2014) Geochemical constraints on the petrogenesis of granitoids in the East Kunlun Orogenic belt, northern Tibetan Plateau: Implications for continental crust growth through syn-collisional felsic magmatism. Chem Geol 370:1–18
57. Huang KN, Opdyke ND, Peng XJ, Li JG (1992) Paleomagnetic results from the Upper Permian of the eastern Qiangtang Terrane of Tibet and their tectonic implications. Earth Planet Sci Lett 111:1–10
58. Ikeda Y, Nagao K, Ishii T, Matsumoto D, Stern RJ, Kagami H, Arima M, Bloomer SH (2016) Contributions of slab fluid and sediment melt components to magmatism in the Mariana Arc-Trough system: Evidence from geochemical compositions and Sr, Nd, and noble gas isotope systematics. Isl Arc 25:253–273
59. Ishizuka O, Tani K, Reagan M, Kanayama K, Umino S, Harigane Y (2014) Izu-Bonin-Mariana forearc crust as a modern ophiolite analogue. Elements 10:115–120
60. Jacobsen SB, Wasserburg GJ (1980) Sm-Nd isotopic evolution of chondrites. Earth Planet Sci Lett 50:139–155
61. Ji Q (1997) Discovery of an Early Cambrian small-shelly fauna in the central sector of the East Kunlun Mountains, Qinghai, and its geological significance. Reg Geol China 4:428–431
62. Jolly WT, Lidiak EG, Dickin AP, Wu T (2001) Secular geochemistry of central Puerto Rican island arc lavas: constraints on Mesozoic tectonism in the eastern Greater Antilles. J Petrol 42:2197–2214
63. Lan CL, Wu J, Li JL, Yu LJ, Li HS, Wang YT (2001) Discovery of early carboniferous radiolarians in Muztag ophiolitic mélange, East Kunlun, Xinjiang. Sci Geol Sin 37:104–106
64. Li CS, Zhang ZW, Li WY, Wang YL, Sun T, Ripley EM (2015) Geochronology, petrology and Hf–S isotope geochemistry of the newly-discovered Xiarihamu magmatic Ni–Cu sulfide deposit in the Qinghai-Tibet plateau, western China. Lithos 216:224–240
65. Li DP, Fan J, Xiao AF, Zhou XK, Li XL, Du SX (2002) Discovery of Early Silurian graptolite fossils in the Qimantag Group in the western sector of the East Kunlun. Geol Bull China 21:136–139
66. Li DP, Li J, Zhang HJ, Li XL, Zhou XK, Du SX (2003) The turbidite of the Silurian Baiganhu formation in the Qimantage Mountain, eastern Kunlun. Geol Shanxi 21:39–44

67. Li DP, Xiao AF (2010) Magmatic sequence and tectonic setting of Bashierxi Granite in W-Sn deposit of Baiganhu Area, the western sector of Qimantage Mountains. Northwest Geol 43:53–61
68. Li GC (2012) The metallogenic model and evolution of structural-fluid study of Baigan lake W-Sn Deposit, Ruoqiang, Xinjiang, Province. Chinese Academy of Geological Sciences, Beijing, pp 1–79
69. Li GC, Feng CY, Wang RJ, Ma SC, Li HM, Zhou AS (2012) SIMS zircon U-Pb age, petrochemistry and tectonic implications of granitoids in northeastern Baiganhu W-Sn orefield, Xinjiang. Acta Geosci Sin 33:216–226
70. Li HK, Lu SN, Xiang ZQ, Zhou HY, Guo H, Song B, Zheng JK, Gu Y (2006) SHRIMP U-Pb zircon age of the granulite from the Qingshuiquan area, Central Eastern Kunlun Suture Zone. Earth Sci Front 13:311
71. Li HM, Shi YD, Liu Z, Wang BJ, Wang ZL (2006) Geological features and origin of the Baigan Lake W-Sn deposit in the Ruoqiang area, East Kunlun Mountains, China. Geol Bull China 25:277–281
72. Li K, Gao YB, Qian B, He SY, Liu YL, Zhang ZW, Zhang JW, Wang YL (2015) Geochronology, geochemical characteristics and Hf isotopic compositions of granite in the Hutouya deposit, Qimantag, East Kunlun. Geol China 42:630–645
73. Li PW, Gao R, Cui JW, Guan Y (2004) Paleomagnetic analysis of eastern Tibet: implications for the collisional and amalgamation history of the Three Rivers region, SW China. J Asian Earth Sci 24:291–310
74. Li SJ, Sun FY, Feng CY, Liu ZH, Zhao JW, Li YC, Wang S (2008) Geochronological study on Yazigou polymetallic deposit in Eastern Kunlun, Qinghai Province. Acta Geol Sin 20:949–955
75. Li SZ, Yang Z, Zhao SJ, Li XY, Suo YH, Guo LL, Yu S, Dai LM, Li SJ, Mu DL (2016) Global Early Paleozoic orogens (II): subduction-accretionary-type orogeny. J Jilin Univ (Earth Sci Ed) 46:968–1014
76. Li W, Neubauer F, Liu YJ, Genser J, Ren SM, Han GQ, Liang CY (2013) Paleozoic evolution of the Qimantagh magmatic arcs, Eastern Kunlun Mountains: Constraints from zircon dating of granitoids and modern river sands. J Asian Earth Sci 77:183–202
77. Li YJ, Jia CZ, Hao J, Wang ZM, Zheng DM, Peng GX (2000) The discovery of radiolarian fauna from Tieshida formation in the East Kunlun. Chin Sci Bull 45:205–208
78. Liao QY (2010) Sedimentary characteristics of carboniferous in margin of the Qaidam Basin and its effect on source rocks. China University of Petroleum, East China, pp 1–91
79. Liu B, Ma CQ, Zhang JY, Xiong FH, Huang J, Jiang HA (2012) Petrogenesis of Early Devonian intrusive rocks in the east part of Eastern Kunlun Orogen and implication for Early Palaeozoic orogenic processes. Acta Petrol Sin 28:1785–1807
80. Liu C (2003) Pb-Sr-Nd-O Isotope Characteristics of Granitoids in East Kunlun Orogenic Belt. Acta Geosci Sin 24:584–588
81. Liu GC, Xu GR (1984) Brachiopod faunae of Upper Devonian in Haerzha region, Qinghai Province. Earth Sci-J Wuhan Coll Geol 26
82. Liu GZ, Zhang YX, Xue JQ, Wu GD, Chen Y (2014) Zircon LA-ICP MS U-Pb dating and geochemistry of basement granites from north Kunlun faults zone, western Qaidam Basin and their geological implications. Acta Petrol Sin 30:1615–1627
83. Liu HT (2005) Petrology, geochemistry and geochronology of late Triassic volcanics, Kunlun orogenic belt, western China: implications for tectonic setting and petrogenesis. Geochem J 39:1–20
84. Liu YC, Ye ZF (1998) A new cognition on high grade-metamorphic rocks in Jinshuikou Area, Eastern Kunlun. Qinghai Geol 18–26
85. Liu YH, Mo XX, Yu XH, Zhang XT, Xu GW (2006) Zircon SHRIMP U-Pb dating of the Jingren granite, Yemaquan region of the east Kunlun and its geological significance. Acta Petrol Sin 22:2457–2463
86. Lu JP, Li J, Qin XF, Xu H, Zhou FS, Li YK, Tang ZH, Tang JH (2005) The Yinieke'agan granite mass in Qimantag, eastern Kunlun and its tectonic significance. Sediment Geol Tethyan Geol 25:46

87. Lu L, Wu ZH, Hu DG, Barosh PJ, Hao S, Zhou CJ (2010) Zircon U-Pb age for rhyolite of the Maoniushan Formation and its tectonic significance in the East Kunlun Mountains. Acta Petrol Sin 26:1150–1158
88. Lu SN, Chen ZH, Li HK, Hao GJ (2005) Two Magmatic belts of the Neoproterozoic in the Qinling orogenic belt. Acta Geol Sin 79:165–173
89. Lu SN, Li HC, Wang HC, Chen ZH, Zheng JK, Xiao ZQ (2009) Detrital zircon population of Proterozoic meta-sedimentary strata in the Qinling-Qilian-Kunlun Orogen. Acta Petrol Sin 25:2195–2208
90. Luo ZH, Deng JF, Cao YQ, Guo ZF, Mo XX (1999) On late Paleozoic-early Mesozoic volcanism and regional tectonic evolution of Eastern Kunlun, Qinghai Province. Geoscience 13:51–56
91. Meng FC, Cui MH, Wu XK, Ren YF (2015) Heishan mafic-ultramafic rocks in the Qimantag area of Eastern Kunlun, NW China: remnants of an Early Paleozoic incipient island arc. Gondwana Res 27:745–759
92. Meng FC, Cui MH, Wu XK, Wu JF, Wang JH (2013) Magmatic and metamorphic events recorded in granitic gneisses from the Qimantag, East Kunlun Mountains, Northwest China. Acta Petrol Sin 29:2107–2122
93. Meng FC, Zhang JX, Cui MH (2013) Discovery of Early Paleozoic eclogite from the East Kunlun, Western China and its tectonic significance. Gondwana Res 23:825–836
94. Metcalfe I (2002) Permian tectonic framework and palaeogeography of SE Asia. J Asian Earth Sci 20:551–566
95. Metcalfe I (2013) Gondwana dispersion and Asian accretion: tectonic and palaeogeographic evolution of eastern Tethys. J Asian Earth Sci 66:1–33
96. Mo X, Luo Z, Deng J, Yu X, Liu C, Yuan W, Bi X (2007) Granitoids and Crustal Growth in the East-Kunlun Orogenic Belt. Geol J China Univ 13:403–414
97. Mo XX, Hou ZQ, Niu YL, Dong GC, Qu XM, Zhao ZD, Yang ZM (2007) Mantle contributions to crustal thickening during continental collision: evidence from Cenozoic igneous rocks in southern Tibet. Lithos 96:225–242
98. Nelson KD (1992) Are crustal thickness variations in old mountain belts like the Appalachians a consequence of lithospheric delamination? Geology 20:498
99. Niu YB, Zhong JH, Duan HL, Yin CM, Wang PJ (2010) Relationship between Carboniferous sedimentary facies and source rock in Qaidam Basin. Acta Sedimentol Sin 28:140–149
100. Niu YL (2003) Initiation of subduction zones as a consequence of lateral compositional buoyancy contrast within the lithosphere: a petrological perspective. J Petrol 44:851–866
101. Niu YL, Liu Y, Xue QQ, Shao FL, Chen S, Duan M, Guo PY, Gong HM, Hu Y, Hu ZX (2015) Exotic origin of the Chinese continental shelf: new insights into the tectonic evolution of the western Pacific and eastern China since the Mesozoic. Sci Bull 60:1598–1616
102. Nohda S (1984) Classification of island arcs by Nd-Sr isotopic data. Geochem J 18:1–9
103. Owens TJ, Zandt G (1997) Implications of crustal property variations for models of Tibetan plateau evolution. Nature 387:37–43
104. Pan GT, Li XZ, Wang LQ, Ding J, Chen ZL (2002) Preliminary division of tectonic units of the Qinghai-Tibet Plateau and its adjacent regions. Geol Bull China 21:701–707
105. Pan GT, Wang LQ, Li RS, Yuan SH, Ji WH, Yin FG, Zhang WP, Wang BD (2012) Tectonic evolution of the Qinghai-Tibet plateau. J Asian Earth Sci 53:3–14
106. Pan YS, Zhou WM, Xu RH (1996) The early Paleozoic geological features and evolutions of the Kunlun Mountains. Sci China (Ser D) 26:302–307
107. Pan YS, Zhou WM, Xu RH, Wang DA, Zhang YQ, Xie YW, Chen TE, Luo H (1996) Geological characteristics and evolution of the Kunlun Mountains region during the early Paleozoic. Sci China (Ser D) 4:337–347
108. Pearce JA (1982) Trace element characteristics of lavas from destructive plate boundaries. Andesites 528–548
109. Pearce JA (2014) Immobile element fingerprinting of ophiolites. Elements 10:101–108
110. Pearce JA, Harris NB, Tindle AG (1984) Trace element discrimination diagrams for the tectonic interpretation of granitic rocks. J Petrol 25:956–983

111. Peng B, Sun F, Li B, Wang G, Li S, Zhao T, Li L, Zhi Y (2016) The geochemistry and geochronology of the Xiarihamu II mafic–ultramafic complex, Eastern Kunlun, Qinghai Province, China: implications for the genesis of magmatic Ni–Cu sulfide deposits. Ore Geol Rev 73:13–28
112. Peng Y, Ma YS, Liu CL, Sun JP, Hu ZY, Mou HL, Zheng C (2016) SHRIMP zircon ages of the Dagangou volcanic rocks in the Eastern Kunlun orogenic belt and their implications. Geol Bull China 35:356–363
113. Polat A, Kerrich R (2001) Magnesian andesites, Nb-enriched basalt-andesites, and adakites from late-Archean 2.7 Ga Wawa greenstone belts, Superior Province, Canada: implications for late Archean subduction zone petrogenetic processes. Contrib Miner Petrol 141:36–52
114. Qi SS (2015) Petrotectonic assemblages and tectonic evolution of the East Kunlun orogenic belt in Qinghai Province. China University of Geosciences, Beijing
115. Qi SS, Deng JF, Ye ZF, Liu R, Wang GL (2013) LA-ICP-MS zircon U-Pb dating of Late Devonian diabase dike swarms in Qimantag area. Geol Bull China 32:1385–1393
116. Qi SS, Song SG, Shi LC, Cai HJ, Hu JC (2014) Discovery and its geological significance of Early Paleozoic eclogite in Xiarihamu-Suhaitu area, western part of the East Kunlun. Acta Petrol Sin 30:3345–3356
117. Qian B, Gao YB, Li K, Zhang ZW, Zhou AS, Wu YS (2015) Zircon U-Pb-Hf isotopes and whole rock geochemistry constraints on the petrogenesis of iron-rare metal mineralization related alkaline granitic intrusive rock in Yugouzi area, eastern Kunlun, Xinjiang. Acta Petrol Sin 31:2508–2520
118. Rapp RP, Shimizu N, Norman MD (2003) Growth of early continental crust by partial melting of eclogite. Nature 425:605–609
119. Rapp RP, Watson EB (1995) Dehydration melting of metabasalt at 8–32 kbar: implications for continental growth and crust-mantle recycling. J Petrol 36:891–931
120. Royden LH, Burchfiel BC, van der Hilst RD (2008) The geological evolution of the Tibetan Plateau. Science 321:1054–1058
121. Rudnick RL, Fountain DM (1995) Nature and composition of the continental crust: a lower crustal perspective. Rev Geophys 33:267–309
122. Sajona FG, Maury RC, Bellon H, Cotten J, Defant M (1996) High Field Strength Element Enrichment of Pliocene-Pleistocene Island Arc Basalts, Zamboanga Peninsula, Western Mindanao (Philippines). J Petrol 37:693–726
123. Santosh M, Teng XM, He XF, Tang L, Yang QY (2016) Discovery of Neoarchean suprasubduction zone ophiolite suite from Yishui Complex in the North China Craton. Gondwana Res 38:1–27
124. Sengör AM (1979) Mid-Mesozoic closure of Permo-Triassic Tethys and its implications. Nature 279:590–593
125. She HQ, Zhang DQ, Jing XY, Guan J, Zhu HP, Feng CY, Li DX (2007) Geological characteristics and genesis of the Ulan Uzhur porphyry copper deposit in Qinghai. Geol China 2:013
126. Shen YC, Yang JZ, Liu TB, Li GM, Zeng QD (2000) Ophiolite found in the Qimantag area, East Kunlun Mountains, Xinjiang. Geol Rev 46:90
127. Shi XD, Gou J, Liao XS (1985) The Devonian strata in the mountains around Qaidam basin, Qinghai Province. Northwest Geol 1–17
128. Song SG (2015) Exotic origin of the Chinese continental shelf. Sci Bull 60:1625
129. Song SG, Niu YL, Su L, Xia XH (2013) Tectonics of the North Qilian orogen, NW China. Gondwana Res 23:1378–1401
130. Song TZ, Zhao HX, Zhang WK, Bai XD, An SW, Yang M (2010) The geological features of Shizigou ophiolites in Qimantage area. Northwestern Geol 43:124–133
131. Stern RJ (1994) Arc assembly and continental collision in the Neoproterozoic East African Orogen: implications for the consolidation of Gondwanaland. Earth Planet Sci 22:319–351
132. Suda Y (2004) Crustal anatexis and evolution of granitoid magma in Permian intra-oceanic island arc, the Asago body of the Yakuno ophiolite, Southwest Japan. J Miner Pet Sci 99:339–356

133. Sun SS, McDonough WF (1989) Chemical and isotopic systematics of oceanic basalts: implications for mantle composition and processes. Geol Soc Lond Spec Publ 42:313–345
134. Tan SX, Bai YS, Chang GH, Tong HK, Bao GP (2004) Discovery and geological significance of metamorphic and intrusive rock (system) of Qimantage region in Jinning epoch. Northwest Geol 1:012
135. Tan SX, Guo TZ, Dong JS, Chang YY, Ma W (2011) Geological characteristics and significance of the peraluminous granite in Late Silurian epoch in Wulanrvuzhuer region of Qinghai. J Qinghai Univ (Nat Sci) 29:36–43
136. Tang LJ, Jin ZJ, Zhang ML, Liu CY, Wu HN, You FB, Zhang BS (2000) An analysis on tectono-paleogeography of the Qaidam Basin, NW China. Earth Sci Front 7:420–429
137. Taylor SR, Mclennan SM (1995) The geochemical evolution of the continental crust. Rev Geophusics 33:241–265
138. Turner S, Hawkesworth C, Liu J, Rogers N, Kelley S, van Calsteren P (1993) Timing of Tibetan uplift constrained by analysis of volcanic rocks. Nature 364:50–54
139. Ueda K, Jacobs J, Thomas RJ, Kosler J, Horstwood MSA, Wartho JA, Jourdan F, Emmel B, Matola R (2012) Postcollisional high-grade metamorphism, orogenic collapse, and differential cooling of the East African Orogen of Northeast Mozambique. J Geol 120:507–530
140. Vroon PZ, Bergen MJV, White WM, Varekamp JC (1993) Sr-Nd-Pb isotope systematics of the Banda Arc, Indonesia: Combined subduction and assimilation of continental material. J Geophys Res Solid Earth 98:22349–22366
141. Wang B, Zhang GW, Li SZ, Yang ZY, Roberts AP, Zhao Q, Wang ZY (2016) Early Carboniferous paleomagnetic results from the northeastern margin of the Qinghai-Tibetan plateau and their implications. Gondwana Res 36:44–51
142. Wang BZ (2011) The study and investigation on the assembly and coupling petrotectonic assemblage during Paleozoic–Mesozoic period at Qimantage geological corridor domain. China University of Geosciences, Beijing, pp 1–230
143. Wang BZ, Luo ZH, Pan T, Song TZ, Xiao PX (2012) Petrotectonic assemblages and LA-ICP-MS zircon U-Pb age of Early Paleozoic volcanic rocks in Qimantag area, Tibetan Plateau. Geol Bull China 31:860–874
144. Wang C, Liu L, Xiao P, Cao Y, Yu H, Meert JG, Liang W (2014) Geochemical and geochronologic constraints for Paleozoic magmatism related to the orogenic collapse in the Qimantagh-South Altyn region, northwestern China. Lithos 202:1–20
145. Wang GL, Ren EF, Bao GP, Li JF, Chang YY (2013) U-Pb age of zircons of Qimantage Late-Devonian intrusive rocks geochemical characteristics and its geological significance, Qinghai. Miner Explor 4:421–434
146. Wang XL, Gao XP, Liu YQ, Li XL, Zhou XK, Cui JG, Dai XY, He LX, Zeng ZC (2010) The characteristics of Heishan structural ophiolite belt in the west part of East Kunlun. Northwestern Geol 43:219–231
147. Wang XL, Gao XP, Liu YQ, Li XL, Zhou XK, Du SX, Cui JG, Dai XY, Zeng ZC (2010) Revision of the Qimantage Group in west part of East Kunlun. Northwestern Geol 43:218–234
148. Wang YB, Yang H (2004) Middle Permian paleobiogeography study in east Kunlun, A'nyêmaqên and Bayan Har. Sci China, Ser D Earth Sci 47:1120–1126
149. Wang ZZ, Han BF, Feng CY, Li GC (2014) Geochronology, geochemistry and tectonic significance of granites in Baiganhu area, Xinjiang. Acta Petrol Mineral 33:597–616
150. Weil AB, Sussman AJ (2004) Classifying curved orogens based on timing relationships between structural development and vertical-axis rotations. Geol Soc Am Spec Pap 383:1–17
151. White WM, Patchett J (1984) Hf-Nd-Sr isotopes and incompatible element abundances in island arcs: implications for magma origins and crust-mantle evolution. Earth Planet Sc Lett 67:167–185
152. Wu C, Yin A, Zuza AV, Zhang JY, Liu WC, Ding L (2016) Pre-Cenozoic geologic history of the central and northern Tibetan Plateau and the role of Wilson cycles in constructing the Tethyan orogenic system. Lithosphere 8
153. Wu HN, Liu CY, Zhang XH, Ren ZL (1997) Paleomagnetic data and their implications for tectonic evolution of the Qaidam block. Sci China (Ser D) 27:9–14

154. Wu HN, Zhu RX, Bai LX, Guo B, Lv JJ (1998) Revised apparent polar wander path of the Yangtze Block and its tectonic implications. Sci China (Ser D) 41:78–90
155. Wu SP, Wu CL, Wang MY, Chen QL, Wooden JL (2008) SHRIMP U-Pb zircon dating of the Tula granite pluton on the south side of the Altun Fault and its geological implications. Acta Geol Sin (Engl Ed) 82:409–414
156. Wu XK, Meng FC (2011) Zircon U-Pb dating, geochemistry and Nd-Hf isotopic compositions of the Maxingdaban Late Triassic granitic pluton from Qimantag in the eastern Kunlun. Acta Petrol Sin 27:3380–3394
157. Xi RG, Xiao PX, Wu YZ (2010) The Geological significances, composition and age of the monzonitic granite in Kendekeke iron mine. Northwestern Geol 43:195–202
158. Xia R, Wang C, Deng J, Carranza EJM, Li W, Qing M (2014) Crustal thickening prior to 220Ma in the East Kunlun Orogenic Belt: Insights from the Late Triassic granitoids in the Xiao-Nuomuhong pluton. J Asian Earth Sci 93:193–210
159. Xiao AF (2005) Yaziquan Silurian volcanic rocks in western Qimantage Mountain of eastern Kunlun. Geol Shaanxi 23:50–60
160. Xiao WJ, Huang BC, Han CM, Sun S, Li JL (2010) A review of the western part of the Altaids: a key to understanding the architecture of accretionary orogens. Gondwana Res 18:253–273
161. Xiao WJ, Kröner A, Windley B (2009) Geodynamic evolution of Central Asia in the Paleozoic and Mesozoic. Int J Earth Sci 98:1185–1188
162. Xiao WJ, Windley BF, Badarch G, Sun S, Li JL, Qin KZ, Wang ZH (2004) Palaeozoic accretionary and convergent tectonics of the southern Altaids: implications for the growth of Central Asia. J Geol Soc Lond 161:339–342
163. Xiao WJ, Windley BF, Yong Y, Yan Z, Yuan C, Liu CZ, Li JL (2009) Early Paleozoic to Devonian multiple-accretionary model for the Qilian Shan, NW China. J Asian Earth Sci 35:323–333
164. Xiao Y, Feng CY, Liu JN, Yu M, Zhou JH, Li DX, Zhao YM (2013) LA-MC-ICP-MS zircon U-Pb dating and sulfur isotope characteristics of Kendekeke Fe-polymetallic deposit, Qinghai Province. Miner Depos 32:177–186
165. Xiong FH, Ma CQ, Wu L, Jiang HA, Liu B (2015) Geochemistry, zircon U-Pb ages and Sr-Nd-Hf isotopes of an Ordovician appinitic pluton in the East Kunlun orogen: new evidence for Proto-Tethyan subduction. J Asian Earth Sci 111:681–697
166. Xu W, Sun ZM, Pei JL, Pan JW, Li HB, Li AZ, Zhao Y, Ren XJ (2011) New Late Permian paleomagnetic results from Qaidam block and tectonic implications. Acta Petrol Sin 27:3479–3486
167. Xu ZQ, Wang Q, Pêcher A, Liang FH, Qi XX, Cai ZH, Li HQ, Zeng LS, Cao H (2013) Orogen-parallel ductile extension and extrusion of the Greater Himalaya in the late Oligocene and Miocene. Tectonics 32:191–215
168. Xu ZQ, Yang JS, Li HB, Zhang JX, Zeng LS, Jiang M (2006) The Qinghai-Tibet plateau and continental dynamics: a review on terrain tectonics, collisional orogenesis, and processes and mechanisms for the rise of the plateau. Geol China 33:221–238
169. Xu ZQ, Yang JS, Li WC, Li HQ, Cai ZH, Yan Z, Ma CQ (2013) Paleo-Tethys system and accretionary orogen in the Tibet Plateau. Acta Petrol Sin 29:1847–1860
170. Xue LM (1986) The Upper-Devonian strata in the Haerzha area, north slope of the Qimantage Mountains. Northwestern Geol 1–6
171. Yang HX, Yu HM, Li PW (1992) Paleomagnetic study of Qaidam plate and its evolution. J Chang Univ Earth Sci 22:420–426
172. Yang JS, Chai YC, Feng BG (1987) The basic volcanic rocks of Wanbaogou Group in the Eastern Kunlun Mountains and their geological significance. Acta Petrol Mineral 6:121–130
173. Yang JZ, Shen YC, Li GM, Liu TB, Zeng QD (1999) Basic features and its tectonic significance of Yaziquan ophiolite belt in eastern Kunlun orogenic belt, Xinjiang. Geoscience 13:309–314
174. Yang JZ, Shen YC, Liu TB (2000) Tectonic environment analysis on volcanic rocks from Qimantage group, East Kunlun orogenic belt, Xinjiang. Xinjiang Geol 18
175. Yang QY, Santosh M, Collins AS, Teng XM (2016) Microblock amalgamation in the North China Craton: Evidence from Neoarchean magmatic suite in the western margin of the Jiaoliao Block. Gondwana Res 31:96–123

176. Yao L (2015) Petrogenesis of the Triassic granitoids and skarn mineralization in the Qimantag area, Qinghai Province, and their geodynamic setting. China University of Geosciences (Beijing), Beijing, pp 1–172
177. Yao L, Lü Z, Yu X, Pang Z, Cai Y, Liu P, Liu C, Wang F (2015) Petrogenesis, geochemistry and zircon U-Pb age of the granite from No. VI section of Hutouya deposit, Qimantag area, Qinghai province, and its geological significance. J Jilin Univ (Earth Sci Ed) 45:743–758
178. Yao L, Lü ZC, Zhao CS, Pang ZS, Yu XF, Zhu XY, Li YS, Liu P, Li ST, Zhang MC (2016) Geochronological study of granitoids from the Niukutou and B section of the Kaerqueka deposits, Qimantag area, Qinghai Province: Implications for Devonian magmatism and mineralization. Geol Bull China 35:1158–1169
179. Yin A, Dang Y, Zhang M, McRivette MW, Burgess WP, Chen X (2007) Cenozoic tectonic evolution of Qaidam basin and its surrounding regions (part 2): wedge tectonics in southern Qaidam basin and the Eastern Kunlun Range. Geol Soc Am Spec Pap 433:369–390
180. Yin A, Harrison TM (2000) Geologic evolution of the Himalayan-Tibetan orogen. Annu Rev Earth Planet Sci 28:211–280
181. Yin JX, Xu JT, Liu CJ, Li H (1988) The Tibetan plateau: regional stratigraphic context and previous work. Philos Trans R Soc Lond A: Math, Phys Eng Sci 327:5–52
182. Yu M, Feng CY, Zhao YM, Li DX (2015) Genesis of post-collisional calc-alkaline and alkaline granitoids in Qiman Tagh, East Kunlun, China. Lithos 239:45–59
183. Yue YJ, Ritts BD, Graham SA (2001) Initiation and long-term slip history of the Altyn Tagh fault. Int Geol Rev 43:1087–1093
184. Zhang AK (2012) Studies on late Paleozoic-early Mesozoic magmatism and mineralization in Yemaquan area, Qinghai Province. China University of Geosciences, Beijing, pp 1–155
185. Zhang B, Sun XC, Guo B, Bao ZJ (2014) Geological characteristics and significance of Yaziquan-Yazidaban tectonic melange in East Kunlun area of Xinjiang. Northwestern Geol 47:95–104
186. Zhang JM, Wang QY, Xu HQ, Yan YF (2012) Discovery and geological significance of metamorphic and intrusive rock (system) of Kayakedengtage region intruded into Jinshuikou Paleo-Proterozoic rock. Northwestern Geol 45:13–19
187. Zhang JX, Meng FC, Wan YS, Yang JS, Dong GA (2003) Early Paleozoic tectono-thermal event of the Jinshuikou Group on the southern margin of Qaidam: Zircon U-Pb SHRIMP age evidence. Geol Bull China 22:397–404
188. Zhang KX, Huang JC, Luo MS, Zhang TP, Bai YS (1999) Sedimentary geochemical features of Animaqing melange zone in eastern Kunlun Mountains. Earth Sci-J China Univ Geosci 24:111–115
189. Zhang KX, Huang JC, Yin HF, Wang GC, Wang YB, Feng QL (2000) Application of radiolarians and other fossils in non-Smith strata. Sci China, Ser D Earth Sci 43:364–374
190. Zhang Q, Jin WJ, Wang YL, Li CD, Wang Y, Jia XQ (2006) Ocean lithosphere delamination and the lower crust delamination: the different mechanism and geological implication: Comment on the delamination model of lower crust and lithosphere mantle. Acta Petrol Sin 22:2631–2638
191. Zhang XF, Li ZM, Jia QZ, Song ZB, Chen XY, Zhang YL, Li DS, Shu XF (2016) Geochronology, geochemistry and geological significance of granite porphyry in Hutouya polymetallic deposit, Qimantage area, Qinghai province. J Jilin Univ 46:749–765
192. Zhang Y, Zhao Y, Heng LI, Zhang S, Zhou Q, Xiong Y (2016) Zircon U-Pb dating and geological significance of monzonitic granite of Kayakedeng zinc polymetallic deposit in Qinghai. Contrib Geol Miner Resour Res 31:47–54
193. Zhang YL, Hu DG, Shi YR, Lu L (2010) SHIRMP zircon U-Pb ages and tectonic significance of Maoniushan Formation volcanic rocks in East Kunlun orogenic belt, China. Geol Bull China 29:1614–1618
194. Zhang ZC, Zhou G, Kusky TM, Yan SH, Chen BL, Zhao L (2009) Late Paleozoic volcanic record of the Eastern Junggar terrane, Xinjiang, Northwestern China: Major and trace element characteristics, Sr-Nd isotopic systematics and implications for tectonic evolution. Gondwana Res 16:201–215

195. Zhang ZW, Tang QY, Li CS, Wang YL, Ripley EM (2016) Sr-Nd-Os-S isotope and PGE geochemistry of the Xiarihamu magmatic sulfide deposit in the Qinghai-Tibet plateau, China. Miner Deposita 52:51–68
196. Zhao ZM, Ma HD, Wang BZ, Bai YS, Li RS, Ji WH (2008) The evidence of intrusive rocks about collision-orogeny during Early Devonian in Eastern Kunlun area. Geol Rev 54:1–11
197. Zheng Z, Chen YJ, Deng XH, Yue SW, Chen HJ (2016) Muscovite $^{40}Ar/^{39}Ar$ dating of the Baiganhu W-Sn orefield, Qimantag, East Kunlun Mountains, and its geological implications. Geol China 43:1341–1352
198. Zhou B, Dong YP, Zhang FF, Yang Z, Sun SS, He DF (2016) Geochemistry and zircon U-Pb geochronology of granitoids in the East Kunlun Orogenic Belt, northern Tibetan Plateau: origin and tectonic implications. J Asian Earth Sci 130:265–281
199. Zhou CJ, Hu DG, Barosh PJ, Wu ZH, Zhang YQ, Geng JZ, Hao S, Ni JY, Zhang YL (2010) Zircon U-Pb dating of the rhyolite-dacite porphyry in the Sandaowan of East Kunlun Mountains and its geological significance. J Geomech 16:28–35
200. Zhou JH, Feng CY, Li DX, Li GC (2015) Geological, geochemical, and geochronological characteristics of Caledonian W-Sn mineralization in the Baiganhu orefield, southeastern Xinjiang, China. Ore Geol Rev 75:125–149
201. Zhou JH, Feng CY, Shen DL, Li D, Hui W, Zhang MY, Ma S (2015) Geochronology, geochemistry and tectonic implications of granodiorite in the northwest of Weibao deposit, Xinjiang Qimantage. Acta Geol Sin 89:473–486
202. Zhu DC, Zhao ZD, Niu YL, Dilek Y, Hou ZQ, Mo XX (2013) The origin and pre-Cenozoic evolution of the Tibetan Plateau. Gondwana Res 23:1429–1454
203. Zhu RK, Xu HX, Deng SH, Li GH (2007) Lithofacies palaeogeography of the Carboniferous in northern China. J Palaeogeogr 9:13–24
204. Zhu YH, Lin QX, Jia CX, Wang GC (2006) SHRIMP zircon U-Pb age and significance of Early Paleozoic volcanic rocks in East Kunlun orogenic belt, Qinghai Province, China. Sci China Ser D 49:88–96

Chapter 2
Skarn Zonation and Mineral Geochemistry of the Galinge Skarn Deposit

The Galinge deposit, the largest skarn Fe-polymetallic deposit in Qinghai province, is located in the center of the Qiman Tagh ore belt in northwestern China (Fig. 2.1) [34]. Qiman Tagh, which extends from Golmud in the east to the Altun strike-slip fault in the west (Fig. 2.1), is one of the most prolific porphyry-skarn metallogenic belts in western China [19, 31, 42]. Many studies of this porphyry-skarn deposit have been conducted during the last decade, suggesting that the single prolonged porphyry-skarn mineralization is closely related to a complex series of Middle- to Late-Triassic (ca. 237–204 Ma) granitic stocks and dikes in Qiman Tagh [3, 21, 42]. Yu et al. [35, 39] distinguished two types of granitic rocks within the Qiman Tagh belt: early emplaced calc-alkaline granodiorite and monzogranite, and later alkaline syenogranite intrusions. Both kinds of intrusions are related to mineralization. Some researchers proposed that mantle-crustal mixing was the primary petrogenetic mechanism for the voluminous Triassic Qiman Tagh intermediate to felsic intrusions [18, 20]. Yu et al. [35, 38] suggested that partial melting of the thickened lower crust may have been a more probable alternative.

Skarn deposits are variably zoned, with a common zonation pattern of magnesian, calcic and manganesian skarn from the proximal to distal zones [1, 23, 41]. Yu et al. [34] showed that the Galinge skarn iron deposit has the typical magnesium-skarn (Mg-SK), calcium-skarn (Ca-SK) and manganese-skarn (Mn-SK) zonation. In general, calcium skarns are associated with Fe-rich plutons intruding into limestone or volcanic rocks, magnesium skarns mostly occur in dolomitic rocks, and manganese-rich skarns are common along the lithologic contacts or structures of carbonate wall rocks distal from the intrusive contact zone and are associated with Pb–Zn–(Ag) mineralization [41, 43].

Magnesian and calcic skarn can be used to describe the dominant composition of the protolith and resulting skarn minerals [25]. The composition and texture of the protolith tend to control the composition and texture of the resulting skarn. Calcic iron skarns are associated with iron-rich plutons intruded into limestone and volcanic wall rocks, consisting dominantly of garnet and pyroxene, with lesser

M. Yu, *Metallogenic Mechanism of the Galinge Polymetallic Iron Skarn Deposit, Qiman Tagh Mountains, Qinghai Province*, Springer Theses,
https://doi.org/10.1007/978-981-10-7907-8_2

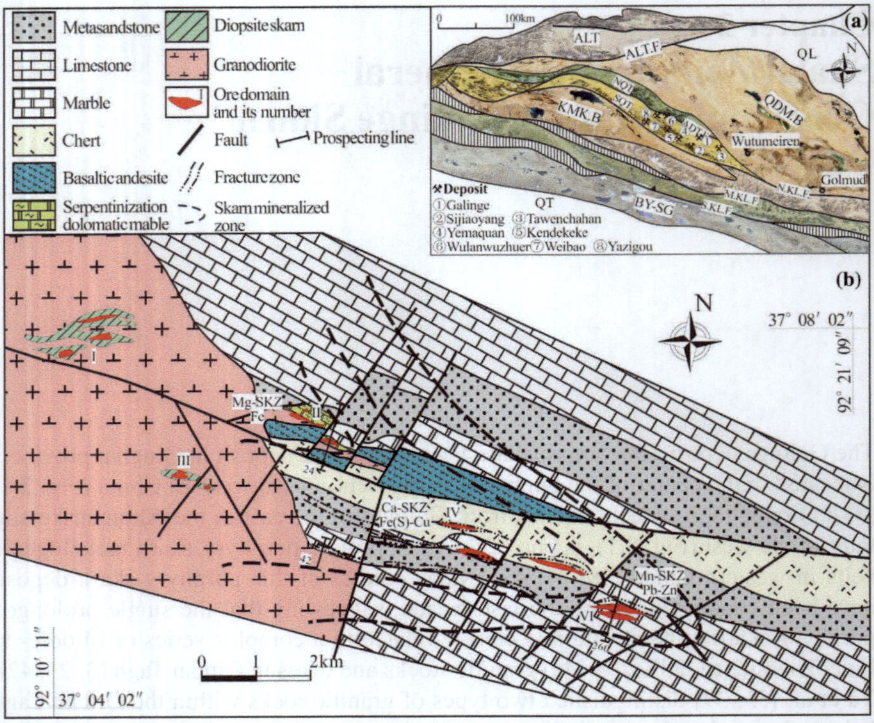

Fig. 2.1 Geological map of the Galinge skarn deposit (after Yu et al. [36, 37, 40]). QL: Qilian terrane; QDM-E.KL: Qaidamu-East Kunlun terrane; NQT: North Qiman Tagh terrane; SQT: South Qiman Tagh terrane; BY-SG: Bayanhar-Songpan-Ganzi terrane; QT: Qiangtang terrane; QDM.B: Qaidamu basin; KMK.B: Kumukuri basin; ADT.F.: Adatan fault; ALT.F.: Altun fault; S.KL.F.: Southern Kunlun fault; M.KL.F.: Middle Kunlun fault; N.KL.F.: Northern Kunlun fault; N.QLT: the north margin of the Northern Qilian thrust fault; MFT: the main Himalaya thrust fault

epidote, ilvaite and actinolite; all are iron rich. In contrast, magnesian iron skarns are associated with diverse plutons in a variety of tectonic settings; the unifying feature is that they all form from dolomitic wall rocks. In magnesian skarns, the main skarn minerals, such as forsterite, diopside, humite, periclase, tremolite and serpentine, do not contain much iron. Thus, the available iron in solution tends to form magnetite rather than andradite or hedenbergite [11]. Over-printing of calcic skarn upon magnesian skarn is reported from many Russian deposits [25]. Zhao et al. [45] summarized the distributed rhythms of magnesian skarn assemblages from unaltered intrusion to dolomitic marble in the most important gold metallogenic belt along the middle and lower Yangtze River Valley and adjacent regions.

In some deposit, the manganese content of pyroxene increase systematically along the fluid path and with the decrease of rock-forming temperature, and thus manganoan skarns can be used to identify proximal and distal skarns or zones [24, 43]. The presence of diverse skarn and skarns does not necessarily indicate a definite geologic setting or a precise protolith composition. Rather, their developments indicate that the combination of temperature, pressure, hydrothermal fluid and host rock composition was within the stability range of the metamorphic reaction [24]. Details of skarn mineralogy and zonation can be used to construct deposit specific exploration models, as well as more general models useful in developing grassroots exploration programs or regional syntheses [24].

2.1 Ore Deposit Geology

The Galinge deposit is located at the center of the transitional zone between the Qaidam Basin and the Qiman Tagh Orogen, at approximately 37° 06′N, 92° 14′E (Fig. 2.1a) [34]. The Galinge deposit is buried under Quaternary sediments with an average surface depth of 200 m. The Lower Tanjianshan Formation, composed of limestone, sandstone and volcanic rocks, forms the basal units of the Ordovician sequence. The limestone, about 750–900 m thick, the main host rock of the Galinge deposit, is composed of calcareous limestone (thickness >300 m) in the southeastern part, and dolomitic limestone in the northwestern part proximal to the pluton (Fig. 2.1b). The limestone is pervasively marmorized and silicified.

Exoskarn bodies at Galinge occur within the metamorphic aureoles of the granodiorite intrusion in the west [34]. Thermal effects extend for approximately 1–1.5 km, and the deposit is divided into six ore domains with increasing distance from the intrusive contacts (Fig. 2.1b). Ore Domains I and III are located within the xenoliths of the granodiorite intrusion, Ore Domain II is located along the intrusive contact between the granodiorite and the dolomitic marble, and Ore Domains IV, V, and VI are distributed along the WNW-trending fracture zone within the exocontact zones. In Ore Domains IV and V, the orebodies are mainly hosted in the limestone and basaltic andesite of the Cambrian-Ordovician Tanjianshan Formation. Magnetite is being mined, and Cu is produced as a byproduct of magnetite from these two ore domains. Ore Domain VI hosts several small NW-trending orebodies, where galena, sphalerite and magnetite in the skarn (developed in the Tanjianshan Formation limestone) are mined.

Skarn zonation at Galinge includes Mg-rich magnetite (Mg-SK), Ca-rich iron sulfide (Ca-SK), and Mn–Ca-rich galena and sphalerite (Mn-SK) [34]. The Mg-SK zonation dominating Ore Domain II, which is hosted by dolomitic rocks nearest the pluton, has lower sulfide content (6 vol.%) and more abundant magnetite (20–35 vol.%). The Mg–Fe skarn forms narrow, zoned envelopes along the intrusive contacts with the dolomitic marble, and is also crosscut by later granodiorite (Fig. 2.2a). It is predominately composed of Mg-rich silicates, including forsterite, diopside, tremolite and phlogopite which are altered by serpentine in different degree, and encompasses a few significant Ca bearing garnet (Fig. 2.2c). The Ca-SK skarn zonation in Ore Domains IV and V occurs distal to the contact, and hosted in the marble and basaltic andesite rocks. It is dominated by pyrrhotite, pyrite and minor chalcopyrite,

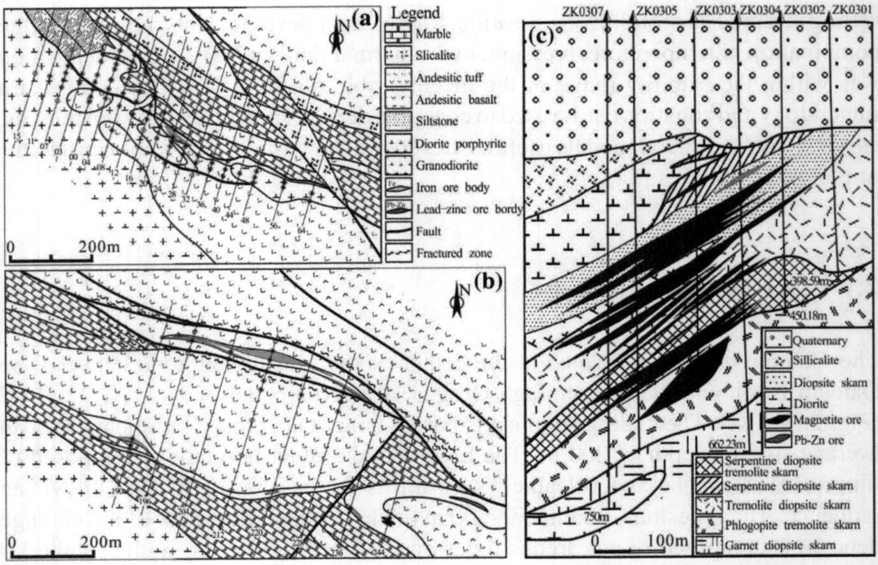

Fig. 2.2 **a**, **b** Geological map of the Ore Domain II and V, showing the skarn geometry, orebody distribution and prospecting lines (**c**). Schematic cross section profiles of prospecting line 03 showing the vertical distribution of the ore bodies and mineral assemblage zonation in the Galinge iron ore, which involves phlogopite-tremolite, diopside-tremolite, serpentine-diopside, serpentine-diopside-tremolite skarn mineral assemblages

and magnetite is disseminated elsewhere. The Mn–Pb–Zn zone encompassing Ore Domain VI occurs farthest from the contact and overlaps the Fe-zone, showing that skarn alteration reached its maximum extent in the distal eastern part of the deposit.

2.1.1 Stage Division of Skarn Evolution

In most large skarn deposits, there is a transition from early/distal metamorphism to later/proximal metasomatism resulting in relatively coarse-grained, ore-bearing skarn. In the Galinge skarn system, the metasomatism process can be divided into the skarn stage, retrograde stage, and quartz-calcite-sulfide stage (Fig. 2.3). In the skarn stage wollastonite, forsterite and hortonolite are formed earlier than garnet, and pyroxene. Notably, the skarn stage terminates with abundant deposition of magnetite. This took place either just prior to, or during, the switch to the retrograde stage and was precipitated by a second boiling event as indicated by the deposition of volatile-rich minerals, such as fluorine-rich phlogopite and chondrodite, chlorite-rich hastingsite, and boron-rich tourmaline [35–40]. The subsequent retrograde alteration also mark out early and later sub-stage, because ilvaite, epidote, tourmaline chondrodite and hastingsite were precipitated before ferro-actinolite, axinite, chlorite,

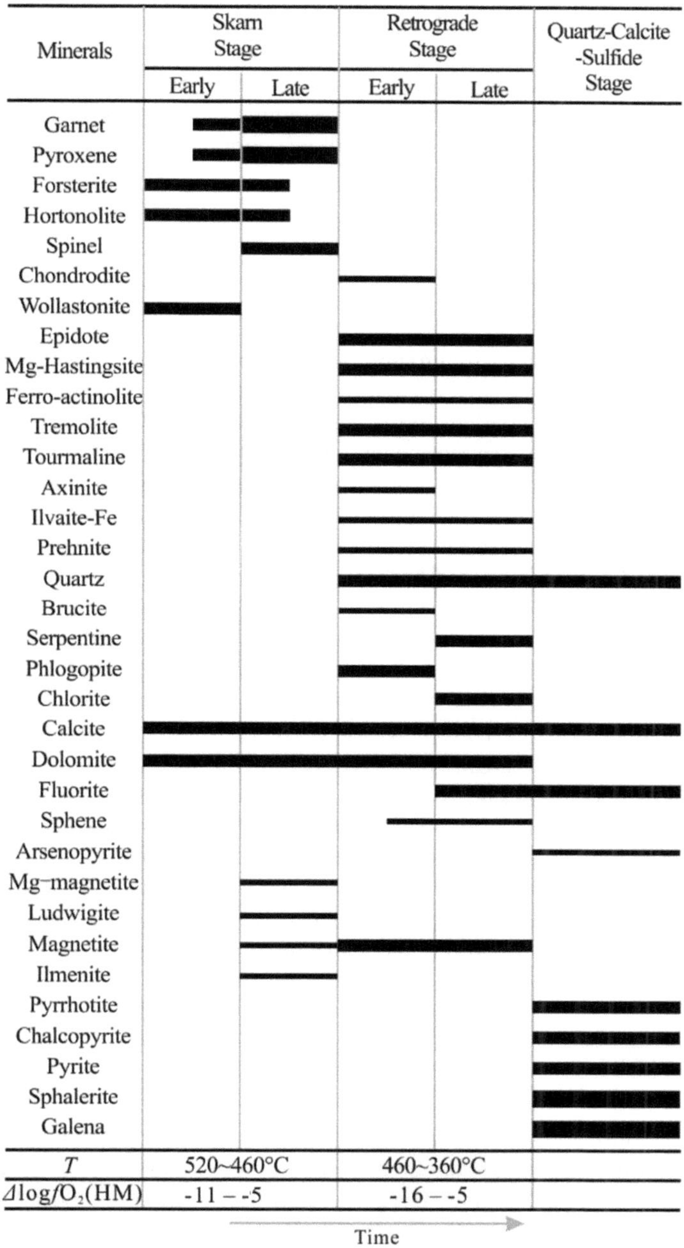

Fig. 2.3 Paragenetic sequence for the major ore- and alteration minerals developed in the three zones at Galinge. The temperature and oxygen fugacity is based on the thermodynamic model discussed next

prehnite, phlogopite and serpentine, but calcite and quartz crystallized throughout the entire retrograde stage. During the sulfide stage mainly low-temperature minerals developed, e.g., calcite, fluorite, pyrrhotite and pyrite. Magnetite was mainly deposited in this stage, ending with the beginning of quartz-calcite-sulfide stage, which is characterized by the deposition of chalcopyrite, pyrrhotite, pyrite, galena, sphalerite, calcite, quartz and fluorite.

2.1.2 Stratagraphy

Based on the generalized stratigraphic sections of this basin summarized in the Fig. 2.4, the Cambrian-Ordovician andesites from the lower Tanjianshan formation are so widespread that interlaid in the variable strata. The Galinge region is completely buried by Quaternary sediments (up to 200 m thick). The based rock is composed of Cambrian-Ordovician Tanjianshan Formation which contains limestone, argillaceous dolomitic limestone, dolomite, chert, sandstone and basaltic andesite [7, 33]. The basaltic andesite interbedded limestone shows a marine sedimentary environment. The carbonate and basaltic andesite rocks are mainly the host of the ore bodies.

Fresh basaltic andesite at Galinge is composed of plagioclase (65–75%), pyroxene (10–15%), amphibole (10–15%) and biotite (5–10%) (Fig. 2.5). Accessory mineral are magnetite, ilmenite, apatite, zircon, and sphene. The small part of euhedral plagioclase phenocryst ($An_{38\text{–}47}$) has reacted to overgrowth margin of plagioclase with a higher An content ($An_{61\text{–}66}$). Clinopyroxene, mainly composed of augite, occurs as corroded crystals and its periphery is altered to amphibole. Abundant igneous amphiboles show edenite composition, and is partly altered to form chlorite with biotite. Magnetite always exsolved ilmenite as a consequences of oxyexsolution of the titaniferous magnetite. Geothermometer and geobarometer applications of the titanomagnetite-ilmenite reliably shows that crystallization temperature of volcanic rocks ranges from 750 to 700 °C. The matrix is substituted by the mafic minerals.

2.1.3 Intrusion

The Galinge intrusions are located in the southern Qaidam Basin margin, adjacent to the central part of the Qiman Tagh Orogen. There is no outcrop of the Galinge intrusions, and thus all samples were collected from drill cores. Based on careful observation on hand specimens and microscopic identification, the Galinge intrusions mainly consist of fine- to medium-grained granodiorite, as well as diorite and diorite porphyry dyke.

Granodiorite: The rocks have a hypidiomorphic-granular texture and are grey to light greyish-green in color with a massive structure (Fig. 2.6a, b). The granodiorites contain plagioclase (40–50%), quartz (15–25%), K-feldspar (15–25%), together

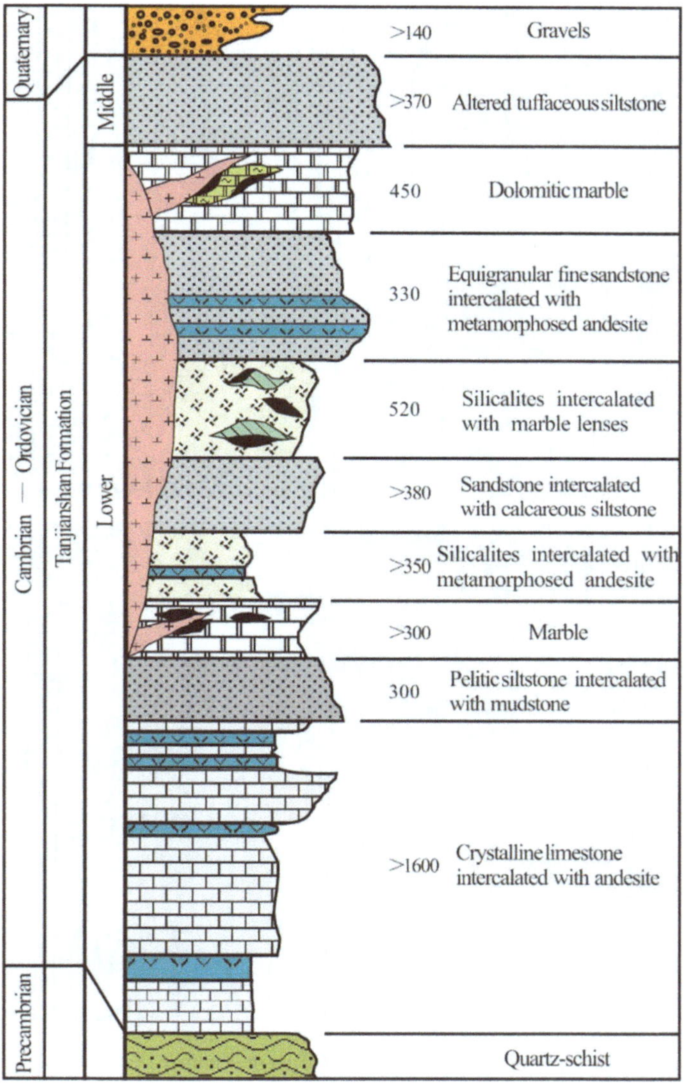

Fig. 2.4 The stratigraphy section of the Galinge skarn deposit

Fig. 2.5 Photomicrographs of the basaltic-andesite from the Galinge skarn deposit

with hornblende and biotite (8–10%). Accessory minerals include apatite, sphene and zircon (Tables 2.1 and 2.2). Microgranular mafic enclaves (MMEs) were found in these plutons. Plagioclase and K-feldspar are subhedral to euhedral, and amphibole and biotite are subhedral. Quartz occurs as an irregularly-shaped interstitial phase. The rocks are commonly altered: Relict clinopyroxenes occur as corroded crystals with its margins altered to amphibole. Biotite is partly altered to chlorite. Interstitial quartz has abundant fluid inclusions containing solid phases. The plagioclase ($An_{40.52\sim64.68}$) contains andesine and labradorite, and the amphibole is dominantly

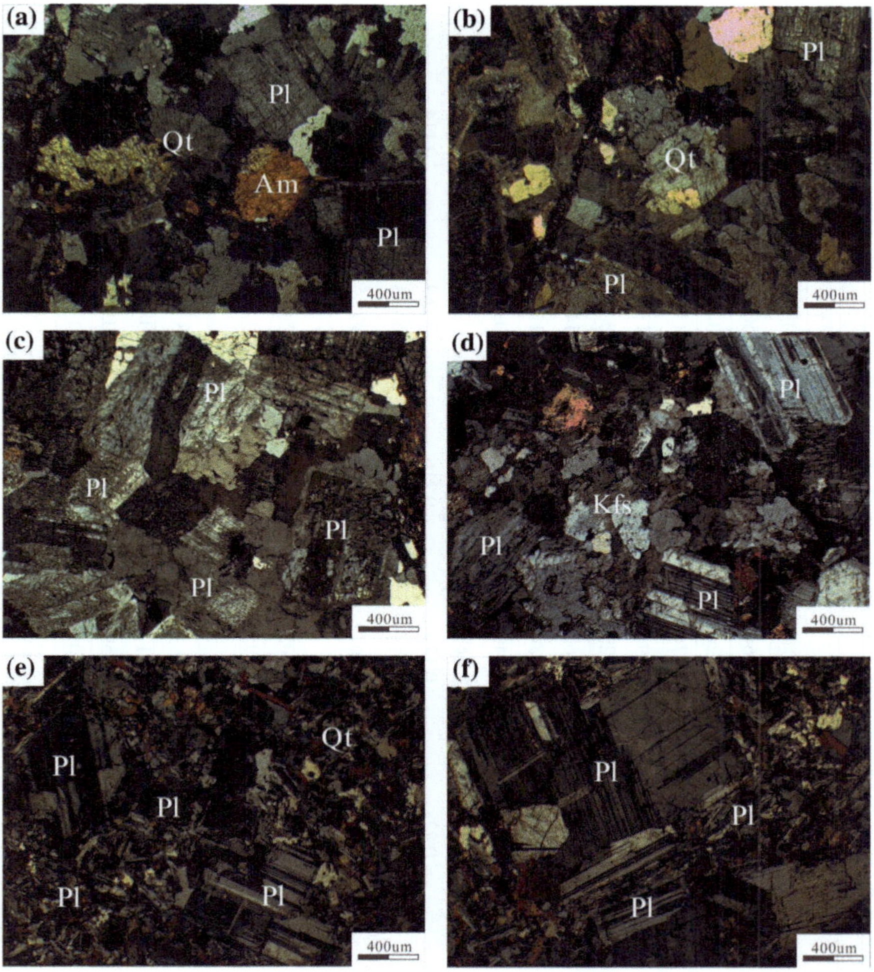

Fig. 2.6 Microphotographs of the Galinge intrusions. **a**, **b** Granodiorite; **c**, **d** Diorites; **e**, **f** Diorite porphyry dikes

composed of high-Si and low-Al actinolitic hornblende, the biotite is significantly Fe rich (Fe** = 0.51). The thermobarometer of amphibole is applied to calculate the crystalized temperature (693–726 °C), pressure (48–57 MPa), oxygen fugacity (log $fO_2 = -13.7$ to -14.7), continental depth (1.8–2.1 km) and water content of melt (3.7–5.0 wt%).

Diorite: The rocks are of porphyraceous texture and massive structure (Fig. 2.6c, d). The diorites contain mainly plagioclase (75–80%), minor K-feldspar (5–10%) and hornblende, clinopyroxene and biotite (10–15%). The feldspar comprises mainly orthoclase ($Or_{92.09}$) and plagioclase ($An_{33.69-55.80}$) (Table 2.2), the amphibole is acti-

Table 2.1 The microprobe data of amphiboles from Galinge intrusions

Sample	Granodiorite		Granodiorite		Diorite		
	16303-14-2-1	16303-14-3-1	2404-35-2-2	2404-35-2-3	0307-76-1-1	0307-76-1-2	0307-76-1-3
SiO_2	50.32	49.849	50.144	51.645	53.805	50.6	50.807
TiO_2	0.173	0.339	0.261	0.524	0.124	0.366	0
Al_2O_3	4.42	3.873	3.657	4.274	1.261	4.022	3.474
FeO	15.629	17.43	15.898	16.94	14.635	13.016	14.091
MnO	0.341	0.27	0.243	0.329	0.302	0.2	0.391
MgO	13.871	12.811	13.298	13.656	14.561	14.772	15.345
CaO	10.225	11.025	10.654	10.848	10.767	12.436	12.012
Na_2O	0.861	0.433	0.872	1.049	0.278	0.115	0.143
K_2O	0.378	0.218	0.4	0.589	0.185	0.371	0
F	0.184	0.102	0.228	0.175	0.216	0.269	0.114
Cl	0.104	0.049	0.133	0.175	0.041	0.109	0.093
Total	96.645	95.616	95.872	96.672	96.291	97.35	96.754
Formula on the basis of 13 cations							
Si	7.29	7.313	7.409	7.305	7.808	7.416	7.414
$Al^{[IV]}$	0.71	0.67	0.591	0.695	0.192	0.584	0.586
Ti	0	0.017	0	0	0	0	0
T site	8	8	8	8	8	8	8
$Al^{[VI]}$	0.044	0	0.046	0.017	0.023	0.111	0.011
Ti	0.019	0.021	0.029	0.056	0.014	0.04	0
Cr	0.006	0.005	0.002	0.005	0	0	0.005

(continued)

Table 2.1 (continued)

Sample	Granodiorite		Granodiorite		Diorite		
	16303-14-2-1	16303-14-3-1	2404-35-2-2	2404-35-2-3	0307-76-1-1	0307-76-1-2	0307-76-1-3
Fe^{3+}	1.137	0.993	0.788	0.881	0.682	0.383	0.953
Mg	2.996	2.802	2.929	2.879	3.15	3.228	3.338
Fe^{2+}	0.757	1.145	1.177	1.123	1.094	1.212	0.645
Mn	0.042	0.034	0.03	0.039	0.037	0.025	0.048
C site	5	5	5	5	5	5	5
Fe^{2+}	0	0	0	0	0	0	0.122
Ca	1.587	1.733	1.687	1.644	1.674	1.953	1.878
Na	0.242	0.123	0.25	0.288	0.078	0.033	0
B site	1.829	1.856	1.936	1.932	1.752	1.986	2
Na	0	0	0	0	0	0	0.04
K	0.07	0.041	0.075	0.106	0.034	0.069	0
A site	0.07	0.041	0.075	0.106	0.034	0.069	0.04
Mg^*	0.798	0.71	0.713	0.719	0.742	0.727	0.813
Name	Actinolitic Hornblende	Actinolitic Hornblende	Actinolitic Hornblende	Actinolitic Hornblende	Actinolitic Hornblende	Actinolitic Hornblende	Actinolitic Hornblende
Physical-chemical conditions							
T (°C)	726	718	693	708	656	718	728
P (MPa)	57	50	48	54	26	52	45
Continental depth (km)	2.1	1.9	1.8	2.0	1.0	2.0	1.7
$\log fO_2$	- 13.7	- 14.1	- 14.7	- 14.4	- 15.4	- 13.8	- 13.2
H_2O_{melt} (wt%)	4.3	5.0	4.2	3.7	3.5	5.0	5.3

Note $Mg^* = Mg/Mg + Fe^{2+}$; Tested Unit: Institute of Mineral Resources, Chinese Academy of Geological Sciences

Table 2.2 The microprobe data of plagioclases from Galinge intrusions

Sample	Granodiorite				Granodiorite		Diorite					Diorite porphyrite
	16303-14-1-1	16303-14-1-2	16303-14-4-1	16303-14-5-1	2404-35-1-1	2404-35-4-1	0307-76-3-2	0307-76-4-2	0307-79-1-2	0307-79-2-1	0307-79-4-1	21605-30-1-2
SiO_2	57.01	57.11	54.5	56.28	51.9	54.39	54.23	64.42	56.47	54.08	59.95	68.83
Al_2O_3	27.78	28.06	28.3	27.49	30.38	28.06	28.66	18.8	27.98	28.97	25.28	22.34
CaO	7.95	6.99	10.27	9.5	12.98	10.51	10.69	0.04	9.67	11.14	6.52	1.94
Na_2O	4.5	4.75	5.14	5.8	3.8	5.02	5.01	0.85	5.17	4.75	6.89	6.24
K_2O	0.34	1.4	0.4	0.33	0.17	0.35	0.29	15.38	0.15	0.19	0.31	0.09
Total	97.58	98.31	98.6	99.39	99.22	98.33	98.88	99.49	99.43	99.12	98.95	99.44
Formula on the basis of 8 oxygen atoms												
Si	2.588	2.584	2.488	2.541	2.37	2.49	2.47	2.983	2.54	2.458	2.688	2.971
Al	1.486	1.496	1.522	1.463	1.635	1.514	1.539	1.026	1.483	1.552	1.336	1.137
Ca	0.386	0.339	0.502	0.46	0.635	0.516	0.522	0.002	0.466	0.542	0.313	0.09
Na	0.396	0.416	0.455	0.508	0.337	0.445	0.443	0.076	0.45	0.419	0.599	0.522
K	0.02	0.081	0.023	0.019	0.01	0.02	0.017	0.909	0.009	0.011	0.018	0.005
An	48.16	40.52	51.27	46.63	64.68	52.54	53.16	0.21	50.37	55.8	33.69	14.56
Ab	49.38	49.83	46.38	51.47	34.31	45.38	45.13	7.7	48.71	43.1	64.4	84.67
Or	2.45	9.65	2.36	1.9	1	2.08	1.71	92.09	0.92	1.1	1.91	0.77

Note An: anorthite; Ab: Albite; Or: Orthoclase; Tested Unit: Institute of Mineral Resources, Chinese Academy of Geological Sciences nolitic hornblende (Table 2.1), the biotite is moderately Mg-rich (Mg** = 0.57–0.62), and the pyroxene is Ca-rich (Wo = 0.44) (Table 2.3). The thermobarometer of amphibole is applied to calculate the crystalized temperature (656–728 °C), pressure (26–52 MPa), oxygen fugacity ($\log fO_2 = -15.4$ to -13.2), continental depth (1.0–2.0 km) and water content of melt (3.5–5.0 wt%).

Diorite porphyry dyke: the rocks are grayish green to light gray and have porphyritic texture. Phenocrysts mainly include abundant plagioclase (70–80%) and minor amount of pyroxene, and the groundmass contains microcrystalline quartz, biotite, plagioclase and hornblende (Fig. 2.6e, f). The plagioclase is mainly composed of basic oligoclase ($An_{14.56}$) (Table 2.2), and the pyroxene is remarkably Ca-rich (Wo = 0.49) (Table 2.3).

2.2 Petrography of Skarns

2.2.1 Magnesian Skarn

Magnesian iron skarns are dominantly composed of prograde minerals of olivine, forsterite, and spinel, and retrograde minerals of chondrodite, phlogopite, serpentine, clinochlore, and chamosite.

Mg-rich silicates in the deposit can be separated into Mg-rich skarn minerals (such as forsterite, diopside and chondrodite) and Mg-rich retrograde skarn minerals (such as phlogopite, tremolite, serpentine, clinochlore and chamosite). Relicts of calcite and/or dolomite are present in all alteration mineral assemblages. The retrograde minerals always overprint the skarn minerals, and yield different mineral zonation (Fig. 2.2c).

The olivine skarn is mainly composed of euhedral-subhedral forsterite-fayalite solid solution (mainly forsterite) crystals with colorless pleochroism. They are paragenetically intergrown with diopside and/or dolomite in the skarn stage (Fig. 2.7a, b), and altered to a mixture of serpentine, chlorite and magnetite as matrix in varying degrees (Fig. 2.7c, d), usually with sulfide concentrations in retrograde stage. This alteration is more intensely developed along the contact between olivine and diopside bands. The crystals of forsterite have been altered around their rims or along their fractures to serpentine, and in the most extreme cases, are preserved only as relics in a matrix of serpentine. In addition, some olivines are completely replaced by magnetite, keeping a pseudomorph of olivines. These observations lead to some questions about the formation of magnetite. What kinds of conditions and reacted processes cause the magnetite precipitation?

Pyroxenes of the diopside–hedenbergite series (mostly diopside) ubiquitously occur in the Mg-SK zone of the Galinge deposit, appearing as anhedral-subhedral fine-coarse grained aggregates or paragenetically intergrown with olivine, spinel and/or chondrodite in skarn stage. They are variably replaced by light green tremolite and phlogopite in retrograde stage. Sometimes pyroxenes with core-rim texture

Table 2.3 The microprobe data of pyroxenes and biotites from Galinge intrusions

Sample	Granodiorite	Diorite porphyry dyke			Diorite			Diorite			Granodiorite
	2404-35-1-2	21605-30-1-1	21605-30-2-1	21605-30-3-1	0307-76-2-1	0307-79-1-1	0307-79-3-1	0307-79-3-2	0307-79-4-2	0307-79-4-3	2404-35-3-1
SiO_2	52.27	52.1	51.17	51.4	51.81	52.14	52.61	38.23	40.35	37.89	34.79
TiO_2	0.29	0.03	0	0.01	0	0.09	0.14	3.87	3.72	2.24	1.79
Al_2O_3	1.51	0.39	0.25	0.41	0.07	0.4	0.64	13.67	11.7	11.84	13.57
FeO	10.83	12.01	14.46	11.65	14.09	27.23	26.49	13.78	15.07	16.05	22.54
MnO	0.25	0.72	0.43	0.65	0.39	0.62	0.48	0.03	0.07	0.11	0.17
MgO	12.26	9.98	8.65	10.31	10.35	17.84	18.36	15.71	14.57	13.29	11.32
CaO	20.94	23.31	22.9	23.17	22.01	1.07	1.07	0.01	0	0.02	0.2
Na_2O	0.35	0.1	0.17	0.18	0.14	0	0.03	0.22	0.15	0.08	0.09
K_2O	0.04	0.01	0	0	0	0	0	9.33	9.46	9.37	5.86
H_2O^+	–	–	–	–	–	–	–	3.6	3.47	3.46	3.37
F	–	–	–	–	–	–	–	0.58	0.75	0.42	0.34
Cl	–	–	–	–	–	–	–	0.16	0.21	0.22	0.22
Total	98.73	98.63	98.03	97.78	98.86	99.39	99.82	99.19	99.52	94.98	94.26
Name	Cpx	Cpx	Cpx	Cpx	Cpx	Cpx	Cpx	Bio	Bio	Bio	Bio
Si	1.985	2.01	2.009	2	2.006	2.004	2.003	2.925	3.113	3.061	2.903
$Al^{[IV]}$	0.015	0	0	0	0.003	0.018	0.029	1.075	0.887	0.939	1.097
$Al^{[VI]}$	0.053	0.018	0.012	0.018	0	0	0	0.157	0.176	0.188	0.237
Ti	0.008	0.001	0	0	0	0.003	0.004	0.223	0.216	0.136	0.112
Fe3+	0	0	0	0	0.008	0	0.002	–	–	–	–
Fe2+	0.345	0.389	0.476	0.379	0.447	0.873	0.84	0.881	0.972	1.084	1.573

(continued)

Table 2.3 (continued)

Sample	Granodiorite	Diorite porphyry dyke			Diorite			Diorite			Granodiorite
	2404-35-1-2	21605-30-1-1	21605-30-2-1	21605-30-3-1	0307-76-2-1	0307-79-1-1	0307-79-3-1	0307-79-3-2	0307-79-4-2	0307-79-4-3	2404-35-3-1
Mn	0.008	0.023	0.014	0.022	0.013	0.02	0.015	0.002	0.005	0.008	0.012
Mg	0.694	0.574	0.506	0.598	0.597	1.022	1.042	1.791	1.675	1.6	1.408
Ca	0.852	0.964	0.963	0.966	0.913	0.044	0.044	0.001	0	0.002	0.018
Na	0.026	0.007	0.013	0.014	0.003	0	0.001	0.032	0.023	0.013	0.015
K	0.002	0	0	0	0	0	0	0.913	0.933	0.968	0.625
F	–	–	–	–	–	–	–	0.141	0.184	0.107	0.091
Cl	–	–	–	–	–	–	–	0.021	0.028	0.03	0.03
OH-	–	–	–	–	–	–	–	1.838	1.788	1.864	1.879
Wo	0.44	0.49	0.49	0.49	0.46	0.02	0.02	–	–	–	–
En	0.36	0.29	0.26	0.3	0.3	0.52	0.54	–	–	–	–
Fs	0.18	0.21	0.25	0.2	0.24	0.46	0.44	–	–	–	–
Ac	0.01	0	0.01	0.01	0	0	0	–	–	–	–
Mg**	–	–	–	–	–	–	–	0.62	0.58	0.57	0.45
Fe**	–	–	–	–	–	–	–	0.3	0.34	0.38	0.51

Mg** = Mg/(Mg + Fe + Ti + Mn); Fe** = Fe/(Mg + Fe + Ti + Mn); Clinopyroxene formula on the basis of 6 oxygens and 4 cations; Biotite formula on the basis of 8 cations; Tested Unit: Institute of Mineral Resources, Chinese Academy of Geological Sciences

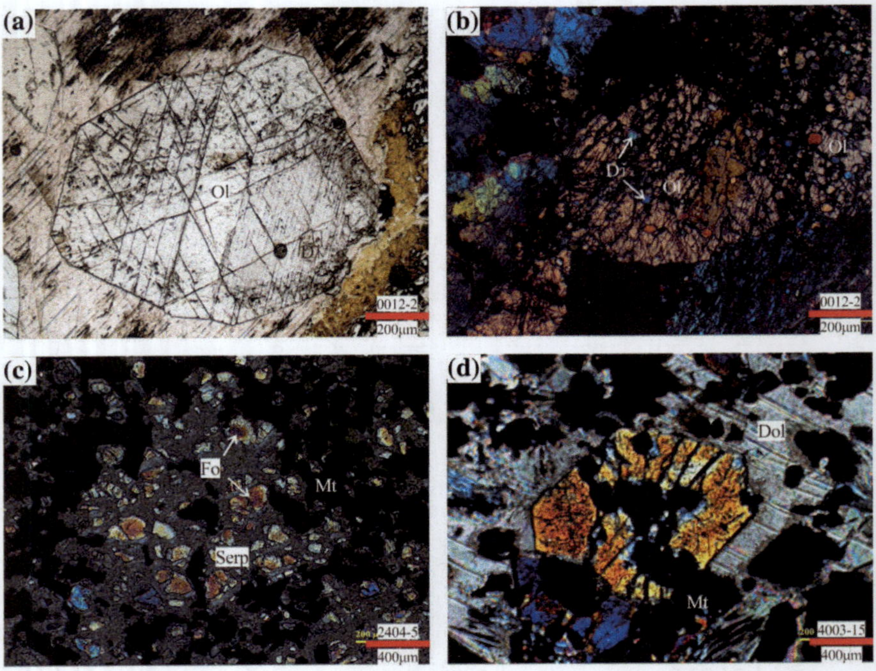

Fig. 2.7 Micrographs of the Galinge olivines. **a** A subhedral olivine crystal in sample 0012 shows cleavage in one direction. The cleavage of {001} is good, and {010} is distinct; **b** subhedral olivines in sample 0012 coexisting with diopside, and enveloping diopside; **c** olivines in sample 2404 were replaced by magnetite and serpentine in later stage; **d** a hexagonal olivine crystal in sample 2404 was replaced by magnetite

are made up of compositionally different two stage medium-crystalline subhedral pyroxene crystals.

The Galinge spinels can be identified as red brown and dark green spinel (Fig. 2.8a). The red brown spinels always appears as subhedral-anhedral granular aggregates in diopside rich rocks, and locally are replaced by dark green spinels (Fig. 2.8a). Most of the spinel occurs as small (0.5 mm or less) isolated grains, although it has been observed as inclusions in diopside. The inclusions do not appear to be systematically different in composition from the isolated grains. In microscopic section, it was usually observed that droplet titanomagnetite exsolved from inner and interstitial of spinel crystals (Fig. 2.8b). The dark green spinels occur as abundant subhedral-euhedral granular aggregates (Fig. 2.8c), in paragenetic growth with diopside. The replacement of spinel-pyroxene skarns by phlogopite or tremolite is very common in the Galinge deposit (Fig. 2.8d).

Chondrodite in the Mg-SK zonation occurs commonly as discrete aggregates of subhedral to euhedral grains paragenetically intergrown with diopside (Fig. 2.9a), but it is also found replacing diopside and altered by serpentine and magnetite.

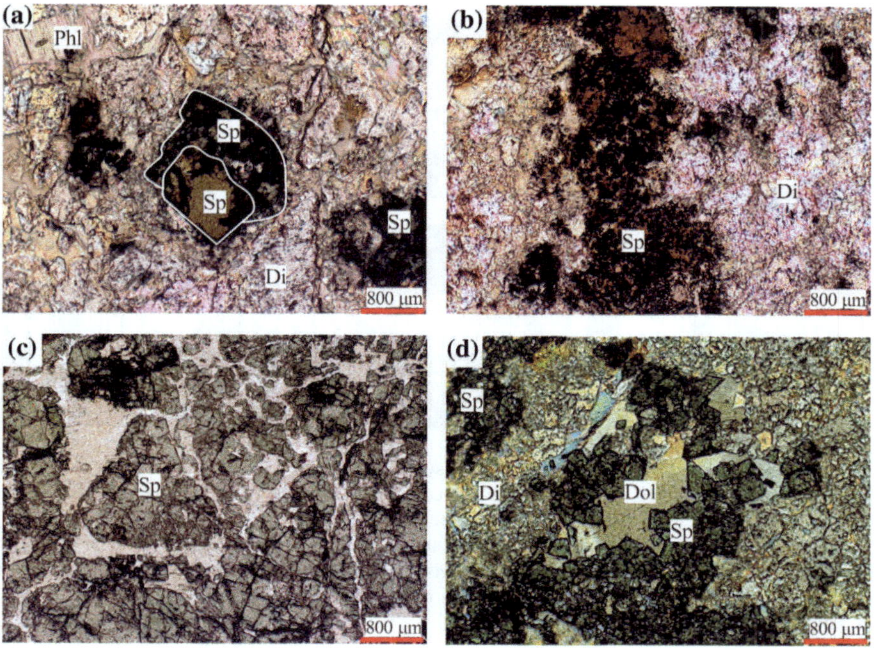

Fig. 2.8 Micrographs of the Galinge spinels. **a** An early spinel with brown color was replaced by a later one with green color; **b** anhedral spinels replaced by diopside; **c** subhedral to euhedral spinels coexisting with calcite; **d** deep green, subhedral to anhedral spinel coexisting with diopside, calcite and dolomite

Euhedral coarse tremolite is commonly associated with sulfide minerals (Fig. 2.9b). Tremolite develops at the expense of diopside, but this alteration is not complete, and relics of diopside are still present (Fig. 2.9c). Locally, tremolite replaces spinel and paragenetically coexists with tourmaline (Fig. 2.9d).

Phlogopite, with variable Fe content, is locally abundant in the Galinge Mg-SK zonation deposit; it is earlier than most retrograde phases, and it has a clear paragenetic relation with tremolite. Phlogopite grain size varies from ~<50 μm to >5 mm, and coarse (~>1 mm). Phlogopite replaces Mg-silicates, forming coarse sheaves (up to 1 cm), with pyroxene relicts located within them and filling interstices (Fig. 2.9e), and locally appears to replace later amphibole. Phlogopite is commonly kinked and hosts Fe sulfides, mostly pyrrhotite along cleavage cracks.

Intensive serpentinisation has affected especially Mg–skarns (with forsterite–diopside-chondrodite) and corresponds to the main stage of mineralisation. Serpentine compositions exhibit variable amounts of iron. There is colorless serpentine with low iron content, and brownish iron-rich serpentine. The serpentines are usually altered to Mg-rich chlorite and talc (Fig. 2.9f).

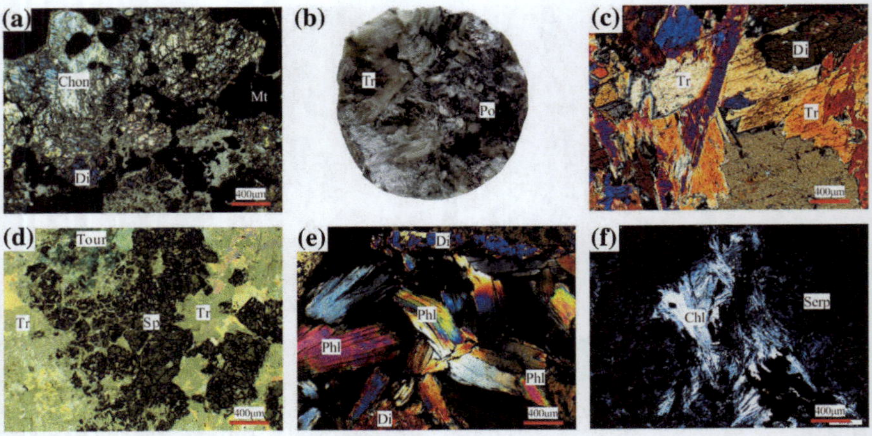

Fig. 2.9 **a** Light yellow chondrodite with polysynthetic twinning coexisting with the diopside is replaced by serpentine and magnetite; **b** hand specimen of tremolite skarn with a metasomatic relic texture; the dark green anhedral diopside is replaced by the light green laminar tremolite coexisting with the pyrrhotite; **c** micrograph of tremolite skarn shows the relic diopside replaced by tremolite; **d** dark green spinel replaced by tremolite coexisting with tourmaline; **e** phlogopite with perfect cleavages, coexisting with diopside; **f** white chlorite coexisting with serpentine. Abbreviations as in Table 2.1

2.2.2 Calcic Skarn

Garnets from different ore domains in the Galinge skarn deposit are generally composed of grossular-andradite-pyralspite solid solution. In the No. I Ore Domain, the prograde garnet-diopside assemblage was developed proximal to the intrusions. Medium- to coarse-grained, euhedral to subhedral garnet associated with pyroxene (Fig. 2.10a). The garnet associated with chalcopyrite is observed that a later stage rim replaces previous inner core (Fig. 2.10b). Garnet occurs in the galena + sphalerite + chalcopyrite assemblage. Euhedral to subhedral pyroxene occurs in anhedral garnet (Fig. 2.10a). Magnetite and clinopyroxene replacing early garnet-dominated skarn are widespread (Fig. 2.10b). In the outside of No. II Ore Domain, the garnet was replace by the Mg-hastingsite. In No. V Ore Domain, the garnet was replaced by retrograde epidote (Fig. 2.10c) in skarn-altered diorites. In No. VI Ore Domain, massive medium-fine grained garnet is related with magnetite. Some coarse-grained, euhedral andraditic garnet was found replaced by fine-medium grained diopside or cut by the later diopside veins along its fractures (Fig. 2.10d). Garnets in altered granodiorite display various colors: Light brown euhedral garnet and light green euhedral garnet (Fig. 2.10e, f). The garnet replaced by Mn-hedenbergite is characterized by compositional zonation.

In No. I Ore Domain, dark green anhedral pyroxene coexisting with garnet occurs locally along skarn-granodiorite dyke contact (Fig. 2.11a, b). Some retrograde actinolite replacing hedenbergite were observed in the altered basaltic andesite. In No.

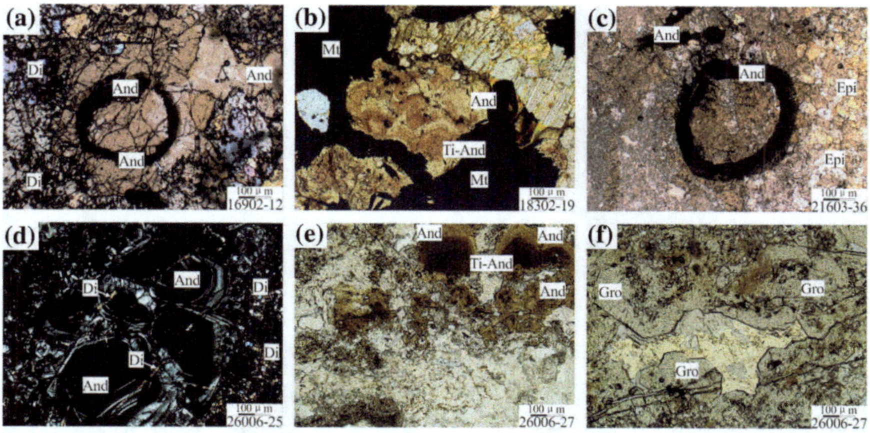

Fig. 2.10 Microphotographs of garnet from the Galinge skarn deposit. **a** Medium- to coarse-grained, euhedral to subhedral andradite minerals replaced by pyroxene. **b** Titanium-rich andradite core and andradite rim. **c** Garnet replaced by retrograde epidote in skarnized diorites shows grossular composition. **d** Garnet replaced by fine-medium grained diopside or cut by the diopside-dominated veins alone its fractures/fissures. **e** Light brown euhedral garnet featured by Ti-rich andradite cores and andradite rims. **f** Light green grossular-rich euhedral garnets

V Ore Domain, pyroxene included in the axinite-dominated skarn was observed (Fig. 2.11c, f), and the diopside + epidote association were observed in magnetite-rich ores (Fig. 2.11e). In No. VI Ore Domain, massive Mn-rich hedenbergite skarn is associated with galena and sphalerite ore (Fig. 2.11d). In addition, the pyroxene generally occurs in altered granodiorite.

Light green anhedral epidote is typically <1 mm (Fig. 2.12a, b). It is found intergrown with magnesio-hastingsite or associated with garnet and zircon. The green to brown pleochroic REE-rich allanite grains or lamellae occur with epidote, garnet and zircon (Fig. 2.12a). Epidote and REE-rich epidote/allanite occur as separate phases adjacent to each other, and as zoned grains with variable REE contents. Greenish subhedral-anhedral granular epidote is enveloped inside axinite (Fig. 2.12e, f).

In the prehnite-bearing sample (Fig. 2.12c, d), prehnite occurs as irregular granular intergrowth with garnet (andradite-grossular) and pyroxene (diopside-hedenbergite), and appears to have been formed after garnet and pyroxene. However, some andradite is zoned with a dark brown (Ti-rich) core and a light brown (Ti-poor) rim. Diopside occurs as fine-grained aggregates appressed to prehnite.

In the axinite-bearing sample (Fig. 2.12e, f), axinite is pleochroic from colorless to pale yellow. Most axinite occurs as aggregates of coarse euhedral crystals. However, some is anhedral to subhedral, and contains inclusions of epidote, pyroxene and anorthite (Fig. 2.12e, f). Some titanite and apatite also occur as euhedral inclusions in axinite or in fractures. Axinite is partially altered to chlorite.

In the Ca-rich skarn zone, mica comprises mainly dark brown to light green color biotite with perfect cleavage (Fig. 2.13a, b). In general, the various colors of biotite

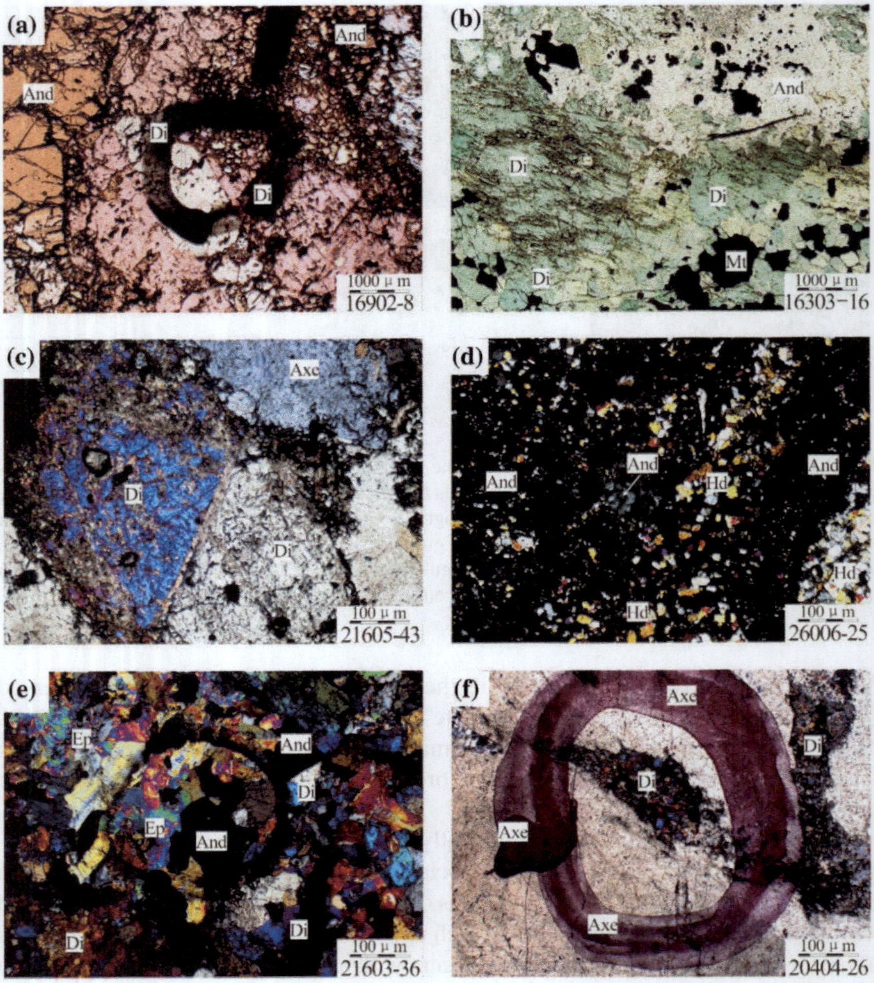

Fig. 2.11 Microphotographs of pyroxenes from Galinge. **a** Subhedral to anhedral diopside replacing andradite. **b** Diopside is green color in granular aggregate. **c** Diopside has simple twin crystalline structure, and replaced by axinite. **d** Subhedral to anhedral Mn-hedenbergite replacing andradite. **e** diopside coexisting with andradite are replaced by epidote. **f** The veins of granular diopside aggregates cut the axinite

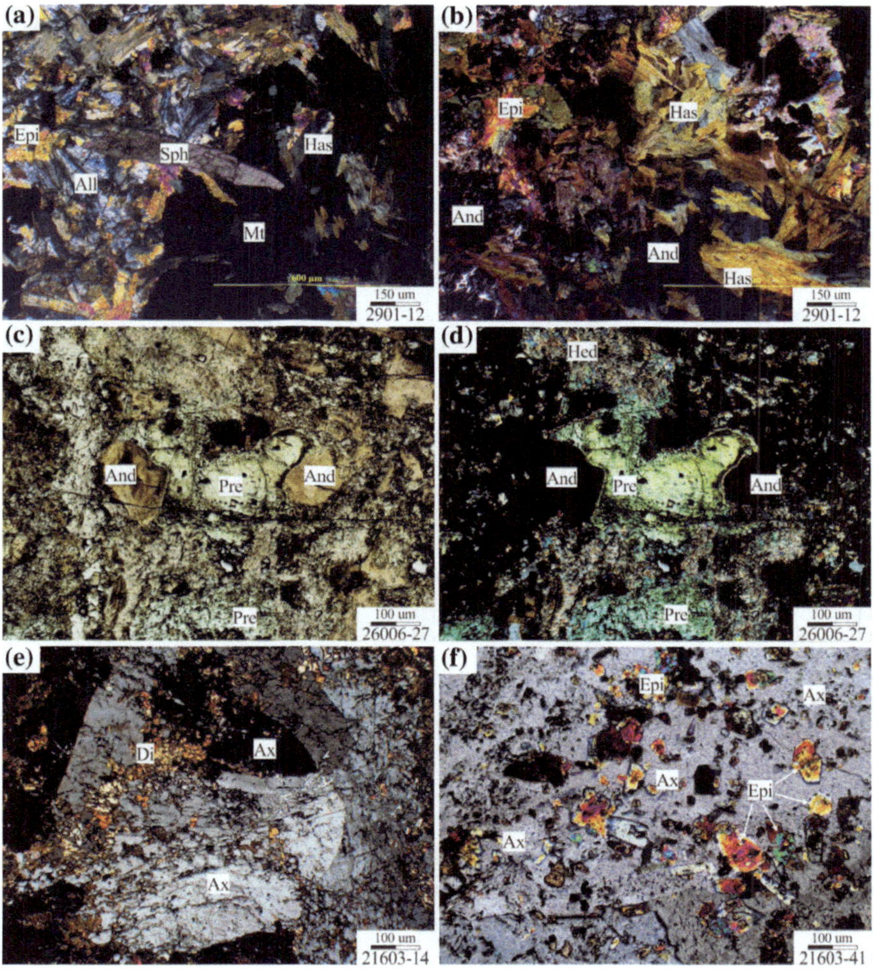

Fig. 2.12 **a** Subhedral to anhedral prismatic epidote and allanite aggregate. **b** Anhedral epidote with andradite and hastingsite. **c**, **d** Anhedral light green anhedral prehnite with andradite and hedenbergite. **e** Colorless euhedral to subhedral axinite replaced by granular pyroxene. **f** Colorless euhedral to subhedral axinite with granular epidote

are due to different Fe and Mg contents, with the brown color intensity increases with increasing Fe content. The evidence that biotite was yielded in retrograde stage by replacing previous pyroxene and garnet with a participation of K_2O and H_2O is very obvious (Fig. 2.13c, d). However, the replacement was not complete and in a very small scale. Fe biotite alteration is always occurred within intrusions or nearest the source of solutions.

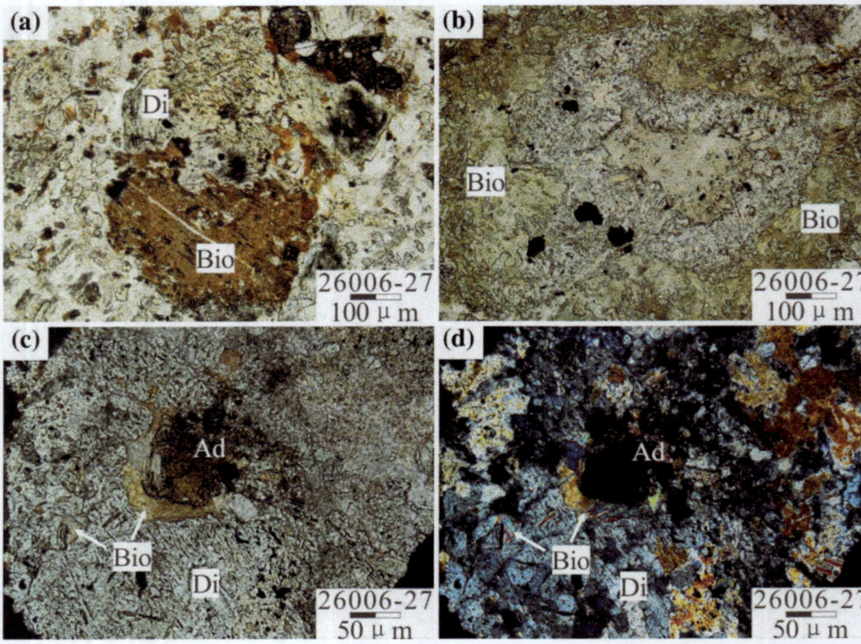

Fig. 2.13 Biotite from the Galinge skarn deposit. **a** Dark brown high-Fe biotite. **b** Light green high-Mg biotite. **c, d** Biotite replaced garnet and pyroxene

Sphene abundantly occurs in altered basaltic andesite throughout retrograde skarn stage. It formed mainly after ilmenite in amphibole-bearing skarn and as large single crystals, and simultaneously formed with ilvaite. The sphene crystals compositionally developed zonation. Back-scattered electron (BSE) imaging of the sphene (Fig. 2.14) shows that it contains a light grey core and a dark grey rim, which suggests that the growth of sphene may have occurred in two stages. Besides, abundant sphene was also found coexisting with magnesian-hastingsite. Magnetite-ilmenite exsolutions were locally observed in their skarn rocks.

2.2.3 *Ore Minerals*

The most widespread ore minerals in the Galinge skarn deposit are magnetite, and a series of sulfides among which galena and sphalerite are most prominent in the VI ore domains, whereas pyrrhotite and chalcopyrite accompanied with magnetite are yielded in IV and V ore domains.

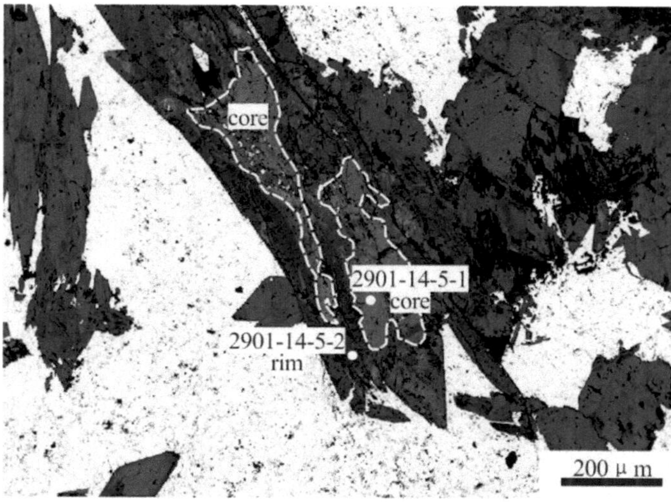

Fig. 2.14 BSE images of sphene, showing light grey cores and dark grey rims. The numbered points show the electron microprobe analyses spots

2.2.3.1 Magnetite

The ore associated spatially with Mg-SK is mainly composed of magnetite, ludwigite, magnesiomagnetite, pyrrhotite and pyrite, and more rarely sphalerite and galena. These minerals occur irregularly disseminated or massive ore in the skarn environment. The main mineralization stage of magnetite generally appears to predate sulfides, and clearly postdates forsterite and diopside. In places olivine replaced by magnetite determines the paragenetic sequence. Magnetite in altered olivine-bearing mineral assemblage is subhedral to anhedral to with a fine- to medium-grained granular texture (Fig. 2.15a, b). Magnetite in rocks that have experienced tremolite-phlogopite alteration generally occurs as euhedral grains (Fig. 2.15c, d), some of which are locally corroded and partially replaced by pyrite and pyrrhotite; magnetite relics within a matrix of pyrite and pyrrhotite display rounded and corroded grain boundaries. This magnetite was simultaneously formed with tremolite and phlogopite in retrograde stage. The ore is richer in the serpentine-diopside-phlogopite and the tremolite-diopside skarn, but poorer in the unaltered diopside skarns. The thermodynamic relationship between the precipitation and retrograde alteration skarn minerals is the main problem about the ore formation in the magnesian skarn system that need to be resolved.

At Galinge, ludwigite, generally coexisting with magnetite, olivine and serpentine in the Mg-SKZ, occurs as irregular bands of randomly oriented elongated grains (Fig. 2.15c, d). They exhibit a texture that obviously results from a metasomatic process, and later truncated by serpentines. Occasionally, ludwigite aggregates usually enclosed by the magnetite mass. Generally, ludwigite in these aggregates forms

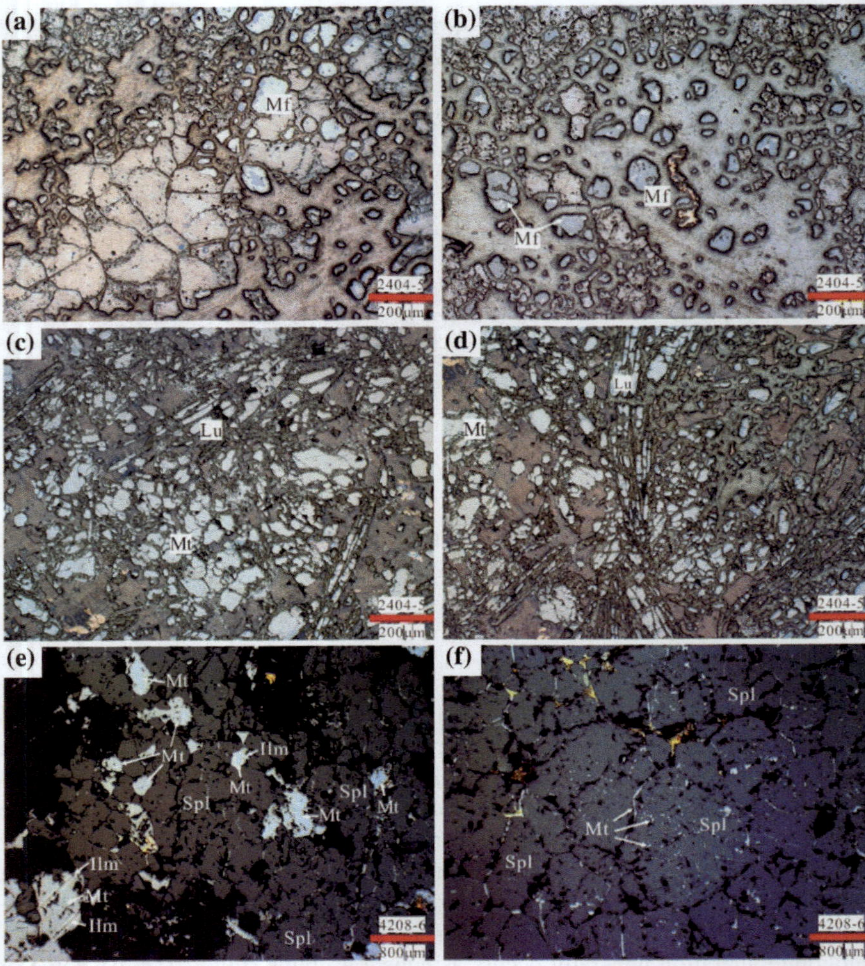

Fig. 2.15 The micrographs of the Galinge iron-oxides; **a**, **b** Granular magnetites replacing olivine, locally as a pseudomorph of olivines; **c**, **d** radial ludwigite coexists with granular magnetites; **e**, **f** the ilmenite exsolved from the magnetite in the spinel skarn

fibrous masses with a silky luster and pitch-black to olive-black color. The component crystals are fibrous and form parallel, radiating or interwoven aggregates, which give to the mass a felt-like appearance. Coarser radiating or fan-like aggregates are scarce; they contain bunches of crystals of up to 35 mm long and up to 2 mm wide.

Polymineralic grains with trellis- or composite-type magnetite-ilmenite intergrowths exsolved from the inner part and interstices of spinel crystals are very common in the Galinge Mg-SK zone (Fig. 2.15e, f). Coexisting Fe–Ti oxides geothermometer and oxygen barometer have been used to postulate the formation conditions of intergrown magnetite-ilmenite in other systems [6, 15]. These postulated conditions are integrated with the thermodynamic model we present below to understand the formation of magnetite in the Galinge deposit.

Magnetite in IV and V ore domains is extensively replaced by later pyrrhotite, pyrite and chalcopyrite. The composition of magnetite from IV and V ore domains has is quite distinct from that of II ore domains proximal to intrusion. Some magnetite is associated with late crosscutting garnet and pyroxene veins. The bulk of the magnetite appears to be later than garnet and pyroxene, as much of the magnetite is associated with ilvaite, epidote, or amphibole alteration of garnet-pyroxene skarn. Furthermore, most of the magnetite in IV and V ore domains is located in massive pods and lenses within skarn-marble or skarn-volcanic rocks. The compositional and physical contrast between prograde and retrograde skarn in addition to a change in fluid properties is responsible for magnetite deposition, e.g. a temperature decline and volatile consumption in the fluid may be involved.

2.2.3.2 Sulfides

The sulfides occurred, in decreasing order of abundance, are: pyrrhotite, pyrite, galena, sphalerite, chalcopyrite, and minor arsenopyrite. Pyrrhotite is the most common sulfide in the Galinge deposit, and generally replace magnetite and pyrite in veinlets or intergranular aggregates (Fig. 2.16a–d, e). However, pyrite also occurs as aggregates or veinlets replacing or crosscutting pyrrhotite sometimes (Fig. 2.16b, f), and occasionally magnetite replacing pyrrhotite are observed (Fig. 2.16g).

Whereas the Fe–Cu association is hosted by calcic skarn in Galinge, the Cu mineralization is significantly related with Mg-hastingsite retrograde skarn in Galinge, and are restricted to the intensely altered mafic volcanic rocks in sedimentary wall rocks. Chalcopyrite is the dominant Cu-bearing phase and is associated with pyrite and pyrrhotite which overprint earlier magnetite. Chalcopyrite is always occurred as minor phase in the pyroxene–garnet-amphibole–magnetite association, and chalcopyrite partially to fully replaced magnetite, pyroxene and amphibole (Fig. 2.16h, i). It also occurs as emulsion droplet exsolved from sphalerite (Fig. 2.16k).

Lead-zinc-silver ores predominately occur in the Mn-bearing skarns in the exocontact zones along lithologic contacts, faults or fractures of carbonate wall rocks, remote from the intrusion, in the form of lenses, veins, stratoid bodies or pockets. In Pb–Zn dominant ore, subhedral sphalerite are surrounded and replaced by pyrrhotite (Fig. 2.16j). Although the chalcopyrite exsolved from sphalerite occurs in Pb–Zn–Ag ore district zone (Fig. 2.16k), the grade of copper is very barren. This Pb–Zn–Ag ore district are generally interpreted as the outermost boundary of mineralization. In some places, arsenopyrite coexisting with pyrrhotite are abundantly observed in Galinge deposit.

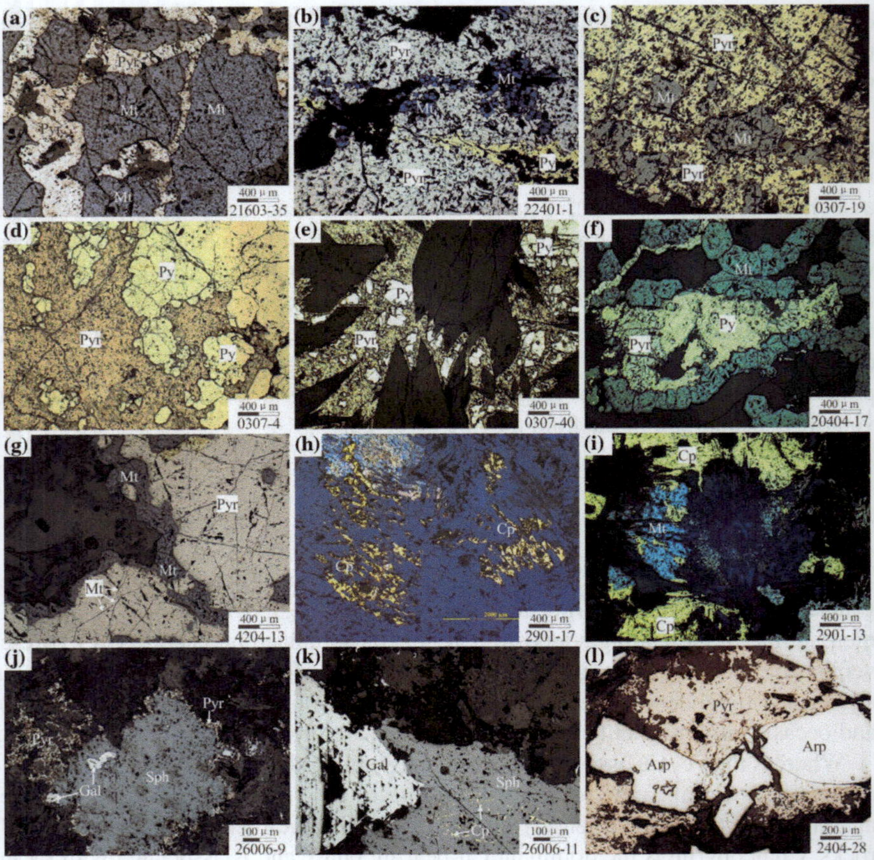

Fig. 2.16 **a** Pyrrhotite (Pyr) veinlets replace and crosscut magnetite (Mt). **b** Magnetite in metasomatic relict surrounded by pyrrhotite, and pyrrhotite replaced and cut by pyrite along granular interstitial fractures. **c** Magnetite replaced and surrounded by pyrrhotite aggregates. **d** Pyrite (Py) in embayed texture and replaced by pyrrhotite. **e** Pyrite grew within granular interstitial fractures of silicates, and replaced by pyrrhotite. **f** Magnetite grew in subhedral grains along the silicate mineral interstice, and replaced by pyrrhotite along the magnetite grain boundaries. **g** Pyrrhotite replaced by magnetite along its granular fractures and boundaries. **h** Chalcopyrite (Cp) grew along the cleavage interstice of Mg-hastingsite. **i** Later stage chalcopyrite superposed early magnetite in skarn. **j** Inclusions of galena (Gal) in sphalerite (Sph), and pyrrhotite crystallization postdates sphalerite. **k** Coeval formation of galena and sphalerite. Chalcopyrite as emulsion droplets exsolved from sphalerite. **l** Rare arsenopyrite associated with pyrrhotite

2.3 Analytical Methods

Spots for detailed microanalyses were selected from optical images of the skarn minerals. Compositional variation of Mg-rich silicates was conducted with wavelength-dispersion spectrometry (WSD) using a JOEL JXA-823V Electron Probe Micro Analyzer (EPMA) at the MLR Key Laboratory of Metallogeny and Mineral Assessment, Institute of Mineral Resources, Chinese Academy of Geological Sciences, Beijing, China. Analytical conditions were: Accelerating potential 15 kV, and beam current 5–10 nA, 3–5 μm focused electron beam. These analyses were calibrated using jadeite (Si, Na, and Al), forsterite (Mg), orthoclase (K), apatite (P), wollastonite (Ca), rutile (Ti), halite (Cl), and synthetic oxide (Cr, Mn, Fe, Ni) as standards.

2.4 Mineral Composition Results

2.4.1 Magnesian Skarn

We found major trends in the Fe and Mg content of olivine the belonging to the hortonolite series (fayalite-forsterite solid solution) (Fig. 2.17a). Fe-olivine is proximal to the intrusion, but only Mg-olivine is replaced by magnetite or serpentine. FeO contents show intensive varieties from 0.96 to 12.66 wt%, and MgO content distinguishably ranges from 43.32 to 55.99 wt% (Table 2.4). The Galinge spinel belongs to the spinel-hercynite-gahnite solid solution, and is compositionally classified into three groups based on the varying MgO, ZnO and FeO contents (Fig. 2.17b). The different colored spinels show a strong relation with the varying ZnO content. The green spinels display a relatively low ZnO content range of 2.26–5.97 wt%, but the red brown ones have a high ZnO content range of 9.28–17.81 wt%. Spinel composition varies from 54.41 to 57.65 mol% (Table 2.5). The calculated formula $(Mg_{0.61}Fe_{0.36}Zn_{0.11})Al_{1.94}O_{4.}$ The chondrodite (F: 7.00–8.72 wt%) and phlogopite (F: 1–4 wt%) both contains high F content, and may be the source of the F anomaly in the skarn-ore concentrates (Table 2.6). The calculated formula of chondrodite is following as $(Mg_{4.852},Fe_{0.148})(Si_{1.995}O_8)(F_{1.523},OH_{1.571})$. Phlogopite is always found with diopside and replaced by later amesite in the retrograde stage, which is MgO-rich (18.83–23.50 wt%) and FeO-poor (2.03–10.55 wt%), and extreme poverty in MnO (<0.23 wt%) (Table 2.6). X_{Mg} varies between 0.68 and 0.79.

The pyroxene coexisting with olivine is diopside-rich ($Di_{97.62-98.29}$) (Fig. 2.18a and Table 2.7), and the pyroxene replaced by serpentine displays a narrow composition ranging from $Di_{96.54}$ to $Di_{97.56}$. Some pyroxenes have compositional zonation with diopside-rich cores ($Di_{55.78-64.86}$) and hedenbergite-rich rims ($Hd_{50.31-56.80}$), suggesting a process of iron increasing in the fluid.

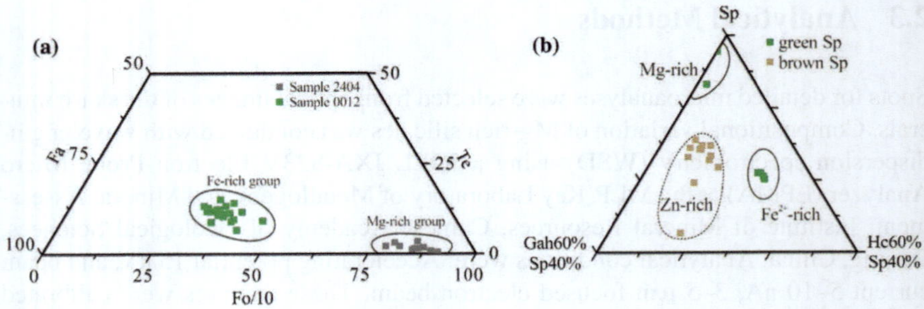

Fig. 2.17 The 3-D plots of three end-member compositions of different minerals from the II ore domains (Mg-skarn zonation). **a** Fo/10–Fa–Te plot. Fo: forsterite (Mg_2SiO_4); Fa: fayalite (Fe_2SiO_4); and Te: tephroite (Mn_2SiO_4). **b** The Sp–Gah–Hc end member plot. It shows the brown spinel is higher in Zn contents than that the green one. Sp-spinel ($MgAl_2O_4$); Hc-hercynite (Fe Al_2O_4); Gah-gahnite ($ZnAl_2O_4$)

Table 2.4 Representative results of electron microprobe analysis of olivine from the Galinge magnesian skarn zonation

Sample (wt%)	0012-2-1	0012-1-2	0012-1-3	0012-1	0012-2	2404/5/4	2404/5/5	2404/5/6
SiO_2	39.95	40.1	40.78	40.81	39.57	42.21	42.32	41.91
TiO_2	0	0.01	0	0	0	0	0.01	0.01
Al_2O_3	0	0	0.01	0	0	0	0.02	0.07
FeOt	12.66	11.61	10.9	8.25	11.79	0.96	0.97	1.13
MnO	1.97	2.24	1.88	1.7	2.43	0.06	0	0.1
MgO	43.32	44.75	45.83	47.94	44.58	55.98	55.99	55.77
CaO	0.13	0.08	0	0	0.01	0	0	0
Total	98.04	98.82	99.4	98.7	98.39	99.24	99.34	99.03
Number of oxygens (4)								
Si	1.019	1.011	1.016	1.011	1.005	1.001	1.002	0.997
Ti	0	0	0	0	0	0	0	0
Al^{IV}	0	0	0	0	0	0	0.001	0.002
Fe^{2+}	0.269	0.244	0.226	0.171	0.25	0.019	0.019	0.022
Mn	0.043	0.048	0.04	0.036	0.052	0.001	0	0.002
Mg	1.647	1.682	1.702	1.771	1.688	1.978	1.976	1.978
Ca	0.004	0.002	0	0	0	0	0	0
$\sum M$	1.963	1.976	1.968	1.978	1.99	1.998	1.995	2.002
Mole percent (mol%)								
Forsterite	84.07	85.2	86.48	89.57	84.83	98.98	99.04	98.79
Tephroite	2.17	2.43	2.01	1.8	2.62	0.06	0	0.1
Fayalite	13.76	12.37	11.51	8.63	12.55	0.95	0.96	1.12
Name	Forsterite	Forsterite	Forsterite	Forsterite	Forsterite	Forsterite	Forsterite	Forsterite

Table 2.5 Representative results of electron microprobe analysis of spinel from the Galinge magnesian skarn zonation

Sample (wt%)	4208-6-1-1	4208-6-2-1	4208-6-3-1	4208-6-4-11	0307-71_1	0307-71_4	0307-71_6	0307-71_7	0307-71_11	0307-67_4	0307-67_5	0307-67_11	0307-67_12	0307-67_3	0307-67_7	0307-68_1	0307-68_3	0307-68_5
Al_2O_3	61.21	61.33	60.82	62.29	58.69	58.89	59.33	59.28	59.09	58.66	60.25	58.04	57.33	57.91	58.28	59.7	58.08	59.57
FeOt	16.45	16.45	17.08	14.77	12.74	12.32	12.28	12.08	13.35	12.39	11.95	12.2	13	12.6	12.44	12.71	14.51	11.73
MnO	0.34	0.33	0.3	0.31	1.29	1.22	0.68	0.82	1.06	1.18	1.25	1.17	1	0.73	1.15	1.19	1.07	1.14
MgO	15.43	15.18	14.92	15.55	16.54	16.62	25.12	21.71	15.88	14.16	17.03	16.58	10.56	16.11	17.21	16.24	15.15	16.96
ZnO	5.6	5.51	5.54	5.97	10.45	10.6	2.26	5.69	10.33	13.38	9.28	11.56	17.81	12.5	10.47	9.88	9.66	10.19
Total	99.36	99.19	99.03	99.23	99.73	99.63	99.66	99.59	99.72	99.76	99.77	99.56	99.71	99.85	99.55	99.72	98.46	99.58
Number of oxygens (4)																		
Al^{IV}	1.927	1.935	1.927	1.95	1.871	1.875	1.806	1.837	1.884	1.894	1.896	1.861	1.904	1.86	1.859	1.893	1.883	1.886
Fe^{3+}	0.103	0.092	0.102	0.07	0.194	0.187	0.265	0.245	0.174	0.16	0.156	0.209	0.144	0.21	0.211	0.161	0.176	0.171
Fe^{2+}	0.266	0.277	0.284	0.259	0.095	0.091	0	0.02	0.129	0.124	0.111	0.068	0.162	0.077	0.07	0.125	0.158	0.093
Mn	0.008	0.007	0.007	0.007	0.03	0.028	0.015	0.018	0.024	0.027	0.028	0.027	0.024	0.017	0.026	0.027	0.025	0.026
Mg	0.616	0.607	0.599	0.617	0.667	0.669	0.967	0.851	0.641	0.578	0.678	0.672	0.444	0.654	0.694	0.651	0.621	0.679
Zn	0.111	0.109	0.11	0.117	0.209	0.211	0.043	0.11	0.206	0.27	0.183	0.232	0.37	0.251	0.209	0.196	0.196	0.202
$\sum M$	1	1	1	1	1	1	1	1	1	1	1	1	1	1	1	1	1	1
Mole percent (mol%)																		
Hercynite	33.41	33.78	34.96	30.72	9.47	9.14	0	2.05	12.86	12.42	11.09	6.84	16.2	7.72	7.01	12.54	15.78	9.28
Spinel	55.85	55.56	54.41	57.65	66.69	66.94	94.34	85.08	64.07	57.8	67.8	67.25	44.36	65.43	69.44	65.12	62.12	67.91
Gahnite	10.04	9.99	10	10.97	20.87	21.14	4.21	11.05	20.64	27.04	18.3	23.21	37.04	25.15	20.91	19.62	19.61	20.21
	spinel	spinel	spinel	spinel	spinel	spinel	spinel	spinel	spinel	spinel	spinel	spinel	spinel	spinel	spinel	spinel	spinel	spinel

Table 2.6 Representative results of electron microprobe analysis of chondrodite, tremolite and phlogopite from the Galinge magnesian skarn zonation

Sample (wt%)	27601-6-4	27601-6-5	27601-6-6	27601-6-7		0307-39-1	0307-61-1	0307-61-2		0307-74	0307-74-2	0307-74-3	0307-74-4	0012-2-1	4208-6-5-1	4208-6-5-2
Location	II	II	II	II		II	II	II		II	II	II	II	II	II	II
SiO_2	33.87	34.14	33.83	33.36		57.93	58.67	56.68		36.35	36.28	39.11	37.66	39.61	40.9	40.58
TiO_2	0.26	0.1	0.13	0.03		0.1	0	0.27		0.1	0.19	0.88	0.22	0.08	0.51	0.82
Al_2O_3	0	0	0.03	0		0.98	0.7	1.24		18.78	18.5	13.24	19.18	16.19	15.12	16.41
FeOt	2.94	3.29	2.69	3.1		3.96	2.42	4.08		7.58	7.31	10.55	7.55	2.03	1.85	1.96
MnO	0.11	0.03	0.07	0		0.08	0	0.12		0.22	0.23	0.02	0.19	0.09	0.07	0.02
MgO	55.23	55.4	55.4	54.56		22.89	23.69	21.04		20.28	19.9	18.83	20.19	23.5	23.11	22.95
CaO	0	0.02	0.04	0.04		13.17	13.76	14.38		0	0.05	0.01	0.03	0.06	0.04	0
Na_2O						0.35	0.2	0.18		0.2	0.33	0.12	0.33	0.17	0.06	0.05
K_2O						0.18	0.08	0.02		11.18	10.15	10.49	10.51	10.49	9.54	9.63
ZnO																
H_2O						2.21	2.23	2.17		3.94	3.32	2.88	3.34	3.92	3.68	3.81
F	8.4	8.49	7	8.72						0.33	1.48	2.13	1.67	0.33	0.69	0.52
Cl										0.08	0.09	0.33	0.11	0.06	0.08	0.06
Total	100.81	101.47	99.18	99.81		99.64	99.52	98.01		99.05	97.83	98.57	100.97	96.53	95.66	96.8
	Based on $\sum Mg + Fe = 5$, $\sum F + OH = 2$					Number of oxygens (24)				Based on $\sum Si + Ti + Al + Ca + Mg + Fe = 8$, $\sum F + Cl + OH = 2$						
Si	1.997	2	1.994	1.987		7.847	7.896	7.833		2.643	2.691	2.953	2.712	2.898	3.043	2.984
Ti	0.012	0.005	0.006	0.001		0.01	0	0.028		0.005	0.011	0.05	0.012	0.004	0.028	0.046
Al^{IV}	0	0	0.002	0		0.153	0.104	0.167		1.357	1.309	1.047	1.288	1.102	0.957	1.016
Al^{VI}						0.003	0.007	0.035		0.253	0.308	0.132	0.339	0.294	0.369	0.405
Fe^{2+}	0.145	0.161	0.132	0.154		0.448	0.272	0.471		0.461	0.453	0.666	0.455	0.124	0.115	0.12

(continued)

Table 2.6 (continued)

Sample (wt%)	27601-6-4	27601-6-5	27601-6-6	27601-6-7		0307-39-1	0307-61-1	0307-61-2		0307-74	0307-74-2	0307-74-3	0307-74-4	0012-2-1	4208-6-5-1	4208-6-5-2
Mn	0.005	0.002	0.003	0		0.009	0	0.014		0.014	0.015	0.001	0.011	0.006	0.005	0.001
Mg	4.855	4.839	4.868	4.846		4.622	4.753	4.335		2.199	2.2	2.12	2.168	2.563	2.564	2.516
Ca	0	0.001	0.003	0.003		1.911	1.984	2.129		0	0.004	0.001	0.002	0.004	0.003	0
Zn																
$\sum M$	5.005	5.003	5.006	5.003		7.003	7.016	7.012		2.927	2.98	2.92	2.975	2.991	3.056	3.042
Na						0.092	0.052	0.048		0.029	0.047	0.017	0.045	0.023	0.009	0.006
K						0.031	0.014	0.004		1.039	0.963	1.012	0.967	0.981	0.907	0.905
F	1.57	1.57	1.31	1.64						0.077	0.348	0.508	0.38	0.077	0.163	0.12
Cl										0.009	0.011	0.042	0.013	0.007	0.01	0.008
OH^-	1.558	1.553	1.631	1.541		2	2	2		1.914	1.641	1.449	1.606	1.915	1.826	1.872
Mole percent (mol%)					Mole percent (mol%)											
Chondrodite	0.999	0.999	0.999	0.999	Tremolite	0.86	0.87	0.88	Mg/(Mg + Fe + Ti + Mn)	0.821	0.821	0.747	0.819	0.95	0.945	0.938
Alleghanyite	0.001	0	0.001	0	Actinolite	0.14	0.13	0.12	Fe/(Mg + Fe + Ti + Mn)	0.172	0.169	0.235	0.172	0.046	0.043	0.045
Tilleyite	0	0	0.001	0.001												
3000	chondrodite	chondrodite	chondrodite	chondrodite		Tremolite	Tremolite	Tremolite		Phl	Phl	Phl	Phl	Phl	Phl	Phl

Table 2.7 Representative results of electron microprobe analysis of pyroxene from the Galinge magnesian skarn zonation

Sample (wt%)	0012-1	0012-1	0012-2	2404-8-1.1	2404-8-1.3	2404-8-2.2	2404/8/6	2404-11-3.1	4003-34 (core)	4003-34 (rim)	4003-34 (core)	4003-34 (rim)	4003-34 (core)	4003-34 (core)	4204-6-1.1	4204-6-1.2	4204-18-1.1	4204-18-2.1
SiO_2	53.68	53.96	53.75	55.26	55.04	55.44	55.31	54.22	52.23	48.01	38.96	51.64	52.51	51.85	49.77	48.08	48.78	48.7
TiO_2	0.02	0.06	0.07	0.11	0.1	0.11	0.02	0	0.1	0.18	0.04	0.05	0.2	0.24	0.07	0	0.11	0
Al_2O_3	0.07	0.1	0.17	0.45	0.41	0.29	0.17	0.2	1.43	2.4	6.86	1.02	0.73	0.68	0.63	0.55	1.91	1.74
FeO	0.53	0.62	0.45	0.93	0.83	0.89	0.94	0.69	16.18	19.47	16.49	19.29	12.07	12.67	23.82	23.02	20.21	22.55
MnO	0.2	0.13	0.14	0.12	0.06	0.08	0.2	0.13	0.37	0.39	0.21	0.24	0.18	0.24	0.41	0.68	0.47	0.52
MgO	17.35	17.1	17.47	17.99	17.89	18.08	17.88	17.58	11.44	7.88	13.41	10.36	11.67	11.06	3.13	3.38	4.96	3.53
CaO	26.49	26.54	26.1	25.71	25.68	24.9	25.85	26	14.7	19.16	10.91	13.97	22.61	23	22.58	23.73	23.3	22.88
Total	98.47	98.51	98.17	100.6	100	99.79	100.4	98.84	96.7	97.77	87.23	96.78	100.5	100.2	100.4	99.51	99.8	99.97
Si	1.982	1.99	1.986	1.989	1.991	2.004	1.996	1.989	2.035	1.925	1.719	2.038	1.981	1.974	1.993	1.957	1.941	1.954
$Al^{(iv)}$	0.003	0.004	0.007	0.008	0.008	0.012	0.003	0.009	0.066	0.039	0.075	0.048	0.014	0.004	0.023	0.026	0.031	0.036
$\sum T$	1.985	1.994	1.993	1.997	1.999	2.016	1.999	1.998	2.101	1.964	1.793	2.086	1.995	1.978	2.016	1.983	1.972	1.99
$Al^{(vi)}$	0	0	0	0.011	0.009	0	0.004	0	0	0.075	0.281	0	0.019	0.026	0.007	0	0.059	0.046
Ti	0	0.002	0.002	0.003	0.003	0.003	0.001	0	0.003	0.005	0.001	0.001	0.006	0.007	0.002	0	0.003	0
Fe^{3+}	0.006	0	0.001	0	0.001	0.001	0	0.001	0.012	0.015	0.02	0.01	0.03	0.027	0	0.003	0.003	0.003
Fe^{2+}	0.01	0.019	0.013	0.028	0.024	0.026	0.028	0.02	0.515	0.636	0.587	0.625	0.35	0.375	0.796	0.779	0.668	0.755
Mn	0.006	0.004	0.004	0.004	0.002	0.003	0.006	0.004	0.012	0.013	0.008	0.008	0.006	0.008	0.014	0.024	0.016	0.018
Mg	0.955	0.94	0.962	0.965	0.965	0.974	0.962	0.961	0.664	0.471	0.882	0.61	0.657	0.627	0.187	0.205	0.295	0.211
Ca	1.048	1.048	1.033	0.991	0.995	0.964	0.999	1.022	0.614	0.823	0.516	0.591	0.914	0.938	0.969	1.035	0.993	0.984
$\sum M$	2.02	2.011	2.013	1.987	1.986	1.967	1.996	2.007	1.805	1.943	1.993	1.833	1.927	1.949	1.965	2.042	1.972	1.967
Hd%	1.08	1.97	1.36	2.77	2.46	2.59	2.84	2.02	43.19	56.8	39.75	50.31	34.57	37.15	79.86	77.31	68.28	76.73
Di%	98.29	97.62	98.2	96.88	97.36	97.16	96.54	97.56	55.78	42.03	59.71	49.06	64.86	62.07	18.75	20.35	30.1	21.46
Jo%	0.63	0.41	0.44	0.35	0.17	0.25	0.62	0.42	1.03	1.18	0.54	0.63	0.57	0.78	1.39	2.33	1.62	1.81
Name	Di	Di	Di	Di	Di	Di	Di	Di	Di	Hd	Di	Hd	Di	Di	Hd	Hd	Hd	Hd

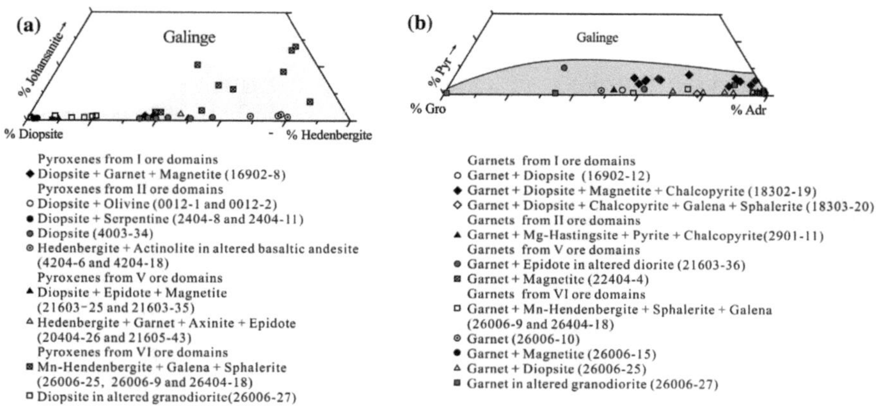

Fig. 2.18 Ternary plots of clinopyroxene (**a**) and garnet (**b**) compositions from the Galinge skarn deposits

2.4.2 Calcic Skarns

In the I ore domain, pyroxenes coexisting with garnet are diopside-rich composition (Di_{60-63}; e.g., sample 16902-8) (Fig. 2.18a and Table 2.8). In V ore domains, the composition of pyroxene grains included in axinite-dominated skarn shows an average composition of $Hd_{47.85}Di_{50.22}Jo_{1.93}$ (Fig. 2.18a), and Fe has been assumed to be all Fe^{2+}. Pyroxenes associated with epidote minerals is diposite rich composition ($Di_{89.73-91.86}$; e.g., 21603-25 and 21603-35).

The Garnet compositions are listed in the Table 2.9. In the I ore domains, garnets associated with pyroxenes are mainly andradite rich ($And_{54.25-98.78}$; e.g., sample 16902-12) (Fig. 2.18b). Nevertheless, the garnet associated with chalcopyrite are compositionally divided into two stages (Fig. 2.18b), e.g., the inner core is Ti-rich andradite (Ti-And: $TiO_2 = 1.26$–7.53 wt%; $And_{55.71-64.15}$; e.g., sample 18302-19), the composition of outer zone is $And_{71.93-93.52}$. Some garnets were found in galena + sphalerite + chalcopyrite ore minerals, and their composition is $And_{78.06}$ (e.g., 18303-20). In the outside of II ore domains, the garnet replaced by the Mg-hastingsite is moderate andradite composition ($And_{51.51}$). In the V domains, the garnet replaced by retrograde epidote in altered diorites shows moderate grossular composition ($Gro_{37.03-57.83}$; e.g., sample 21603-36) (Fig. 2.18b). The garnet associated with magnetite is andradite rich ($And_{92.59}$; e.g., sample 22404-4). In the VI ore domains, the garnet related with high grade magnetite ore is andradite rich composition ($And_{98.18}$; e.g., sample 26006-15). However, the massive medium-fine grained garnet skarn associated with low grade magnetite ore consist an intermediate composition ($Gro_{50.54}$; e.g., sample 26006-10). Some well-developed euhedral and coarse-grained garnets replaced by fine-medium grained diopsite or cut by the later diopsite-dominated veins consist of andradite with a large compositional range ($Ad_{69.89-96.76}$; e.g., sample 26006-25). In altered granodiorite, light brown garnets display compositional zonation patterns

Table 2.8 Representative results of electron microprobe analysis of pyroxene from the I and V ore domains in the Galinge deposit

Sample (wt%)	16902-8	16902-8	20404-26	21603-25	21603-35-2	21605-43	21605-87-6-4
Location	I	I	V	V	V	V	V
SiO_2	51.7	51.13	51.13	50.41	53.48	49.86	51.72
TiO_2	0	0.04	0.05	0.29	0.02	0.01	0
Al_2O_3	0.49	0.38	0.23	2.71	0.48	0.17	1.23
FeO	10.94	11.92	14.39	2.24	2.85	14.93	19.53
MnO	0.5	0.54	0.8	0.16	0.24	0.35	3.96
MgO	10.73	10.36	8.79	15.15	14.72	8.22	9.04
CaO	25.06	24.95	24.95	25.28	25.71	23.95	12
Na_2O	0.01	0.02	0.05	0	0.05	0.14	0.04
Total	99.43	99.35	100.42	96.24	97.55	97.66	97.53
Si	1.98	1.971	1.975	1.918	2.006	1.983	2.046
$Al^{(iv)}$	0.002	0.017	0.01	0.039	0.021	0.008	0.057
$\sum T$	1.982	1.989	1.985	1.957	2.027	1.991	2.104
$Al^{(vi)}$	0.02	0	0	0.082	0	0	0
Ti	0	0.001	0.001	0.008	0.001	0	0
Fe^{3+}	0	0.001	0.004	0	0.003	0.009	0.003
Fe^{2+}	0.349	0.382	0.46	0.071	0.087	0.487	0.642
Mn	0.016	0.018	0.026	0.005	0.008	0.012	0.133
Mg	0.613	0.596	0.506	0.859	0.823	0.488	0.533
Ca	1.028	1.031	1.032	1.03	1.033	1.02	0.509
$\sum M$	2.007	2.027	2.03	2.056	1.954	2.016	1.82
Na	0	0	0.001	0	0.001	0.003	0.001
Hd%	35.7	38.4	46.35	7.58	9.44	49.35	49.09
Di%	62.63	59.83	51	91.86	89.73	49.44	40.76
Jo%	1.67	1.76	2.64	0.56	0.83	1.21	10.15
Name	Di	Di	Hd	Di	Di	Hd	Mn-Hd

characterized by Ti-rich andradite cores ($And_{87.98}$; e.g., 26006-27) up to 7.54 wt% Ti, and generally the TiO_2 content of the core is higher than the rim (Fig. 2.18b). Light green euhedral garnets shows grossular-rich composition range from $Gro_{65.41}$ to $Gro_{99.10}$ (Fig. 2.18b). The garnet replaced by Mn-hedenbergite are characterized by compositional zonation of which the core is $And_{94.89}$, the inner zone is $And_{58.17}$, and the outer zone is $And_{97.80}$ (e.g., sample 26006-9) (Fig. 2.18b).

The vast majority of Galinge epidotes belongs to the Al-Fe^{3+} solid solution series with Fe^{3+} contents between about $X_{Epi} = 0.14$ and $X_{Epi} = 0.28$ (Table 2.10). The Galinge epidotes is Fe^{3+}-rich varying from 0.406 to 0.811, with Fe^{3+} the only sig-

Table 2.9 Representative results of electron microprobe analysis of garnet from the Galinge deposit

Sample (wt%)	16902-12	16902-12	16902-12	16902-12	18302-19(core)	18302-19(rim)	18302-19(core)	18302-19(rim)	18302-19-3-1(core)	18302-19-3-2(core)
Location	I	I	I	I	I	I	I	I	I	I
SiO_2	35.42	34.4	33.66	36.14	33.84	34.57	33.48	34.66	34.86	35.22
TiO_2	0.04	0.03	0.01	0.11	5.24	0.3	5.49	0.1	3.06	3.35
Al_2O_3	0.01	0	0.08	9.59	6.19	0.67	6.06	4.47	6.88	7.34
FeO	28.19	28.11	27.21	16.71	16	27.19	15.85	22.08	16.47	15.2
MnO	0.09	0.25	0.13	0.66	0.39	0.78	0.38	3.38	0.45	0.39
MgO	0.1	0.15	0.12	0.09	1.06	0.49	0.93	0.02	0.93	1.22
CaO	33.84	33.72	32.89	34.08	34.31	31.2	34.16	30.92	34.41	34.44
Total	97.69	96.65	94.1	97.37	97.03	95.19	96.35	95.62	97.05	97.16
Si	3.246	3.204	3.212	3.085	2.932	3.232	2.922	3.155	3.004	3.015
$Al^{[IV]}$	0	0	0	0	0.068	0	0.078	0	0	0
$\sum$	3.246	3.204	3.212	3.085	3	3.232	3	3.155	3.004	3.015
$Al^{[VI]}$	0.001	0	0.009	0.965	0.564	0.073	0.545	0.479	0.698	0.741
Ti	0.003	0.002	0.001	0.007	0.341	0.021	0.36	0.007	0.198	0.216
Fe^{3+}	2.156	2.184	2.167	1.191	1.157	2.121	1.154	1.677	1.184	1.086
$\sum$	2.16	2.187	2.176	2.163	2.061	2.215	2.059	2.163	2.081	2.042
Fe^{2+}	0	0	0	0	0	0	0	0	0	0
Mn	0.007	0.019	0.01	0.048	0.028	0.062	0.028	0.26	0.033	0.028
Mg	0.014	0.021	0.017	0.011	0.137	0.068	0.122	0.002	0.119	0.156
Ca	3.323	3.364	3.363	3.117	3.184	3.125	3.194	3.016	3.176	3.158
$\sum$	3.343	3.404	3.39	3.176	3.349	3.255	3.344	3.278	3.328	3.342
And%	99.31	98.79	98.78	54.25	63.19	92.98	64.15	71.93	59.69	55.71

(continued)

Table 2.9 (continued)

Sample (wt%)	16902-12	16902-12	16902-12	16902-12	18302-19(core)	18302-19(rim)	18302-19(core)	18302-19(rim)	18302-19-3-1(core)	18302-19-3-2(core)
Location	I	I	I	I	I	I	I	I	I	I
Gro%	0.05	0	0.4	43.96	30.79	3.22	30.29	20.56	35.2	38.01
Pyr%	0.63	1.21	0.82	1.79	6.01	3.8	5.55	7.51	5.11	6.29
Name	And	And	And	And	Ti-Gt	And	Ti-And	And	Ti-And	Ti-And
Sample (wt%)	18302-19-3-3(core)	18302-19-4-1(core)	18302-19-4-2(rim)	18302-19-4-3(rim)	18302-19-4-4(core)	18302-19-4-5(rim)	18303-20	2901-11-1	2901-14-1-3	2901-14-2-1
Location	I	I	I	I	I	I	I	II	II	II
SiO_2	34.99	34.87	34.77	34.98	36.11	34.98	34.43	37.23	36.93	35.48
TiO_2	0.1	4.02	0.12	0.18	1.59	0	0	0.29	0.33	0.42
Al_2O_3	1.26	6.73	0.2	2.23	7.53	1.5	4.43	9.76	8.74	8.65
FeO	27.23	16.05	28.51	26.46	16.41	27.31	22.51	15.29	16.69	17.43
MnO	2.3	0.24	2.48	1.39	0.31	2.83	0.09	0.58	0.62	0.48
MgO	0.02	1.11	0.02	0.05	0.79	0.03	0.01	0.19	0.18	0.10
CaO	31.07	34.57	31.03	32.19	34.89	30.92	35.06	35.48	34.67	36.34
Total	96.95	97.59	97.12	97.48	97.63	97.56	96.53	98.81	98.17	98.89
Si	3.22	2.989	3.223	3.178	3.085	3.206	3.103	3.109	3.127	3.021
$Al^{[IV]}$	0	0.011	0	0	0	0	0	0	0.000	0.000
$\sum$	3.22	3	3.223	3.178	3.085	3.206	3.103	3.109	3.127	3.021
$Al^{[VI]}$	0.136	0.668	0.021	0.239	0.758	0.162	0.471	0.96	0.872	0.868
Ti	0.007	0.259	0.008	0.013	0.102	0	0	0.018	0.021	0.027
Fe^{3+}	2.091	1.148	2.206	2.006	1.17	2.088	1.693	1.065	1.179	1.239
$\sum$	2.234	2.076	2.235	2.257	2.03	2.25	2.164	2.044	2.072	2.133
Fe^{2+}	0	0	0	0	0	0	0	0	0.000	0.000

(continued)

Table 2.9 (continued)

Sample (wt%)	18302-19-3-3(core)	18302-19-4-1(core)	18302-19-4-2(rim)	18302-19-4-3(rim)	18302-19-4-4(core)	18302-19-4-5(rim)	18303-20	2901-11-1	2901-14-1-3	2901-14-2-1
Location	I	I	I	I	I	I	I	II	II	II
Mn	0.179	0.017	0.195	0.107	0.022	0.22	0.007	0.041	0.045	0.034
Mg	0.003	0.141	0.002	0.007	0.101	0.003	0.001	0.023	0.023	0.012
Ca	3.064	3.175	3.082	3.132	3.193	3.035	3.386	3.174	3.145	3.315
$\sum$	3.246	3.334	3.279	3.246	3.316	3.258	3.393	3.239	3.213	3.362
And%	89.04	59.73	93.52	86.44	58.19	87.05	78.06	51.51	56.24	57.95
Gro%	5.81	34.76	0.91	10.29	37.73	6.75	21.7	46.41	41.60	40.60
Pyr%	5.15	5.51	5.57	3.27	4.08	6.2	0.24	2.09	2.16	1.45
Name	And	Ti-And	And	And	Ti-And	And	And	And	And	And

Sample (wt%)	21603-36	21603-36	22404-4-1	26006-25-2	26006-25-3.1	26006-25-3.2	26006-25-3.3	26006-25-3.4	26006-25-3.5	26006-25-3.6	26006-27-5
Location	V	V	V	VI	VI	VI	VI	VI	VI	VI	VI
SiO_2	36.8	35.62	35.36	36.36	36.11	35.64	35.44	36.2	36.64	35.74	33.82
TiO_2	0.46	0.44	0.01	0.81	0	0.02	0	0.03	0	0	7.54
Al_2O_3	8.16	13.98	0.15	3.92	0.45	2.06	0.78	0.47	6.02	2.57	1.31
FeO	18.93	15.38	26.66	22.73	27.82	25.84	27.28	27.68	20.82	24.87	19.8
MnO	0.69	0.38	0.19	0.48	0.3	0.33	0.48	0.52	0.55	0.37	0.18
MgO	0.17	0.19	0.04	0.04	0.08	0.09	0.09	0.09	0.08	0.13	0.63
CaO	33.61	29	33.02	33.54	32.07	32.7	32.38	32.06	33.35	32.49	33.36
Total	98.82	94.99	95.42	97.88	96.83	96.68	96.43	97.03	97.47	96.17	96.65
Si	3.124	3.023	3.293	3.201	3.308	3.241	3.265	3.309	3.195	3.247	2.998
$Al^{[IV]}$	0	0	0	0	0	0	0	0	0	0	0.002

(continued)

Table 2.9 (continued)

Sample (wt%)	21603-36	21603-36	22404-4-1	26006-25-2	26006-25-3.1	26006-25-3.2	26006-25-3.3	26006-25-3.4	26006-25-3.5	26006-25-3.6	26006-27-5
Location	V	V	V	VI	VI	VI	VI	VI	VI	VI	VI
$\sum$	3.124	3.023	3.293	3.201	3.308	3.241	3.265	3.309	3.195	3.247	3
$Al^{[VI]}$	0.816	1.399	0.016	0.406	0.048	0.221	0.084	0.05	0.619	0.275	0.135
Ti	0.029	0.028	0.001	0.054	0	0.002	0	0.002	0	0	0.503
Fe^{3+}	1.341	0.778	2.071	1.67	2.126	1.961	2.097	2.111	1.515	1.885	1.464
$\sum$	2.187	2.205	2.088	2.131	2.174	2.183	2.181	2.163	2.134	2.16	2.103
Fe^{2+}	0	0.311	0	0	0	0	0	0	0	0	0
Mn	0.049	0.027	0.015	0.036	0.024	0.025	0.037	0.04	0.041	0.028	0.014
Mg	0.022	0.024	0.006	0.005	0.011	0.012	0.012	0.012	0.01	0.018	0.084
Ca	3.056	2.638	3.294	3.164	3.147	3.186	3.196	3.14	3.115	3.162	3.168
$\sum$	3.128	3	3.314	3.205	3.182	3.223	3.246	3.191	3.166	3.209	3.265
And%	60.81	32.18	98.59	79.39	96.76	88.89	94.71	96.14	69.89	86.03	87.98
Gro%	37.03	57.83	0.77	19.32	2.19	10	3.8	2.28	28.56	12.56	8.12
Pyr%	2.15	9.99	0.64	1.29	1.05	1.11	1.48	1.57	1.56	1.41	3.89
Name	And	Gro	And	And	And	And	And	And	And	And	Ti-And

Sample (wt%)	26006-27-6.1	26006-27-6.2	26006-9-2.2	26006-9-4.1(core)	26006-9-4.2(rim)	26006-9-4.3(rim)	26006-9-10.1	26006-10-2.1	26006-15-1	26404-18
Location	VI	VI	VI	VI	VI	VI	VI	VI	VI	VI
SiO_2	37.71	38.02	36.11	35.93	37	35.97	35.61	37.12	35.97	34.19
TiO_2	1.3	0.11	0.02	0	0	0.01	0.02	0.49	0.01	0.46
Al_2O_3	13.44	21.82	0.59	0.8	8.86	0.24	0.37	10.41	0	4.7
FeO	9.93	0.24	27.89	27.23	17.66	28.46	28.4	13.9	28.07	20.85

(continued)

Table 2.9 (continued)

Sample (wt%)	26006-27-6.1	26006-27-6.2	26006-9-2.2	26006-9-4.1(core)	26006-9-4.2(rim)	26006-9-4.3(rim)	26006-9-10.1	26006-10-2.1	26006-15-1	26404-18
Location	VI	VI	VI	VI	VI	VI	VI	VI	VI	VI
MnO	0.15	0.04	0.27	0.35	0.28	0.44	0.37	0.28	0.45	0.72
MgO	0	0	0.01	0.09	0.01	0	0.03	0.24	0.18	0.13
CaO	35.76	37.3	33.5	33.41	33.99	33.25	32.89	34.99	33.28	33.56
Total	98.29	97.53	98.39	97.8	97.79	98.37	97.69	97.42	97.95	94.61
Si	3.06	2.943	3.265	3.261	3.148	3.267	3.257	3.114	3.278	3.119
$Al^{[IV]}$	0	0.057	0	0	0	0	0	0	0	0
$\sum$	3.06	3	3.265	3.261	3.148	3.267	3.257	3.114	3.278	3.119
$Al^{[VI]}$	1.285	1.934	0.063	0.085	0.888	0.026	0.04	1.029	0	0.505
Ti	0.079	0.006	0.001	0	0	0.001	0.002	0.031	0.001	0.031
Fe^{3+}	0.673	0.016	2.104	2.062	1.254	2.157	2.167	0.973	2.134	1.588
$\sum$	2.037	1.956	2.169	2.148	2.142	2.183	2.208	2.033	2.136	2.124
Fe^{2+}	0	0	0	0	0	0	0	0	0	0
Mn	0.01	0.003	0.02	0.027	0.02	0.034	0.029	0.02	0.034	0.056
Mg	0	0	0.001	0.012	0.001	0	0.004	0.03	0.024	0.017
Ca	3.109	3.093	3.245	3.248	3.098	3.235	3.222	3.144	3.249	3.28
$\sum$	3.119	3.096	3.267	3.287	3.119	3.269	3.255	3.194	3.308	3.353
And%	34.24	0.81	96.45	94.89	58.17	97.8	97.22	47.81	98.18	74.15
Gro%	65.41	99.1	2.89	3.93	41.19	1.19	1.79	50.54	0.02	23.58
Pyr%	0.35	0.09	0.65	1.19	0.63	1.02	0.99	1.64	1.8	2.27
Name	Gro	Gro	And	And	And	And	And	Gro	And	And

nificant substituent for Al. Epidote resulting from late replacement of grossular-rich garnet shows X_{Epi} ratio (mean 0.21; e.g., sample 21603-36) than that which replaces axinite skarn (mean 0.255; e.g., sample 21603-41 and 21605-43). Epidote in axinite-bearing assemblage is relatively richer in FeO_{tot} and poorer in Al_2O_3 than in the prehnite-bearing one, and X_{Epi} vary between 0.61 and 0.85 (mean $X_{Epi} = 0.72$; Table 2.10). The mean X_{Epi} ratio of epidote coexisted with hematite (e.g., sample 26006-8 and 26006-7) is 0.22.

The Galinge prehnites are generally regarded to be $Ca_{1.97}Al_{1.88}Fe_{0.01}(Si_3O_{10})(OH)_2$ (Table 2.11), showing a largely pure Al-prehnite component. The X_{Fe} ($Fe^{3+}/(Fe^{3+}$+Al)) range of 0.001–0.11, which is abundantly Al-rich and Fe^{3+}-poor. In the calcium-rich skarn zonation, mica is compositionally classified into biotite which is distinctive by having a characteristic dark brown to light green color. Generally, the varying colors are caused by the different Fe and Mg content, e.g. the intensity of brown color increase with the Fe concentration. Fe biotite contains up to 14.08 wt% FeO (total) and to 19.88 wt% MgO (inferred) (Table 2.11).

The Galinge axinites commonly occur with epidote and pyroxene, which are very close to the Fe end-member representing the Fe-dominant species axinite-(Fe)in the (Mn, Fe^{2+}, Mg)-composition ternary, with Fe/(Fe + Mn + Mg) = X_{Fe} = 0.719-0.818 (Table 2.12). B_2O_3 content calculated based on 2 apfu boron ranges from 5.81 to 5.84%. Hydrogen content ranges from 1.944 apfu to 1.999 apfu. OH content plays a crucial role in charge-balance relations of which coupled mechanism of hydroxyl deficiency is proposed by $SiO(BOH)_{-1}$.

The compositions of sphene cores vary greatly from the sphene rims, e.g., the Al_2O_3 and SiO_2 contents of cores are lower than rims (Table 2.11). Titanite in the axinite-bearing assemblage contains relatively high TiO_2 (mean = 35.16 wt%). It is note that "high Ti content" may actually mean "low Al content" as a consequence of the $(TiO)(Al(OH,F))_{-1}$ substitution mechanism.

2.4.3 Manganesian-Calcic Skarns

In No. VI Ore Domain, massive Mn-rich hedenbergite skarn is associated with galena and sphalerite ore. The MnO content in such pyroxenes is generally high and varies between 0.15 and 7.84 (Fig. 2.18a and Table 2.13). In addition, the pyroxene in altered granodiorite displays a wide diopside composition ($Di_{44.08–98.03}$).

2.4.4 Magnetite Composition

Step-wise provenance analysis found that TiO_2, Cr_2O_3, MnO, MgO, V_2O_3, and Al_2O_3 could appropriately discriminate the igneous magnetite from hydrothermal magnetite [9]. In the ternary Fig. 2.19, the magnetites from the II ore domains are

Table 2.10 Representative results of electron microprobe analysis of epidote from the V and VI ore domains

Sample (wt%)	21603-14	21603-36	21603-36	21603-36	21603-36	21603-36	21603-36	21603-36	21603-36	21603-36	21603-41	21603-41	21605-43	26006-6-1	26006-7-1	26006-8-1	26006-8-2
Location	V	V	V	V	V	V	V	V	V	V	V	V	V	VI	VI	VI	VI
SiO_2	37.8	37.56	38.2	36.58	36.72	37.76	36.87	32.7	35.11	39.32	37.54	36.94	41.36	39.27	39.03	38.18	38.23
TiO_2	0.05	0.05	0	0.03	0.03	0	0.01	0.05	0.05	0.06	0.11	0.08	0.06	0.05	0.02	0.01	0.01
Al_2O_3	25.19	26.56	24.88	23.71	22.71	25.26	22.8	20	18.05	25.27	24.27	22.42	17.11	26.74	24.56	24.94	24.19
FeO	8.51	6.09	8.91	7.94	9.26	8.92	8.72	9.16	9.36	8.68	9.66	12.01	8.4	7.68	10.79	8.8	9.76
MnO	0.03	0.2	0.77	0.6	0.53	0.56	0.68	0.72	0.73	0.64	0.82	0	2.16	0.64	0.14	0.04	0.34
MgO	0.01	0.06	0.12	0.54	1.33	0.04	0.87	0.99	2.58	0.56	0.01	0	0.98	0.03	0	0.08	0
CaO	24.35	24.22	22.04	21.32	20.38	23.04	20.82	19.3	16.86	21.87	23.12	23.7	19.95	23.39	23.72	23.68	23.05
Total	95.97	94.76	94.95	90.76	91.05	95.59	90.85	82.99	82.83	96.47	95.54	95.14	90.04	97.8	98.25	95.74	95.58
Si	3	2.998	3.072	3.067	3.07	3.014	3.094	3.012	3.231	3.105	3.009	2.99	3.547	3.054	3.046	3.041	3.062
Ti	0.003	0.003	0	0.002	0.002	0	0.001	0.003	0.003	0.004	0.006	0.005	0.004	0.003	0.001	0.001	0.001
Al	2.356	2.499	2.358	2.343	2.238	2.376	2.255	2.171	1.957	2.351	2.293	2.139	1.729	2.45	2.259	2.341	2.284
Fe^{3+}	0.563	0.406	0.598	0.556	0.646	0.594	0.611	0.704	0.719	0.572	0.646	0.811	0.601	0.498	0.702	0.585	0.652
Mn	0.002	0.013	0.052	0.043	0.038	0.038	0.048	0.056	0.056	0.043	0.056	0	0.157	0.042	0.01	0.003	0.023
Mg	0.001	0.008	0.014	0.068	0.166	0.004	0.109	0.136	0.354	0.066	0.001	0	0.125	0.004	0	0.009	0
$\sum M$	2.926	2.929	3.023	3.011	3.089	3.013	3.024	3.071	3.089	3.036	3.003	2.955	2.616	2.997	2.971	2.939	2.96
Ca	2.07	2.071	1.899	1.916	1.826	1.971	1.871	1.905	1.662	1.85	1.986	2.055	1.833	1.949	1.983	2.02	1.978
X_{Epi}	0.19	0.14	0.2	0.19	0.22	0.2	0.21	0.24	0.27	0.2	0.22	0.28	0.26	0.17	0.24	0.2	0.22
Name	Epi	Epi	Epi	Epi	Epi	Epi	Epi	Al	Al	Epi	Epi	Epi	Epi	Epi	Epi	Epi	Epi

Note A + M + Si = 8; M + Si = 8; Hypothetically there is no Fe^{2+} in M site for epidote formula; $X_{Epi} = Fe^{3+}/(Fe^{3+} + Al)$

Table 2.11 Representative results of electron microprobe analysis of biotite, prehnite and sphene from the Galinge deposit

Sample (wt%)	26006-27-3.3	26006-27-5.2	26006-27-5.4	26006-27-7.1	26006-27-6.3	26006-27-7.3	26006-27-1.1	2901-14-7-1	2901-14-7-2	2901-14-7-3	2901-14-7-4	2901-14-5-1	2901-14-5-2	2901-14-6-1	2901-14-6-2
SiO_2	41.52	41.26	39.11	41.08	44.48	44.39	43.96	30.85	29.41	31.50	30.83	30.12	31.48	31.00	30.94
TiO_2	3.04	2.86	2.81	1.03	0.03	0.04	0.03	28.54	29.69	29.29	30.87	27.93	32.03	32.56	28.40
Al_2O_3	9.63	12.71	12.94	13.52	22.86	22.54	23.14	3.65	1.69	4.63	2.72	3.37	3.58	3.13	2.97
FeO	14.08	9.04	9.33	12.64	0.22	0.36	0.04	0.53	0.75	0.98	0.82	0.63	0.43	0.60	0.95
MnO	0.18	0.17	0.17	0.23	0.00	0.04	0.00	0.00	0.00	0.03	0.04	0.00	0.02	0.02	0.02
MgO	17.23	19.88	18.53	18.61	0.02	0.00	0.00	0.01	0.04	0.03	0.01	0.00	0.01	0.00	0.00
CaO	1.69	0.30	0.99	0.05	26.43	25.87	26.73	28.41	27.62	28.63	28.27	27.68	28.94	28.96	28.24
Na_2O	0.01	0.09	0.04	0.06	0.06	0.00	0.05	0.01	0.02	0.00	0.00	0.04	0.00	0.03	0.00
K_2O	10.15	10.64	8.98	10.33	0.00	0.00	0.01	0.00	0.00	0.00	0.00	0.00	0.00	0.00	0.00
F	0.52	0.52	0.52	0.52	–		–	0.73	0.05	0.90	0.44	0.76	0.88	0.49	0.85
H_2O	3.77	3.85	3.66	3.85	4.30	4.27	4.29	–	–	–	–	–	–	–	–
Total	101.9	101.4	97.15	101.9	98.40	97.52	98.24	92.72	89.28	95.99	94.00	90.52	97.37	96.78	92.37
Si	3.087	3.006	2.989	2.997	3.097	3.115	3.068	1.086	1.075	1.073	1.071	1.088	1.057	1.047	1.098
Ti	0.17	0.157	0.162	0.056	0.002	0.002	0.001	0.756	0.817	0.750	0.807	0.759	0.809	0.827	0.758
Al^{IV}	0.844	0.994	1.011	1.003											
Al^{VI}	0	0.096	0.155	0.159	1.876	1.864	1.904	0.151	0.073	0.186	0.111	0.143	0.142	0.124	0.124

(continued)

Table 2.11 (continued)

Sample (wt%)	26006-27-3.3	26006-27-5.2	26006-27-5.4	26006-27-7.1	26006-27-6.3	26006-27-7.3	26006-27-1.1	2901-14-7-1	2901-14-7-2	2901-14-7-3	2901-14-7-4	2901-14-5-1	2901-14-5-2	2901-14-6-1	2901-14-6-2
Fe^{3+}					0.013	0.021	0.002	–	–	–	–	–	–	–	–
Fe^{2+}	0.876	0.551	0.596	0.771	–	–	–	0.015	0.023	0.028	0.024	0.019	0.012	0.017	0.028
Mn	0.011	0.01	0.011	0.014	0	0.002	0	0.000	0.000	0.001	0.001	0.000	0.001	0.001	0.001
Mg	1.911	2.159	2.112	2.024	0.002	0	0	0.001	0.002	0.002	0.000	0.000	0.000	0.000	0.000
$\sum M$	2.968	2.973	3.035	3.025	1.89	1.888	1.906	–	–	–	–	–	–	–	–
Ca	0.135	0.023	0.081	0.004	1.971	1.945	1.999	1.072	1.082	1.045	1.052	1.071	1.041	1.048	1.073
Na	0.002	0.012	0.006	0.008	0.008	0	0.007	0.001	0.002	0.000	0.000	0.003	0.000	0.002	0.000
K	0.965	0.991	0.877	0.963	0	0	0.001	0.000	0.000	0.000	0.000	0.000	0.000	0.000	0.000
$\sum A$	1.101	1.027	0.965	0.975	1.979	1.945	2.006	1.072	1.083	1.045	1.052	1.074	1.041	1.049	1.073
F	0.122	0.119	0.125	0.12	–	–	–	0.082	0.006	0.096	0.049	0.086	0.094	0.053	0.096
OH^-	1.87	1.873	1.866	1.873	2	2	2	–	–	–	–	–	–	–	–
X_{Mg}	0.683	0.794	0.777	0.721	1	0	0	0.033	0.089	0.057	0.017	0	0.023	0	0.002
Fe^{2+}/Fe^{3+}															
Name	Bio	Bio	Bio	Bio	Pre	Pre	Pre	Spn	Spn	Spn	Spn	Spn	Spn	Spn	Spn

Table 2.12 Representative results of electron microprobe analysis of axinite from the Galinge deposit

Sample (wt%)	20404-26	21603-14	21603-41	21605-77	21605-43	21603-14_1	21603-41_2	21605-43_1	21605-77_1
SiO_2	41.43	41.40	40.91	41.32	41.36	38.20	37.07	35.83	41.57
TiO_2	0.13	0.03	0.01	0.06	0.06	0.17	0.04	0.01	0.01
Al_2O_3	16.52	16.49	16.45	16.40	17.11	18.82	18.64	20.84	16.89
FeO	9.67	9.17	9.91	9.37	8.40	8.65	9.79	9.82	9.21
MnO	1.94	2.01	1.78	2.16	2.16	2.67	2.63	2.95	2.65
MgO	0.94	0.00	0.61	0.82	0.98	1.12	0.67	1.21	0.87
CaO	19.41	20.47	20.11	19.68	19.95	22.50	23.25	24.15	19.36
Na_2O	0.00	0.08	0.02	0.00	0.02	0.02	0.00	0.01	0.00
K_2O	0.00	0.05	0.02	0.02	0.02	0.00	0.00	0.00	0.00
B_2O_3	7.63	6.88	8.43	7.86	6.97	5.35	5.42	2.67	6.94
F	0.00	0.00	0.05	0.06	0.00	0.00	0.00	0.00	0.00
Cl	0.00	0.05	0.08	0.01	0.00	0.00	0.00	0.00	0.00
H_2O	1.59	1.55	1.57	1.57	1.58	1.76	1.81	1.92	1.60
Total	99.27	98.19	99.94	99.32	98.62	99.26	99.31	99.42	99.10
Based on $Ca + Fe + Mn + Mg = 3$									
Si	3.894	3.969	3.808	3.879	3.918	3.249	3.064	2.801	3.901

(continued)

Table 2.12 (continued)

Sample (wt%)	20404-26	21603-14	21603-41	21605-77	21605-43	21603-14_1	21603-41_2	21605-43_1	21605-77_1
Ti	0.009	0.002	0.001	0.004	0.004	0.011	0.002	0.000	0.000
$Al^{[IV]}$	0.097	0.029	0.191	0.117	0.078	0.740	0.934	1.199	0.098
$\sum$	4.000	4.000	4.000	4.000	4.000	4.000	4.000	4.000	4.000
$Al^{[VI]}$	1.733	1.834	1.613	1.696	1.833	1.147	0.881	0.721	1.769
B	1.267	1.166	1.387	1.304	1.167	1.853	2.119	2.279	1.231
$\sum$	3.000	3.000	3.000	3.000	3.000	3.000	3.000	3.000	3.000
Fe2+	0.759	0.734	0.770	0.734	0.664	0.614	0.675	0.641	0.721
Mn	0.154	0.164	0.140	0.172	0.173	0.192	0.184	0.196	0.211
Mg	0.132	0.000	0.085	0.115	0.138	0.143	0.082	0.141	0.122
Ca	1.955	2.103	2.005	1.979	2.024	2.051	2.059	2.023	1.946
$\sum$	3.000	3.000	3.000	3.000	3.000	3.000	3.000	3.000	3.000
Na	0.000	0.014	0.004	0.000	0.004	0.004	0.000	0.001	0.000
K	0.000	0.006	0.002	0.003	0.003	0.000	0.000	0.000	0.000
F	0.000	0.000	0.015	0.018	0.000	0.000	0.000	0.000	0.000
Cl	0.001	0.009	0.012	0.001	0.000	0.000	0.000	0.000	0.000
OH^-	0.999	0.991	0.973	0.981	1.000	1.000	1.000	1.000	1.000
X_{Fe}	25.28	24.46	25.65	24.47	22.14	20.48	22.50	21.36	24.04
Name	Fax	Fax	Fax	Fax	Fax	Fax	Fax	Fax	Fax

Table 2.13 Representative results of electron microprobe analysis of pyroxene from the VI ore domain

Sample (wt%)	26006-25-4	26006-25-5	26006-25-6	26006-25-7	26006-27-2.1	26006-27-3.1	26006-27-3.4	26006-27-5.3	26006-27-7.2	26006-9-2.1	26006-9-3	26006-9-13	26006-26-2	26404-18	26404-18	26404-18-1	26404-18-2
SiO_2	51.97	48.97	50.8	52.13	54.28	54.14	51.65	54.66	54.78	51.06	50.7	50.66	50.78	52.28	48.32	49.63	49.6
Al_2O_3	0.02	0.01	0.17	0	0.1	0.8	1.12	0.14	0.13	0	0.02	0.01	6	0.35	0.06	0.03	0.05
FeO	11.84	24.46	16	12.51	9.31	9.33	6.56	8	5.49	12.99	16.14	17.6	2.4	2.4	20.21	20.19	21.65
MnO	0.93	2.02	1.11	0.92	0.45	0.53	0.15	0.34	0.32	6.19	3.92	3.43	0.03	0.5	7.84	7.51	5.48
MgO	10.14	1.45	7.54	9.9	13.47	13.67	10.62	14.61	16.06	6.31	5.85	5.16	15.68	15.24	0.51	0.84	2.02
CaO	24.33	22.82	24.07	24.24	20.38	20.54	23.21	21.04	22.73	23.38	23.66	23.84	25.49	26	23.03	23.08	21.52
Total	99.26	99.73	99.72	99.76	99.98	100.9	99.4	100.3	100.1	99.97	100.3	100.7	100.8	96.79	100.0	101.3	100.4
Si	2.001	2.003	1.989	2.002	2.026	2.002	1.96	2.02	2.01	2.006	1.997	1.997	1.84	1.983	1.989	2.005	2.01
$Al^{(iv)}$	0.001	0	0.008	0	0.004	0.035	0.01	0.006	0.006	0	0.001	0.001	0.097	0.015	0.003	0.001	0.002
$\sum T$	2.002	2.003	1.997	2.002	2.03	2.037	1.971	2.027	2.016	2.006	1.998	1.998	1.937	1.998	1.992	2.006	2.012
$Al^{(vi)}$	0	0	0	0	0	0	0.04	0	0	0	0	0	0.16	0	0	0	0
Fe^{3+}	0.001	0	0.001	0.003	0.105	0.088	0.202	0.077	0.03	0.001	0.001	0.002	0	0	0.004	0.003	0.002
Fe^{2+}	0.379	0.835	0.521	0.398	0.185	0.2	0.005	0.17	0.138	0.425	0.53	0.577	0.072	0.076	0.691	0.677	0.73
Mn	0.03	0.07	0.037	0.03	0.014	0.016	0.005	0.011	0.01	0.206	0.131	0.114	0.001	0.016	0.273	0.257	0.188
Mg	0.582	0.089	0.44	0.567	0.749	0.753	0.601	0.805	0.879	0.37	0.343	0.303	0.847	0.862	0.031	0.051	0.122
Ca	1.003	1	1.01	0.997	0.815	0.814	0.944	0.833	0.894	0.984	0.998	1.007	0.99	1.057	1.016	0.998	0.934
$\sum M$	1.995	1.993	2.008	1.992	1.763	1.783	1.555	1.819	1.92	1.985	2.002	2.002	1.911	2.01	2.011	1.983	1.974
Hd %	38.26	84.05	52.23	40.04	19.54	20.59	0.89	17.24	13.43	42.49	52.75	58.02	7.87	7.94	69.4	68.79	70.20
Di %	58.67	8.92	44.08	56.95	78.98	77.71	98.33	81.69	85.58	36.93	34.21	30.47	92.02	90.37	3.13	5.13	11.72
Jo%	3.07	7.03	3.69	3.01	1.48	1.7	0.78	1.07	0.98	20.58	13.04	11.51	0.11	1.69	27.46	26.07	18.07
Name	Di	Mn-Hd	Hd	Di	Di	Di	Di	Di	Di	Mn-Hd	Mn-Hd	Mn-Hd	Di	Di	Mn-Hd	Mn-Hd	Mn-Hd

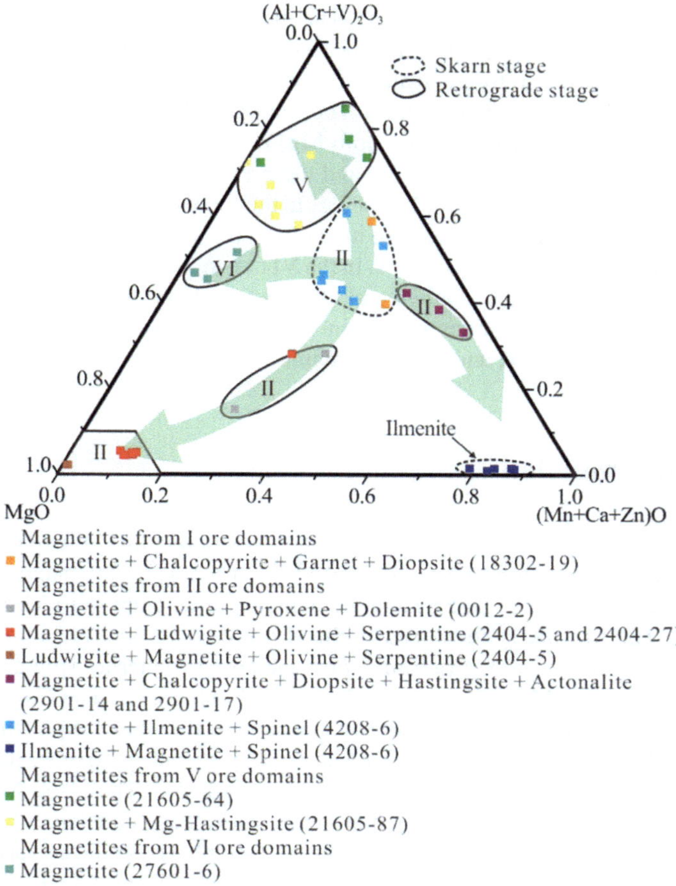

Fig. 2.19 The various content of minor elements in the magnetite, ludwigite and ilmenite from the II, V, and VI ore domains are plotted

characterized by MgO-rich. Magnetite coexisting with olivine replaced by serpentine is featured by its high MgO content, indicating that composition of magnetite is overwhelmingly controlled by host rocks that have been attacked by magmatic-derived fluids. However, the magnetite coexisting with diopside in the I and II ore domains contains relatively low MgO content (Fig. 2.19).

In addition, coexisting ilmenite-magnetite exsolved from spinel skarns in equilibrium texture (ilmenite lamellae parallel to the (111) planes of the magnetite octahedron) are very common [36, 37, 40]. Based on its composition, including intermediate MgO contents up to 6 wt% (Fig. 2.19 and Table 2.14) and fields of exsolved skeletal spinels, they are probably formed in the skarn stage and associated with the early hydrothermal fluid evolved from magma. The magnetite are distinctively higher on content of $V_2O_3 + Cr_2O_3 + Al_2O_3$ than the intergrowth ilmenite (Fig. 2.19).

From the skarn to the retrograde alteration stage, the MgO contents are gradually increased in the II ore domain (Fig. 2.19). The Galinge ludwigite is completely closer to the "pure" magnesian (up to 2.364 apfu Mg) end-member (Fig. 2.19). It almost contains no vonsenite in solid solution, and the presumed B_2O_3 compositions take weight percentages ranging of 19.82–20.87%. In the V ore domain, the $V_2O_3 + Cr_2O_3 + Al_2O_3$ contents of magnetite increase much from the skarn stage to the retrograde stage, and identically higher than others. In the VI ore domain, the magnetites contain very low MnO + CaO + ZnO content and intermediate $V_2O_3 + Cr_2O_3 + Al_2O_3$.

Figure 2.19 The various content of minor elements in the magnetite, ludwigite and ilmenite from the II, V, and VI ore domains are plotted.

2.5 Skarn Zonation

Based on the mineral assemblages and their compositions, the Galinge deposit can be roughly divided into three proximal to distal skarn zones; i.e. the magnesian skarn (Mg-SK), the calcic skarn (Ca-SK) and the manganese-calcium skarn (Mn–Ca-SK) zones [34]. The Mg-SK is developed in the II ore domain, and is featured by the widespread occurrence of Mg-rich minerals, including forsterite, spinel, diopside, tremolite, serpentine, Mg-chlorite, chondrodite and phlogopite. The Ca-SK, occurring in the IV and V ore domains, is associated with the metamorphism of mafic volcanic rocks and marbles, and is characterized by garnet, pyroxene, tourmaline, axinite, magnesian-hastingsite, ferro-actinolite and epidote. The distal Mn–Ca-SK zone of the VI ore domain contains johannsenite, galena and sphalerite.

Different calc–magnesium–silicate mineral assemblage reflects variations in magma chemistry, wall rock composition, temperature of formation, and oxidation state [26]. The main sedimentary rocks associated with ore deposition in the Tanjianshan Formation involve recrystallized marble, dolomitic marble and basaltic andesite. As the reason of diverse wall rock composition, Galinge skarn deposit is compositionally divided into three zonation. Magnesian skarn, developed in the dolomitic marble strata proximal to granodioritic intrusions mainly in the II ore domain, is probably coeval with initial emplacement of plutonic rocks, and the related ore association characterized by magnetite, suggests that the bulk of the magnesian skarns were formed under oxidizing conditions. In oxidized environments, Mg-rich pyroxene and garnet close to andradite end-member are the most stable skarn silicates [10]. In Ca-SK zonation developed in IV and V ore blocks associated with basaltic andesite and marble strata. Wollastonite, garnet and hedenbergite are abundantly formed during early metasomatic alteration stage at relatively high temperatures and oxidizing conditions. Successively, the prograde minerals were replaced by retrograde calcic assemblages far away the intrusive contacts. In addition with the decrease of oxygen fugacity and temperature, magnetite was replaced by the pyrrhotite and pyrite, and chalcopyrite was occurred in disseminated and massive structure. While the Mn-SK

Table 2.14 Representative results of electron microprobe analysis of magnetite and ludwigite from the Galinge deposit

Sample (wt%)	18302-19-3-4	18302-19-4-6	0012-2	0012-2	2404-5-1	2404-5-2	2404-5-3	2404-5_4	2404-5_5	2404-5_11	2404-5_9	2404-5_10	2404-27-4.1	2901-17-5-1	2901-14-2-1	2901-14-3-1	4208-6-3-2
Location	I		II														
SiO_2	0.30	1.91	0.66	0.08	0.00	0.04	0.03	0.23	0.14	0.10	0.11	0.22	0.08	0.60	0.95	2.01	0.00
TiO_2	0.12	0.06	0.06	0.51	0.00	0.00	0.01	0.02	0.02	0.00	0.02	0.03	0.00	0.00	0.03	0.00	1.49
Al_2O_3	0.05	0.21	0.00	0.44	0.40	0.39	0.47	0.49	0.47	0.00	0.74	0.75	0.02	0.08	0.12	0.11	0.80
FeO(T)	93.85	91.38	87.55	91.75	84.39	84.86	84.18	90.72	91.10	99.95	44.24	45.46	89.14	93.38	89.89	89.39	90.26
MnO	0.06	0.14	3.34	0.68	1.00	0.96	0.85	1.05	1.05	0.00	0.26	0.19	1.31	0.11	0.12	0.06	0.36
MgO	0.05	0.06	0.99	0.63	7.88	7.76	7.75	8.10	7.69	0.00	34.51	34.43	0.43	0.02	0.03	0.00	0.50
CaO	0.00	0.00	0.11	0.00	0.00	0.00	0.00	0.00	0.00	0.00	0.07	0.05	0.01	0.00	0.00	0.17	0.00
Cr_2O_3	0.07	0.15	0.01	0.07	0.01	0.01	0.02						0.01	0.02	0.00	0.02	0.09
ZnO	0.07	0.05	0.00	0.00	0.01	0.00	0.01	0.10	0.13	0.00	0.02	0.02	0.03	0.08	0.01	0.12	0.19
V_2O_3	0.00	0.00	0.01	0.00	0.00	0.01	0.00						0.03	0.00	0.00	0.00	0.03
B_2O_3											20.87	19.82					
Sum:	94.57	94.03	92.78	94.18	93.69	94.02	93.31	100.72	100.61	100.05	79.98	81.16	91.05	94.30	91.27	91.97	93.72
N(O)	4	4	4	4	4	4	4	4	4	4	5	5	4	4	4	4	4
Si	0.012	0.073	0.025	0.003	0.000	0.001	0.001	0.008	0.005	0.004	0.005	0.010	0.003	0.023	0.037	0.078	0.000
Ti	0.003	0.002	0.002	0.015	0.000	0.000	0.000	0.001	0.001	0.000	0.001	0.001	0.000	0.000	0.001	0.000	0.043
Al	0.002	0.009	0.000	0.020	0.017	0.016	0.020	0.019	0.019	0.000	0.040	0.040	0.001	0.004	0.005	0.005	0.036
Fe^{3+}	1.966	1.837	1.945	1.943	1.983	1.981	1.977	1.964	1.971	1.993	1.700	1.722	1.992	1.950	1.916	1.837	1.875
Fe^{2+}	1.008	1.064	0.857	0.960	0.548	0.558	0.559	0.572	0.587	1.004	0.000	0.000	0.933	1.016	1.031	1.064	0.997
Mn	0.002	0.005	0.108	0.022	0.030	0.029	0.026	0.030	0.030	0.000	0.010	0.007	0.044	0.004	0.004	0.002	0.012
Mg	0.003	0.004	0.057	0.035	0.421	0.414	0.416	0.404	0.385	0.000	2.364	2.324	0.025	0.001	0.002	0.000	0.029

(continued)

Table 2.14 (continued)

Sample (wt%)	18302-19-3-4	18302-19-4-6	0012-2	0012-2	2404-5-1	2404-5-2	2404-5-3	2404-5_4	2404-5_5	2404-5_11	2404-5_9	2404-5_10	2404-27-4.1	2901-17-5-1	2901-14-2-1	2901-14-3-1	4208-6-3-2
Location	I		II														
Ca	0.000	0.000	0.005	0.000	0.000	0.000	0.000	0.000	0.000	0.000	0.003	0.003	0.000	0.000	0.000	0.007	0.000
Cr	0.002	0.005	0.000	0.002	0.000	0.000	0.001	0.000	0.000	0.000	0.000	0.000	0.000	0.001	0.000	0.001	0.003
Ba	0.000	0.000	0.000	0.000	0.000	0.000	0.000	0.000	0.000	0.000	0.000	0.000	0.000	0.000	0.000	0.000	0.000
Zn	0.002	0.001	0.000	0.000	0.000	0.000	0.000	0.002	0.003	0.000	0.001	0.001	0.001	0.002	0.000	0.003	0.005
V	0.000	0.000	0.000	0.000	0.000	0.000	0.000	0.000	0.000	0.000	0.000	0.000	0.001	0.000	0.000	0.000	0.001
Total	3.000	3.003	3.002	3.001	3.000	3.000	3.000	3.000	3.000	3.000	4.124	4.108	3.000	3.000	3.006	3.005	3.000
Fe_2O_3	68.94	64.30	67.53	68.25	73.48	73.57	72.93	78.09	78.02	73.87	49.16	50.52	67.45	68.23	64.96	62.89	65.49
FeO	31.82	33.53	26.78	30.34	18.28	18.66	18.56	20.46	20.90	33.48	0.00	0.00	28.44	31.99	31.44	32.79	31.33
Total	100.75	97.82	94.32	98.59	91.75	92.23	91.49	98.55	98.91	107.35	49.16	50.52	95.90	100.22	96.40	95.69	96.82
	Mt	Mt	Mt	Mt	Mt	Mt	Mt	Mt	Mt	Mt	Lug	Lug	Mt	Mt	Mt	Mt	Mt

Sample (wt%)	21605-64-7-6	21605-64-7-8	21605-64-7-9	21605-64-7-1	21605-87-6-1	21605-87-7-12	21605-87-7-13	21605-87-7-14	21605-87-7-15	21605-87-7-16	21605-87-7-17	27601-6-1	27601-6-2	27601-6-3
Location	V	VI												
SiO_2	0.03	0.00	0.00	1.62	1.07	1.52	1.70	1.22	1.94	1.75	1.75	0.00	0.01	0.06
TiO_2	0.08	0.11	0.19	0.06	0.11	0.18	0.05	0.02	0.02	0.14	0.11	0.25	0.12	0.38
Al_2O_3	0.21	0.19	0.21	2.12	0.70	0.92	0.93	0.76	1.28	0.97	0.99	0.33	2.61	2.53
FeO(T)	90.03	92.51	93.40	82.07	91.11	91.98	90.23	91.36	88.93	90.50	90.64	91.39	89.66	89.07
MnO	0.05	0.03	0.01	0.00	0.00	0.08	0.05	0.04	0.00	0.08	0.07	0.35	0.13	0.33
MgO	0.01	0.01	0.00	0.76	0.28	0.18	0.46	0.36	0.52	0.48	0.47	1.86	2.85	2.77
CaO	0.00	0.00	0.00	0.09	0.00	0.06	0.06	0.00	0.15	0.12	0.09	0.00	0.00	0.00
Cr_2O_3	0.07	0.03	0.00	0.07	0.01	0.00	0.03	0.04	0.07	0.01	0.06	0.05	0.02	0.01

Table 2.14 (continued)

Sample (wt%)	21605-64-7-6	21605-64-7-8	21605-64-7-9	21605-64-7-1	21605-87-6-1	21605-87-7-12	21605-87-7-13	21605-87-7-14	21605-87-7-15	21605-87-7-16	21605-87-7-17	27601-6-1	27601-6-2	27601-6-3
Location	V	VI												
ZnO	0.00	0.05	0.04		0.00	0.00	0.00	0.11	0.00	0.00	0.17	0.06	0.04	0.02
V_2O_3	0.07	0.04	0.05		0.04	0.00	0.00	0.03	0.00	0.03	0.05	0.02	0.01	0.05
Sum:	90.64	93.16	93.91	86.85	93.42	95.05	93.71	94.06	93.08	94.23	94.63	94.35	95.45	95.22
N(O)	4	4	4	4	4	4	4	4	4	4	4	4	4	4
Si	0.001	0.000	0.000	0.066	0.041	0.057	0.064	0.046	0.074	0.066	0.066	0.000	0.000	0.002
Ti	0.002	0.003	0.006	0.002	0.003	0.005	0.001	0.000	0.000	0.004	0.003	0.007	0.003	0.010
Al	0.010	0.008	0.009	0.101	0.031	0.041	0.041	0.034	0.057	0.043	0.044	0.015	0.112	0.109
Fe^{3+}	1.977	1.980	1.978	1.761	1.878	1.834	1.823	1.869	1.789	1.814	1.812	1.968	1.880	1.865
Fe^{2+}	1.001	0.999	1.004	1.017	1.027	1.046	1.034	1.021	1.037	1.034	1.030	0.890	0.845	0.851
Mn	0.002	0.001	0.000	0.000	0.000	0.003	0.002	0.001	0.000	0.002	0.002	0.011	0.004	0.010
Mg	0.001	0.001	0.000	0.046	0.016	0.010	0.026	0.020	0.029	0.027	0.026	0.104	0.154	0.151
Ca	0.000	0.000	0.000	0.004	0.000	0.003	0.003	0.000	0.006	0.005	0.004	0.000	0.000	0.000
Cr	0.002	0.001	0.000	0.002	0.000	0.000	0.001	0.001	0.002	0.000	0.002	0.002	0.001	0.000
Ba	0.000	0.000	0.000	0.000	0.000	0.000	0.000	0.000	0.000	0.000	0.000	0.000	0.000	0.000
Zn	0.000	0.001	0.001	0.000	0.000	0.000	0.000	0.003	0.000	0.000	0.005	0.002	0.001	0.000
V	0.002	0.001	0.001	0.000	0.001	0.000	0.000	0.001	0.000	0.001	0.002	0.001	0.000	0.001
Total	3.003	3.008	3.000	3.004	3.004	3.005	3.009	3.005	3.008	3.006	3.009	3.002	3.000	3.000
Fe_2O_3	66.43	68.34	68.85	57.81	65.45	65.10	63.99	65.65	62.56	64.06	64.22	69.93	68.74	67.96
FeO	30.26	31.02	31.44	30.05	32.21	33.41	32.65	32.28	32.63	32.86	32.85	28.46	27.80	27.92
Total	96.69	99.36	100.30	87.86	97.67	98.51	96.64	97.94	95.19	96.92	97.08	98.39	96.54	95.88
	Mt	Mt	Mt	Mt	Mt	Mt	Mt	Mt	Mt	Mt	Mt	Mt	Mt	Mt

zonation are developed in the VI ore domains, manganoan hedenbergite is volumetrically the most important mineral, and galena and sphalerite accompanying magnetite are developed in the absence of copper.

2.5.1 Mg-SK Zonation

This skarn forms narrow, zoned envelopes along contacts of granodiorite intruded into the dolomite marbles, but is also crosscut by granodiorite. The magnesian skarns exhibits forsterite, diopsite, spinel, phlogopite, chondrodite, tremolite, and serpentine. Usually individual magnesian skarns were replaced by serpentine to varying degrees. Relicts of calcite and/or dolomite are present in all zones.

From unaltered granodiorite to dolomitic marble, the Mg-skarn zonation is roughly divided into as follows: unaltered granodiorite → K-feldspathized granodiorite → grossularitized and epidotized granodiorite → forsterite skarn → forsterite-diopside skarn → diopside skarn → diopside-tremolite skarn → diopside-clinohumite (chondrodite) skarn → serpentinized marble → dolomitic marble. The Mg-rich magnetite from the II ore domain is mostly associated with serpentine which is an alteration product of forsterite, tremolite and diopside [5, 44]. Serpentine replacing olivine increases MgO content of the magnetite. After hydration of olivine to produce chrysotile, more Fe would be liberated to form more magnetite [27]. Crystalline phase association for the olivine, chondrodite, tremolite, and diopsite composition in the Ca–Mg–Si–C–O–H system are plotted in Fig. 2.20. Some coexisting assemblages constrained the valid bulk composition. The bulk composition of the system from the skarn stage to the retrograde stage varies extensively as an influence of fluid composition. In the retrograde stage, the amounts of CaO, Al_2O_3 and SiO_2 increase, and that of MgO decreases.

2.5.2 Ca-SK Zonation

With successive decrease in fO_2, bulk composition of garnet moves toward the grossular end-member with simultaneously increasing almandine component (Fig. 2.21) [16]. In the Galinge deposit, the garnet associated with high grade magnetite ore shows intensively andradite-rich composition ranging from $And_{92.59}$ to $And_{98.18}$, however, massive grossular-rich garnet skarn is associated with low grade magnetite ore. In the sulfide ore association, the garnet moderately consist of andradite with a wide compositional range ($Ad_{51.51–78.06}$). From magnetite to sulfide ore association, the decreased almandine component of garnet implies the decreased oxygen fugacity. The garnet replaced by retrograde epidote in skarnized diorites shows moderate grossular composition. Generally, the well-developed euhedral to subhedral garnet minerals replaced or cut by later pyroxene contains a wide andradite composition ($And_{54.25–98.78}$). The garnet replaced by Mn-hedenbergite which is

Fig. 2.20 The 3-D plot of the CaO–MgO–Al_2O_3–SiO_2 (+K_2O + H_2O) system showing positions of minerals considered for stability calculations

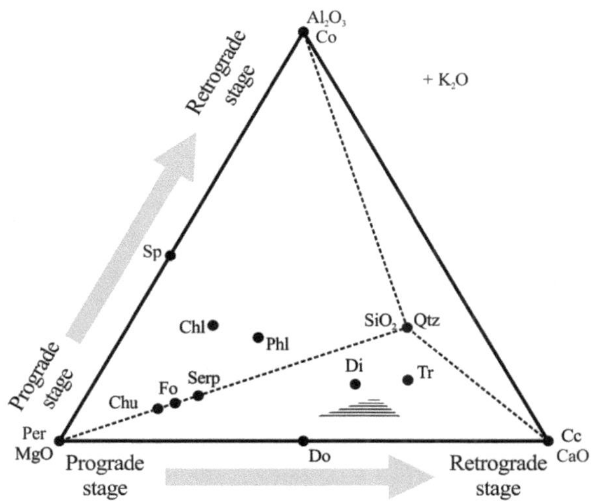

commonly related with galena and sphalerite mineralization are characterized by compositional zonation of which the core is $And_{94.89}$ composition, the inner zone is $And_{58.17}$, and the outer zone is $And_{97.80}$. The large composition variation is extensively dependent on the variation of oxygen fugacity. In another word, galena and sphalerite mineralization obviously related with the decrease of oxygen fugacity.

Crystalline phases for the epidote and prehnite composition in the Ca–Fe–Al–Si–C–O–H system are plotted in Fig. 2.21. The Galinge prehnite clearly shows extremely low X_{Fe} values. The X_{Fe} in the prehnite octahedral site varies as a function of fO_2 and temperature, and becomes more aluminous with decreasing fO_2 and increasing temperature [17].

Liou [16] pointed out maximum Fe^{3+} content of epidotes varies as a function of fO_2. Epidote is most iron-rich at high (HM and CCO) oxygen buffers and becomes progressively more aluminous with decreasing fO_2 values and temperatures. Therefore, Fe rich epidotes cannot be found in basaltic-andesitic igneous rocks, where fO_2 is generally low. Enrichment of Fe (relative to Mg) in an igneous rock with high Ca and Al contents is typical for granodiorite-granite, which supply a proper chemical condition for epidote minerals formation. Galinge epidotes belongs to the Al–Fe^{3+} solid solution series with high Fe^{3+} contents between $X_{Epi} = 0.14$ and $X_{Epi} = 0.28$. From deep epidote-axinite-diopsite association to shallow epidote-grossular association, the composition of epidote show a decrease in the values of X_{Epi}. The very common occurrence of all types of epidote minerals, Fe-poor zoisite to Fe-rich epidote, to such fact that Al as a major component must be transported via a fluid phase and cannot be considered as immobile. The formation of Fe rich epidotes requires in addition to the enrichment in Ca and Al enrichment of Fe and a relatively high fO_2. In summary, metasomatic reactions that form epidote minerals can be explained as mass transport of all major elements essential for epidote, Ca, Al, Fe and Si [4].

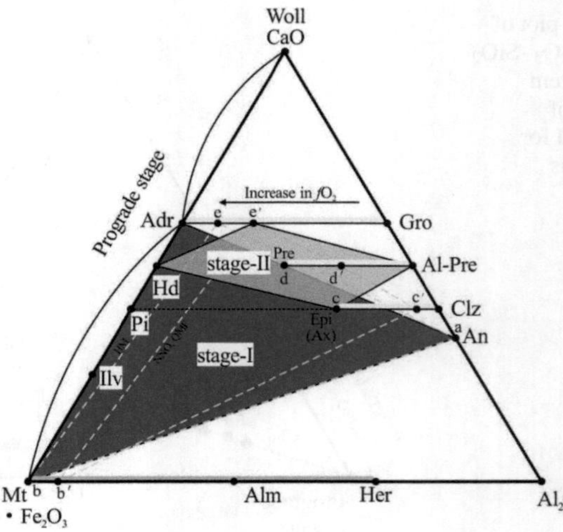

Fig. 2.21 Ca–Fe–Al ternary plot showing the composition of crystalline phases in the Ca–Fe–Al–Si–C–O–H system considered for stability calculations. The fO_2 conditions defined by the HM buffers in the trilateral abc, abd and abe, and controlled by the NNO and QMF buffer in the tetrahedron b′c′d′e′. Garnet composition moves towards the grossular end-member with decreasing fO_2. The Woll + Adr + Mt assemblage was formed at the prograde stage. The light gray fields stands for the Adr-Epi-Mt assemblage; the dark gray fields stands for Hd-Adr-Pre assemblage

The axinite-group minerals are complex phases typically occurring in low-to medium-grade metamorphic environments [8]. Axinite formation is not only broadly associated with the host rock components [28], but also reflects the metasomatic introduction of B, Fe and Mn [29]. Axinite, in most cases, is a late-stage mineral in skarn and other contact metamorphic rocks [2], typically formed by replacing epidote, amphibole, garnet, clinopyroxene and feldspar [29]. Axinite varies largely in composition even within a single deposit [32]. Also, axinite is believed to be alternative to tourmaline as a B-rich phase in Al-poor assemblages [30], as it is the only borosilicate in some regionally metamorphosed rocks. At Galinge, axinite-(Fe) commonly coexists with hedenbergite, anorthite and epidote. Although the Galinge deposit develops large amounts of the tourmaline + epidote assemblage [36, 37, 40], the tourmaline + axinite assemblage is rare. In this case, tourmaline + epidote acts a barrier to restrict axinite to rocks richer in B and Ca [8]. Such a temperature range corresponds to the Stage I temperature range in retrograde alteration. Fluid inclusion measurements for the epidote-quartz-calcite-axinite skarn assemblage from the JC tin skarn deposit, southern Yukon Territory, yielded a high formation temperature of ca. 410–500 °C [13, 14]. The formation of the hedenbergite-axinite assemblage required temperatures <500 °C [12].

2.5.3 Mn–Ca-SK Zonation

In the VI ore domains, the MnO content in pyroxenes are generally high and varies between 0.15 and 7.84. Massive Mn-rich hedenbergite skarn associated with high grade galena and sphalerite ore can be considered to have formed from fluids which are distal in the overall zonation sequence. In contrast, pyroxene formed in proximal locations is enriched in magnesium. The general sequence of pyroxene-forming fluids is depletion first in Mg, then in Fe, and finally in Mn.

In the Mn–Ca skarn zonation, sphalerite and galena mineralization occurred during the sulfide stage at lower temperatures. This could indicate multiple stages of fluid infiltration during the skarn genesis. Meinert [22]suggests that the manganese-rich skarn forming hydrothermal systems can also evolve from initially iron-rich and manganese-poor intrusions, which request fluid travel long distances prior to reaction with carbonate rock cause unusually high manganese contents in pyroxene (up to 27 mol% johannsenite). Studies of fluid inclusions of the Mn-bearing skarn minerals show that the formation temperatures range from 220 to 420 °C, corresponding with the formation temperatures of the associated sphalerite (230–380 °C) [43]. The temperature obtained from a sphalerite–galena pair ranges from 263 to 748 °C, which could be the lower temperature limit for ore precipitation.

References

1. Atkinson WW, Einaudi MT (1978) Skarn formation and mineralization in the contact aureole at Carr Fork, Bingham, Utah. Econ Geol 73(7):1326–1365
2. Deer WA, Howie RA, Zussman J (1986) Rock-forming minerals, Disilicates and ring silicates, v. 1B. Longman Scientific and Technical, London
3. Feng CY, Li DS, Wu ZS et al (2010) Major types, time-space distribution and metallogenesis of polymetallic deposits in the Qimantage metallogenic belt, eastern Kunlun area. Northwestern Geol 43(4):10–17
4. Franz G, Liebscher A (2004) Physical and chemical properties of the epidote minerals—an introduction. Rev Mineral Geochem 56(1):1–81
5. Frietsch R (1984) Formation of Mg-bearing magnetite and serpentine in skarn iron ores in northern Sweden. GFF 106(3):219–230
6. Frost BR, Lindsley DH (1992) Equilibria among Fe-Ti oxides, pyroxenes, olivine, and quartz: part II. Application. Am Mineral 77:1004
7. Gao YB, Li WY, Ma XG et al (2012) Genesis, geochronology and Hf isotopic compositions of the magmatic rocks in Galinge iron deposit, eastern Kunlun. J Lanzhou Univ (Nat Sci) 48:36–47
8. Grew ES (1996) Borosilicates (exclusive of tourmaline) and boron in rock-forming minerals in metamorphic environments. Rev Mineral Geochem 33(1):387–502
9. Grigsby JD (1990) Detrital magnetite as a provenance indicator. J Sediment Res 60(6):940–951
10. Gustafson WI (1974) The stability of andradite, hedenbergite, and related minerals in the system Ca–Fe–Si–O–H. J Petrol 15(3):455–496
11. Hall DL, Cohen LH, Schiffman P (1988) Hydrothermal alteration associated with the iron hat iron skarn deposit, eastern Mojave Desert, San Bernardino County, California. Econ Geol 83(3):568–587

12. Kurshakova LD, Tikhomirova VI (1974) Axinite and its paragenesis with hedenbergite. Int Geol Rev 16(12):1360–1369
13. Layne GD, Longstaffe FJ, Spooner E (1991) The JC tin skarn deposit, southern Yukon Territory; II, A carbon, oxygen, hydrogen, and sulfur stable isotope study. Econ Geol 86(1):48–65
14. Layne GD, Spooner E (1991) The JC tin skarn deposit, southern Yukon Territory; I, Geology, paragenesis, and fluid inclusion microthermometry. Econ Geol 86(1):29–47
15. Lindsley DH, Frost BR (1992) Equilibria among Fe-Ti oxides, pyroxenes, olivine, and quartz; Part I, Theory. Am Mineral 77(9–10):987–1003
16. Liou JG (1973) Synthesis and stability relations of epidote, $Ca_2Al_2FeSi_3O_{12}(OH)$. J Petrol 14(3):381–413
17. Liou JG, Kim HS, Maruyama S (1983) Prehnite–epidote equilibria and their petrologic applications. J Petrol 24(4):321–342
18. Liu CD, Mo XX, Luo ZH et al (2004) Mixing events between the crust-and mantle-derived magmas in Eastern Kunlun: evidence from zircon SHRIMP II chronology. Chin Sci Bull 49(8):828–834
19. Liu JN, Feng CY, Zhao YM et al (2013) Characteristics of intrusive rock, metasomatites, mineralization and alteration in Yemaquan skarn Fe-Zn polymetallic deposit, Qinghai Province. Miner Depos 1:008
20. Luo ZH, Ke S, Cao YQ et al (2002) Late Indosinian mantle-derived magmatism in the East Kunlun. Geol Bull China 21(6):292–297
21. Mao JW, Zhou ZH, Feng CY et al (2012) A preliminary study of the Triassic large-scale mineralization in China and its geodynamic setting. Geol China 39(6):1437–1471
22. Meinert LD (1984) Mineralogy and petrology of iron skarns in western British Columbia, Canada. Econ Geol 79(5):869–882
23. Meinert LD (1987) Skarn zonation and fluid evolution in the Groundhog mine, Central mining district, New Mexico. Econ Geol 82(3):523–545
24. Meinert LD (1992) Skarns and skarn deposits. Geosci Can 19(4):145–162
25. Meinert LD (1993) Igneous petrogenesis and skarn deposits. Miner Depos Model 40:569–583
26. Meinert LD, Hefton KK, Mayes D et al (1997) Geology, zonation, and fluid evolution of the Big Gossan Cu-Au skarn deposit, Ertsberg district, Irian Jaya. Econ Geol 92(5):509–534
27. O'Hanley DS, Robin O (1992) Characterization of multiple sepernization, Woodsreef, New South Wales. Can Mineral 30:1113–1126
28. Ozaki M (1972) Chemical composition and occurrence of axinite. Kumamoto J Sci Geol 9(2):1–34
29. Pertsev NN (1971) Parageneses of boron minerals in magnesian skarns. Nauka, Moscow
30. Pringle IJ, Kawachi Y (1980) Axinite mineral group in low-grade regionally metamorphosed rocks in southern New Zealand. Am Mineral 65(11–12):1119–1129
31. She HQ, Zhang DQ, Jing XY et al (2007) Geological characteristics and genesis of the Ulan Uzhur porphyry copper deposit in Qinghai. Geol China 2:013
32. Sonnet PM, Verkaeren J (1989) Scheelite-, malayaite-, and axinite-bearing skarns from El Hammam, central Morocco. Econ Geol 84(3):575–590
33. Yao L (2015) Petrogenesis of the Triassic granitoids and skarn mineralization in the Qimantag area, Qinghai Province, and their geodynamic setting. China University of Geosciences (Beijing), Beijing, pp 1–172
34. Yu M, Feng CY, Bao GY et al (2013) Characteristics and zonation of skarn minerals in Galinge iron deposit, Qinghai Province. Miner Depos 32(1):55–76
35. Yu M, Feng CY, Liu HC et al (2015) The significance of mineralization and geochemistry of the Cl-rich amphiboles from the Galinge skarn iron deposit in Qinghai Province. Acta Petrol Mineral 34(5):721–740
36. Yu M, Feng CY, Liu HC et al (2016) Mineralogy, element geochemistry and genesis of tourmaline from Galinge skarn deposit, Qinghai Province. Miner Depos 35(1):69–84
37. Yu M, Feng CY, Liu HC et al (2016) The Iron-Titanium Oxides in the Galinge Iron Polymetallic Skarn Deposit of Qinghai Province and their thermodynamic significance. Acta Geosci Sin 37(2):204–214

38. Yu M, Feng CY, Mao JW et al (2017) Multistage skarn-related tourmaline from the Galinge deposit, Qiman Tagh, Western China: a fluid evolution perspective. Can Mineral 55(1):3–19
39. Yu M, Feng CY, Zhao YM et al (2015) Genesis of post-collisional calc-alkaline and alkaline granitoids in Qiman Tagh, East Kunlun, China. Lithos 239:45–59
40. Yu M, Feng CY, Zhu YF et al (2016) Multistage amphiboles from the Galinge iron skarn deposit in Qiman Tagh, western China: evidence of igneous rocks replacement. Miner Petrol 1–17
41. Zhao YM, Dong YG, Li DX et al (2003) Geology, mineralogy, geochemistry, and zonation of the Bajiazi dolostone-hosted Zn–Pb–Ag skarn deposit, Liaoning Province, China. Ore Geol Rev 23(3):153–182
42. Zhao YM, Feng CY, Li DX et al (2013) Metallogenic setting and mineralization-alteration characteristics of major skarn Fe-polymetallic deposits in Qimantag area, western Qinghai Province. Miner Depos 32(1):1–19
43. Zhao YM, Li DX (2004) Pb-Zn-Ag-bearing manganoan skarns of China. Acta Geol Sin (Engl Ed) 78(2):524–528
44. Zhao YM, Zhang YN, Bi CS et al (1998) The discovery of magnesioferrite from Au (Fe, Cu) magnesian skarn deposits and study of the magnesioferrite-magnesiomagnetite Series. Acta Geol Sin (Engl Ed) 72(4):382–391
45. Zhao YM, Zhang YN, Bi CS (1999) Geology of gold-bearing skarn deposits in the middle and lower Yangtze River Valley and adjacent regions. Ore Geol Rev 14(3):227–249

Chapter 3
Thermodynamic Model for the Galinge Fe Skarn Deposit in Qinghai

All reactions in the skarn system involve hydration and carbonation, which are inevitably driven by chemical disequilibrium between rock and flowing fluid [10]. Generally, the skarn system is not equilibrium in a large scale, but local equilibrium to a small scale. With lowering of temperature of the post-magmatic solutions, the mineral assemblages occurring in the skarn stage are transformed into retrograde alteration hydrous mineral assemblages. In the meantime, a sequence of the metasomatically altered zones will be yielded in the metasomatic section of magnesian skarns, which depends upon: (a) concentrations and migrations of the active constituents (Si, Al, Ca, Mg, K, and Na) in the fluids; (b) continuously varying temperature and mole fractions of CO_2 of the fluid. For the thermodynamic model, a quartz-poor protolith, typical for the host-rock dolomites, was assumed. As found in the compositional analysis, from the skarn alteration assemblages to the retrograde alteration assemblages, the bulk composition of the system experienced a change from host rock to fluid control, in which the CaO, SiO_2, and Al_2O_3 contents are elevated.

To determine the general environments of skarn formation, it is necessary to consider the effects of temperature (T), activities of variable elements in the system and fluid component containing CO_2 and H_2O. Stable reactions in the system were identified using winTWQ [2], then were normalized to the hematite-magnetite buffer and plotted using CHNOSZ [5]. In the model, the pressure range based on the calculation of amphibole geobarometer is 57–26 Mpa (Table 3.1). Therefore, we set a ideal pressure of 50 Mpa for the thermodynamic model. As documented the water content of most shallow (~2 km) granitic magmas is almost 3 wt% at a pressure of 0.5 Kbar, with lesser amounts of volatiles such as CO_2, fluorine and boron [23]. All the activities of different minerals used in the calculations hinge upon the chemistry data of various minerals formed in the same stages, in which equilibrium is reached or as replacement of one by another(s). In skarn environments, stoping of carbonate rocks and related decarbonation reactions would greatly elevate the CO_2 content of

M. Yu, *Metallogenic Mechanism of the Galinge Polymetallic Iron Skarn Deposit, Qiman Tagh Mountains, Qinghai Province*, Springer Theses,
https://doi.org/10.1007/978-981-10-7907-8_3

Table 3.1 The geochemical and geophysical conditions of amphiboles from the granodiorites and diorites in the Galinge deposit

Label	Granodiorite	Granodiorite	Granodiorite	Granodiorite	Diorite	Diorite	Diorite
Spot (wt%)	16303-14-2-1	16303-14-3-1	2404-35-2-2	2404-35-2-3	0307-76-1-1	0307-76-1-2	0307-76-1-3
SiO_2	50.32	49.85	50.14	51.65	53.81	50.60	50.81
TiO_2	0.17	0.34	0.26	0.52	0.12	0.37	0.00
Al_2O_3	4.42	3.87	3.66	4.27	1.26	4.02	3.47
FeO	15.63	17.43	15.90	16.94	14.64	13.02	14.09
MnO	0.34	0.27	0.24	0.33	0.30	0.20	0.39
MgO	13.87	12.81	13.30	13.66	14.56	14.77	15.35
CaO	10.23	11.03	10.65	10.85	10.77	12.44	12.01
Na_2O	0.86	0.43	0.87	1.05	0.28	0.12	0.14
K_2O	0.38	0.22	0.40	0.59	0.19	0.37	0.00
F	0.18	0.10	0.23	0.18	0.22	0.27	0.11
Cl	0.10	0.05	0.13	0.18	0.04	0.11	0.09
Physical-chemical conditions							
T (°C)	726	718	693	708	656	718	728
P (MPa)	57	50	48	54	26	52	45
Continental depth (km)	2.1	1.9	1.8	2.0	1.0	2.0	1.7
$\log fO_2$	−13.7	−14.1	−14.7	−14.4	−15.4	−13.8	−13.2
H_2O_{melt} (wt%)	4.3	5.0	4.2	3.7	3.5	5.0	5.3

silicate melts, and thus magma is easier to be water saturation. The hydrothermal fluid will release more CO_2. Early metamorphism and continued metasomatism is generally occurred at relatively high temperature, and retrograde alteration follows as temperatures decline and fluids evolve or undergo phase separation [24]. In the skarn stage, the $X(CO_2)$ of the rock (approximate 0.3) is much higher than that of the fluid ($X(CO_2) = 0.01$) in the retrograde stage [22].

3.1 Thermodynamic Model for Magnesian Skarn

The Galinge magnesian-skarn associations were formed at the contact between the dolomitic limestones of the Tanjianshan Formation and the Triassic granodiorites. However, such strata and similar plutons are very common in the Qiman Tagh metallogenic belt, raising questions about why Mg-SK have only been found in the Galinge deposit, what conditions are sufficient for formation of Mg-SK, and whether other iron deposits associated with Mg-SK could be prospected in this belt. In addition, lots of various mineral assemblages were developed in the magnesian skarn zonation through different equilibrated reactions (Table 3.2). Another question is raised that these reactions reflect what kind of thermodynamic processes. These questions will be answered progressively, taking advantage of the unique opportunity to study the complete magnesian-skarn zonation in the Galinge skarn deposit and its relation to the associated pluton, in order to describe the forming condition for this zonal pattern within a thermodynamic model.

In the skarn stage of the Galinge deposit, a stable assemblage diopside + forsterite + calcite widespreadly developed, and its formation temperature is constrained between 500 °C at $X(CO_2) = 0.3$ through the reaction of Di + 3 Do = 2 Fo + 4 Cc + 2 CO_2, accompanying the disappearance of Cc (Fig. 3.1a). The reaction theoretically control the formation and disappearance of diopside in the skarn stage [11]. In addition, the diopside can be abundantly produced by replacing dolomite through the reaction of Do + 2 Qtz = Di + 2 CO_2, leaving a stable assemblage of Di + Do after the consumption of quartz. Generally, this reaction significantly indicates a reacting temperature of 475 °C at $X(CO_2) = 0.3$ (Fig. 3.1a). In the retrograde stage, diopside is usually altered to various minerals including tremolite, phlogopite, serpentine, clinohumite and clinochlore. The diopside is stable at intermediate activities of MgO and CaO in the fluid, and constrained between the saturation boundaries of magnesite and brucite in the fluid (Fig. 3.3). With increasing Al_2O_3 activity in the fluid, diopside will be altered to phlogopite.

Due to the lack of monticellite and fayalite in the Galinge Mg-SK zonation, the appearance of forsterite plus calcite in the dolomitic marbles is suggested by the reaction of 2 Do + Qtz = Fo + 2 Cc + 2 CO_2, accompanying with the disappearance of quartz, and leaving a stable assemblage forsterite + calcite + dolomite under a temperature of 490 °C at $X(CO_2) = 0.3$ (Fig. 3.1a). Actually, the lower limits of the skarn stage are consistent with the stability fields of the forsterite + calcite + dolomite assemblage (Fig. 3.1a), which are evidently higher than 500 °C at relatively high $X(CO_2)$ ranges ($X(CO_2)>0.2$) [15, 36]. Although the effects of CO_2 on temperature are maximal in a relatively low CO_2 environment, the earlier decarbonation

Table 3.2 The stable magnesian skarn assemblages in the Galinge deposit

Stage	Mineral assemblages		The inferred equilibrium reactions
Skarn stage	a	forsterite + calcite + dolomite	2 Do + Qtz = Fo + 2 Cc + 2 CO_2
	b	spinel + calcite + dolomite	Do + Al_2O_3 = Sp + Cc + CO_2
	c	spinel + diopside + calcite + quartz	Di + Al_2O_3 + CO_2 = Cc + Qtz + Sp
	d	diopside + dolomite	Do + 2 Qtz = Di + 2 CO_2
	e	diopside + forsterite + dolomite	Di + 3 Do = 2 Fo + 4 Cc + 2 CO_2
Retrograde stage	f	diopside + clinohumite (chondrodite) + calcite	2Di + 7 Do + H_2O = Chu + 9 Cc + 5 CO_2
	g	tremolite + diopside + dolomite	3 Cc + Tr = Do + 4 Di + H_2O + CO_2
	h	chrysotile + diopside + calcite	3Cc + Chr + CO_2 = 2Do + Di + 2 H_2O
	i	chrysotile + forsterite + diopside + calcite	2 Chr + 2 Cc = 2 Fo + Do + Di + 4 H_2O
	j	chrysotile + forsterite	2 Chr = Qtz + Fo + 4 H_2O
	k	chrysotile + diopside + calcite	3 Cc + Chr + 4 Qtz = 3 Di + 2 H_2O + 3 CO_2
	l	phlogopite + diopside + calcite	3 Cc + Phl + 6 Qtz = Kfs + 3 Di + H_2O + 3 CO_2
	m	tremolite + phlogopite + diopside	4 Di + Phl + 4 CO_2 = Tr + Kfs + 2Do
	n	chrysotile ± calcite ± dolomite	3 Cc + Chr + 3 CO_2 = 2 Qtz + 3 Do + 2 H_2O

Cc-calcite; Chu-clinohumite; Chr-chrysotile; Di-diopside; Do-dolomite; Fo-forsterite; Kfs-K feldspar; Phl-phlogopite; Qtz-quartz; Serp-serpentine; Tr-tremolite

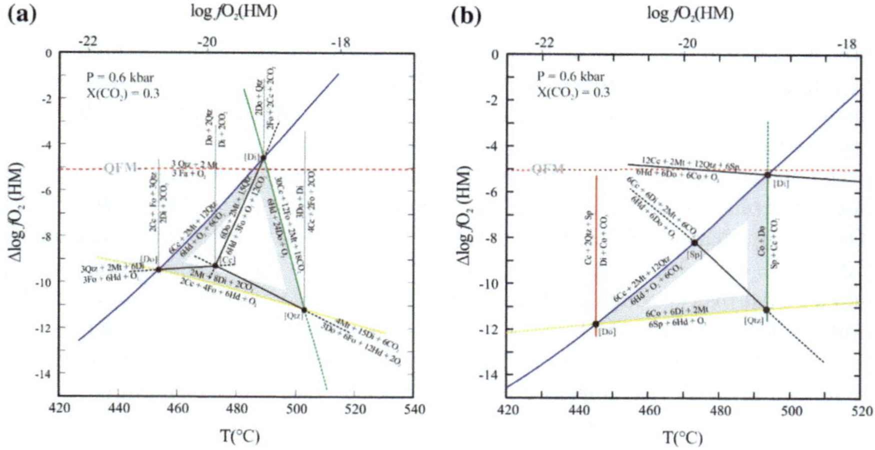

Fig. 3.1 $\Delta \log f O_2$ (HM)-T (°C) diagram for the reactions in the skarn stage assuming P = 600 bar and $X(CO_2)$ = 0.3. **a** The mineral phases of forsterite-diopside-hedenbergite-dolomite-calcite-quartz-magnetite in the system of Ca–Mg–Fe–Si–O–C–H. The shadow area represents a stable zone of forsterite + diopside + hedenbergite + magnetite + calcite assemblage; **b** the mineral phases of spinel-diopside-hedenbergite-dolomite-calcite-corundum-quartz-magnetite in the system of Ca–Mg–Fe–Al–Si–O–C–H. The shadow area represents a stable zone of diopside + hedenbergite + dolomite + magnetite + calcite + corundum assemblage. The curve marked with QFM refers to the quartz-fayalite-magnetite oxygen buffer under conditions of $P_{total} = P_{H_2O}$ = 600 bar [1]. The spinel and forsterite activities are both assumed to equal 0.8 based on their chemical compositions

will produce large amounts of carbon dioxide to limit the reaction at very high temperature. If fluids infiltrating the dolomite are Qtz-saturated, phase equilibria require that development of forsterite from dolomite during contact metamorphism must be that where most of the silica initially in the fluid has been removed after formation of Di in the flow systems [12]. The corresponding mineral assemblage of Fo + Di + Do suggests a upstream reaction (Fig. 3.1a). With decreasing temperature, the reverse of reaction Fo + 2 Cc + 3 Qtz = 2 Di + 2 CO_2 will form forsterite at a relatively low temperature range of 450 °C and $X(CO_2) = 0.3$ (Fig. 3.1a).

The formation of the spinel + dolomite + calcite assemblage limits the upper temperature of 506 °C in the $\Delta \log f O_2$(HM) interval of −5.5 to −9.8 at $X(CO_2) =$ 0.8 (Fig. 3.1b), through the reaction of Do + Al_2O_3 = Sp + Cc + CO_2, accompanying with the disappearance of Al_2O_3 [36]. Moreover, the spinel + diopside + calcite + quartz assemblage constraints a lower temperature of 472 °C in the $\Delta \log f O_2$(HM) interval of −5 to −10.2 (Fig. 3.1b), via the reaction of Di + Al_2O_3 + CO_2 = Cc + Qtz + Sp, accompanying with the consumption of Al_2O_3. The spinel formation also requires a high activity of Al_2O_3 and an intermediate activity of MgO (Fig. 3.3). The model temperature range is approximately consistent with the geothermometer and oxygen barometer of the Fe–Ti oxides exsolved from the spinel skarns. The computational temperature of magnetite-ilmenite exsolution ranges 432–565 °C (average 503 °C) [41], based on ILMAT (ilmenite-magnetite thermometry) [17, 19]. We therefore assume a average temperature of 503 °C for spinel crystallization.

We estimate a relatively wide temperature range of the retrograde stage (460~ to 360~ °C) by using paragenetic mineral assemblages, involving clinohumite, tremolite, phlogopite, clinochlore and serpentine. Because large amounts of carbonates are consumed under decarbonation in the skarn stage, the mole percent of CO_2 becomes very low in the retrograde alteration stage.

The presence of tremolite implies the onset of the retrograde stage, and it is produced by replacing diopside through the reaction of 3 Cc + Tr = Do + 4 Di + H_2O + CO_2, leaving an assemblage of diopside + tremolite + calcite after complete expense of dolomite (Fig. 3.2a) [34]. Generally, tremolite has a wide temperature stability and can exist over a wide range of $X(CO_2)$ [33]. It is easy to be replaced by phlogopite under a relatively high Al_2O_3 ($a(Al_2O_3) > -20$) activity of fluid environment (Fig. 3.3). The upper temperature limited by the genesis of tremolite + diopside + calcite equilibrated assemblages are approximately at 470 °C which corresponds to the tremolite-in reaction. In addition, fluids for the genesis of tremolite + diopside equilibrated assemblages are characterized by high concentrations of Ca relative to Mg and increasing Ca/Mg ratios with decreasing temperatures [20]. With decreasing temperature, reaction of Tr + 3 Cc + 7 CO_2 = 8 Qtz + 5 Do + H_2O, causing tremolite to react out (tremolite-out) and leaving calcite-quartz-dolomite assemblage [38, 39], appears at about 375 °C with $X(CO_2) = 0.01$ (Fig. 3.2a).

The clinohumite replacing diopside in the stable assemblage of Chu + Di $\pm$ Cc is found locally in Mg-SK zonation, through the reaction: 2 Di + 7 Do + H_2O = Chu + 9 Cc + 5 CO_2, accompanying with the disappearance of dolomite, which defines a Chu-in reaction in the retrograde alteration stage at an upper temperature of 450 °C with an $X(CO_2) = 0.01$ (Fig. 3.2b). Moreover, the reaction of 9 Cc + 14 Qtz + Chu = 9 Di + H_2O + 9 CO_2 limits a lower temperature at ca. 340 °C for the formation of clinohumite. The formation of clinohumite needs a very high activity of MgO, and it is more stable than tremolite and phlogopite under varying fluid activity (Fig. 3.3). Liang and Qiao [18] suggest that the principal conditions for the synthetized humite group are that the Mg/Si (mole number ratio) in the starting H_2O-saturated materials should be higher than 3:1, under appropriate *T*-*P*, neutral-weakly alkaline solution and oxidation environment. The effect of fluorine substitution is such as to greatly expand the stability field of clinohumite in rocks of siliceous dolomitic composition [30]. The formation of Cc + Chu needs a water-rich fluid condition irrespective of solid solution [40].

The stability field of phlogopite as a retrograde mineral is almost overlapped by those of tremolite with decreasing temperature (Fig. 3.2c). Phlogopite in the Galinge deposit coexists with diopside via the reaction as followed: 3 Cc + Phl + 6 Qtz = Kfs + 3 Di + H_2O + 3 CO_2. This metamorphic process considered as phl-in reaction takes place at ca. 420 °C and $X(CO_2)$ values of 0.01 at 0.6 kb in aqueous fluid phase (Fig. 3.2c). In addition, the assemblage of diopside + tremolite + phlogopite $\pm$ dolomite are locally observed in the Galinge deposit, in which the diopside and phlogopite are altered by the tremolite through the reaction of Tr + Kfs + 2 Do = 4 Di + Phl + 4 CO_2. The lowest temperature limits of phlogopite appearance is constrained by the reaction of 3 Cc + Phl + 3 CO_2 = Kfs + 3 Do + H_2O, which approximately happens at 360 °C at $X(CO_2) = 0.01$ (Fig. 3.2c).

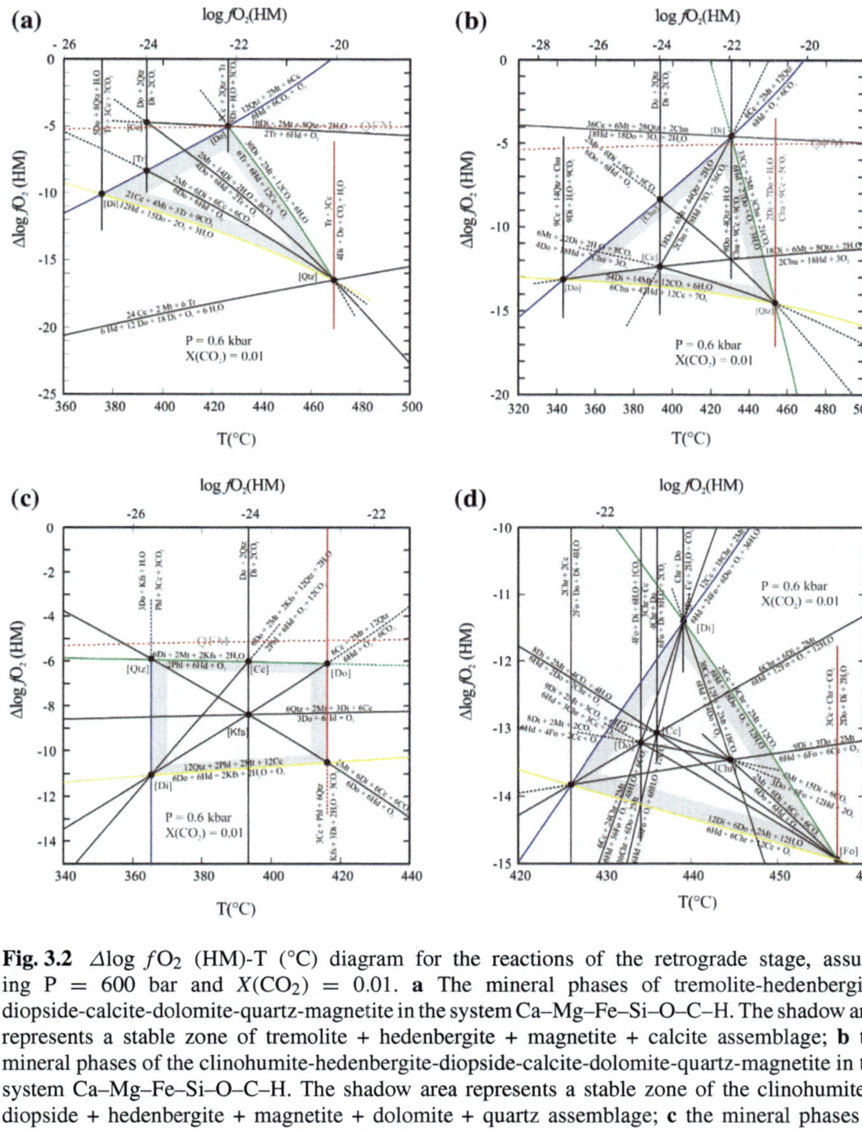

Fig. 3.2 Δlog fO_2 (HM)-T (°C) diagram for the reactions of the retrograde stage, assuming P = 600 bar and $X(CO_2)$ = 0.01. **a** The mineral phases of tremolite-hedenbergite-diopside-calcite-dolomite-quartz-magnetite in the system Ca–Mg–Fe–Si–O–C–H. The shadow area represents a stable zone of tremolite + hedenbergite + magnetite + calcite assemblage; **b** the mineral phases of the clinohumite-hedenbergite-diopside-calcite-dolomite-quartz-magnetite in the system Ca–Mg–Fe–Si–O–C–H. The shadow area represents a stable zone of the clinohumite + diopside + hedenbergite + magnetite + dolomite + quartz assemblage; **c** the mineral phases of phlogopite-potassic feldspar-hedenbergite-diopside-calcite-dolomite-quartz-magnetite in the system K–Ca–Mg–Fe–Al–Si–O–C–H. The shadow area represents a stable zone of the phlogopite + diopside + hedenbergite + magnetite + dolomite + quartz assemblage; **d** the mineral phases of chrysotile-forsterite-hedenbergite-diopside-calcite-dolomite -magnetite in the system Ca–Mg–Fe–Si–O–C–H. The shadow area represents a stable zone of the chrysotile + diopside + forsterite + hedenbergite + magnetite + dolomite assemblage; the curve marked with QFM refers to the quartz-fayalite-magnetite oxygen buffer under conditions of $P_{total} = P_{H_2O}$ = 0.6 kbars [1]

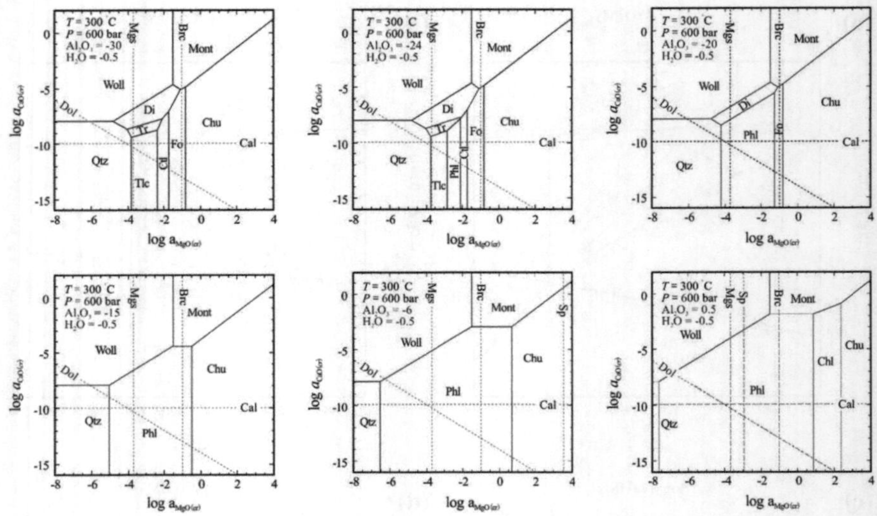

Fig. 3.3 Log a_{MgO} versus Log a_{CaO} Assuming T = 300 °C and P = 600 bar. The solid line shows the stability field of various minerals with elevating activity of Al_2O_3, and the dashed lines represent the saturation line of magnesite, brucite, calcite, dolomite and spinel in the hydrothermal fluid

The polymorphs of serpentine, lizardite-chrysotile, occur in nature in virtually identical ranges of temperature and pressure, from surficial or near-surficial environments to temperatures perhaps as high as 400 °C [9]. Laboratory evidence indicates that lizardite is the more stable at low temperatures. Above about 300 °C, antigorite + brucite is more stable than both; in other words, chrysotile is nowhere the most stable. Mazurov et al. [21] used the method of mineralogical thermometry to study that mineral assemblage of clinochlore + serpentine + pyrrhotite were formed at 450 °C. At the Galinge deposit, chrysotile are ubiquitously developed via replacing diopside, olivine, and clinohumite. Chrysoltile replacing diopside are usually considered as Chr-in reaction of 3 Cc + Chr + CO_2 = 2 Do + Di + 2 H_2O, leaving a stable assemblage of Di + Chr + Cc after the complete consumption of dolomite. The geodynamic data of chrysotile formation give a calculated stability temperature field less than ca. 460 °C (Fig. 3.2d). In addition, the chrysotile replacing the forsterite + diopside + dolomite assemblages formed in the skarn stage are significantly observed in the Galinge deposit, which corresponds to the reactions 2 Chr + 2 Cc = 2 Fo + Do + Di + 4 H_2O occurred at a temperature of ca. 425 °C (Fig. 3.2d).

To summarize the findings of the thermodynamic model: The skarn and retrograde stages are characterized by temperatures of 520–460 and 460–380 °C, respectively. The predominant activities of fluid components, represented by CaO, MgO and Al_2O_3, are constrained by reactions that generate forsterite, diopside, clinohumite, tremolite, phlogopite and chrysotile (Fig. 3.3). The high Al_2O_3 mainly comes from magmatic hydrothermal fluid, and it increase from skarn to retrograde stage as the presence of phlogopite and disappearance of diopside. The reactions considered above are not sensitive to oxygen fugacity, but this is a key constraint on the formation conditions of magnetite, which is discussed next.

3.2 Thermodynamic Model for Calcic Skarn

In the calcic skarn zonation, prehnite occurs as anhedral crystals intergrown with wollastonite, andradite, hedenbergite, which reflects three alteration stages: (1) wollastonite formation in the early skarn stage, via the reaction: Qtz + Cc = Wo + CO_2 (Table 3.3), almost occurred at a minimum temperature of >520 °C (Fig. 3.4); (2) andradite and hedenbergite formation in the late skarn stage, via the reaction: 18 Wo + 4 Mt + O_2 = 6 Adr; 2 Mt + 12 Wo + 6 CO_2 = 6 Hd + 6 Cc + O_2; 18 Cc + 4 Mt +18 Qtz + O_2 = 6 Adr + 18 CO_2; 6 Cc + 2 Mt +12 Qtz = 6 Hd + O_2 + 6 CO_2. These reactions accompany a disappearance of magnetite and leave a stable zone of andradite + hedenbergite + quartz + calcite which are constrained at a temperature range of ca. 500–580 °C and $\Delta \log fO_2$(HM) range of ca. −10 to 0 (Fig. 3.4); (3) prehnite formation. At this stage, prehnite is likely to be formed through the reaction with abundant magnetite precipitations: 12 Qtz + 2 Mt + 6 Grs + 6 H_2O = 6 Hd + 6 Prh + O_2 and 18 Qtz + 4 Mt + 18 Grs + O_2 + 18 H_2O = 6 Adr + 18 Prh (Fig. 3.5a), which are occurred at a successive fO_2 values of −5.5 to −3.5 (log unit below hematite-magnetite buffer) and a temperature range of 440–465 °C (Fig. 3.5a). Therefore, most magnetite are precipitated in the retrograde stage rather than skarn stage as most iron in the fluid are occupied by andradites and hedenbergites.

Retrograde alteration-related epidote was also considered as an essential mineral along with hedenbergite and/or grandite in skarns. As illustrated in Fig. 3.5b, the upper stability limit of epidote is controlled by the reactions: 8 Adr + 4 Grs + 16 CO_2 + 2 H_2O = 4 Ep + 12 Hd + 16 Cc + 3 O_2, and lower limit is controlled by the reaction of 4 Mt + 12 Grs + O_2 + 6 H_2O + 12 CO_2 = 12 Cc + 12 Ep at successive fO_2 values of −14 to −4 (log unit below hematite-magnetite buffer) and a temperature range of 465–485 °C (Fig. 3.5b). Therefore, the epidote are always formed before prehnite as its higher forming temperature.

In the skarn stage, iron in the fluid are most incorporated by hedenbergite and andradite, therefore, a few magnetites are precipitated. However, in the retrograde stage, abundant magnetite precipitated with the decomposition of hedenbergite and andradite and crystallization of retrograde skarn minerals. In addition, the decrease of temperature and increase of magnetite are the main cause for the magnetite precipitation.

3.3 The Precipitation of Magnetite

To reach water saturation, 10% crystallization would be required for the melt under a pressure of 0.5 Kbar. Therefore, too many metal elements such as Fe and Cu can be involved in the fluid before largely crystallized at the Galinge deposit. The primary oxygen fugacity condition of magma has important effects on magnetite precipitation. Magnetite may form readily in magnesian skarns since magnesian-silicates (for example forsterite, talc and serpentine) do not take up much iron in

Table 3.3 The stable calcic skarn assemblages in the Galinge deposit

Stage	Mineral assemblages		The inferred equilibrium reactions
Skarn stage	a	wollastonite + calcite	$Qtz + Cc = Wo + CO_2$
	b	andradite + magnetite	$18\ Woll + 4\ Mt + O_2 = 6\ Adr$
	c	hedenbergite + magnetite + calcite	$2\ Mt + 12\ Wo + 6\ CO_2 = 6\ Hd + 6\ Cc + O_2$ $6\ Cc + 2\ Mt + 12\ Qtz = 6\ Hd + O_2 + 6\ CO_2$ $12\ Adr + 18\ CO_2 = 18\ Cc + 18\ Hd + 2\ Mt + 5\ O_2$
	d	andradite + hedenbergite + calcite ± magnetite	$2\ Adr + 2\ Wo + 4\ CO_2 = 9\ Hd + 4\ Cc + O_2$ $2\ Adr + 2\ Qtz + 2\ CO_2 = 4\ Hd + 2\ Cc + O_2$
	e	Grossular + quartz + wollastonite	$Wo + An = Qtz + Grs$
Retrograde stage	f	epidote + andradite + hedenbergite + calcite	$8\ Adr + 4\ Grs + 16\ CO_2 + 2\ H_2O = 4\ Ep + 12\ Hd + 16\ Cc + 3O_2$
	g	epidote + magnetite + calcite	$4\ Mt + 12\ Grs + O_2 + 6\ H_2O + 12\ CO_2 = 12\ Cc + 12\ Ep$
	h	prehnite + hedenbergite + magnetite + grossular	$12\ Qtz + 2\ Mt + 6\ Grs + 6\ H_2O = 6\ Hd + 6\ Prh + O_2$
	i	prehnite + andradite + magnetite + grossular	$18\ Qtz + 4\ Mt + 18\ Grs + O_2 + 18\ H_2O = 6\ Adr + 18\ Prh$

Adr: andradite; Cc-calcite; Ep: epidote; Grs: grossular; Hd: hedenbergite; Mt: magnetite; Prh: prehnite; Qtz-quartz; Wo: wollastonite

solid solution at the prevailing oxidation states [3]. We investigate the thermodynamic relations of the replacement reactions responsible for magnetite precipitation in order to track the compositional and physical contrast between skarn and retrograde stage minerals, in addition to changes in fluid properties ($fO_2, T, X(CO_2)$).

Whereas magnetite are probably from the only one magmatic source, too many reasons including temperature changes, varying composition of host rocks and fluid could cause the subtle identification on the magnetite component of TiO_2, Cr_2O_3, MnO, MgO, V_2O_3 and Al_2O_3 [14]. In the skarn stage of the Galinge deposit, the magnetite contains very little MgO contents, but is relatively enriched in Al_2O_3 and CaO. Generally, higher Al and Ti in magnetite reflects higher temperatures [28]. This thermodynamic principle also applies to hydrothermal magnetite [26]. However, the Galinge hydrothermal magnetite formed in the retrograde alteration stage is characterized by high MgO and low TiO_2 and V_2O_3, which likely reflect extensive fluid-rock interaction [27]. Al and Ti are typically immobile at sub-magmatic temperatures, and their solubilities in hydrothermal fluids are considered to be strongly temperature dependent [35, 37], which is for immiscibility gaps in both the (Fe,

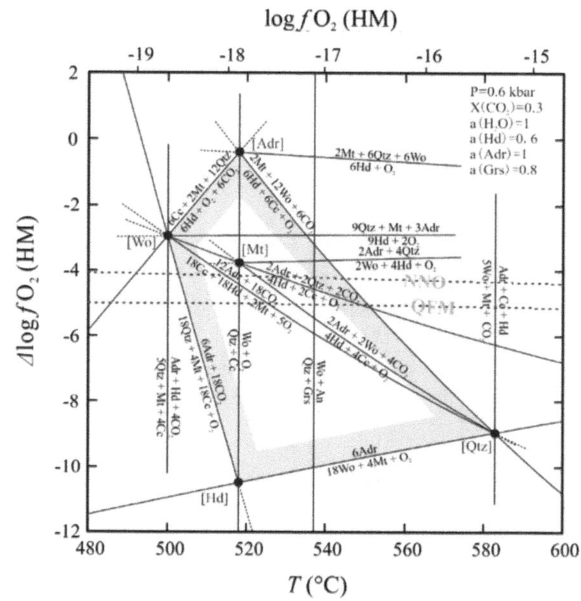

Fig. 3.4 Δlog fO_2(HM)-T diagram for several reactions in the system Ca–Fe–Si–O–H–C at a pressure of 0.6 Kbar. Mole fractions of CO_2 in C–O–H fluids are set to 0.3. The HM defined the hematite-magnetite oxygen buffer [25, 31]; While those curves marked with QFM and NNO referred to the quartz-fayalite-magnetite and busenite-nickel-oxygen assemblage oxygen buffers under a pressure condition of 0.6 kbar [1]. The shadow area is a stable zone of andradite + hedenbergite + wollastonite + calcite + quartz assemblage

Mg)Al_2O_4–(Fe, Mg)Fe_2O_4, and the Fe_3O_4–Fe_2TiO_4 series [8, 13, 32]. Our data are consistent with this model, as Al and Ti are on average more concentrated in the magnetite coexisting with spinel in the skarn stage than that coexisting with tremolite and phlogopite in the retrograde stage. It is reasonable to suggest that the temperature of the magnetite precipitation is between ca. 430 and 560 °C based on the Fe–Ti oxides geothermometer and oxygen barometer, which almost match the temperature range of 440–500 °C of the spinel + magnetite formation in our thermodynamic model (Fig. 3.1b).

In the early metasomatism stage (skarn stage), the magnetite genesis is associated with the consumption of forsterite-fayalite and diopside-hedenbergite solid solution containing much Fe. Structure and stability of iron-bearing olivines were considered as a function of temperature, forsterite composition, oxygen fugacity and crystallographic orientation [4, 6, 7, 16, 29]. The FeO content of the Galinge olivines shows a reverse relation with the magnetite mineralization degrees in magnesian skarn systems because the Fe in solution tends to form fayalite series. Thus olivine with high FeO content (fayalite series) is commonly unrelated with magnetite precipitation, but nearly pure forsterite shows strong relationship with magnetite via 2 Cc + 4 Fo + 6 Hd + O_2 = 2 Mt + 8 Di + 2 CO_2 and 6 Hd + 24 Do + +O_2 = 30 Cc + 12 Fo + 2

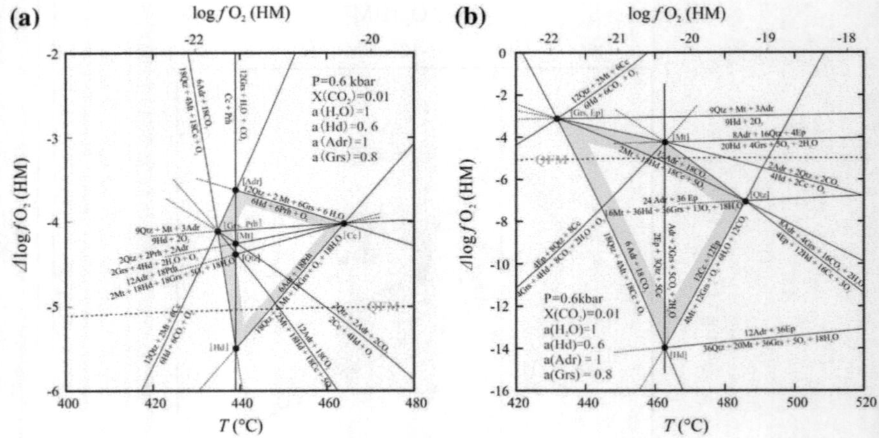

Fig. 3.5 Schematic $\Delta\log fO_2$(HM)-T relations in the Ca–Fe–Al–Si–C–O–H system with excess SiO_2 and fluids, including epidote + grossular + andradite + hedenbergite + magnetite + quartz + calcite phases (**a**) and prehnite + grossular + andradite + hedenbergite + magnetite + quartz + calcite (**b**), respectively. The shadow area in A stands for a stable zone of prehnite + andradite + hedenbergite + magnetite + calcite assemblage, but the shadow area in (**b**) represents a stable zone of epidote + andradite + hedenbergite + calcite + magnetite assemblage. HM: the hematite-magnetite oxygen buffer [25, 31]; QFM: quartz-fayalite-magnetite buffer [1]

Mt + 18 CO_2, accompanying with the disappearances of hedenbergite, and leaving a stable assemblage of forsterite + diopside + magnetite ± calcite and forsterite + magnetite + calcite ± dolomite, respectively. The thermodynamic model indicates a stability field of magnetite below the reaction of 6 Cc + 2 Mt + 12 Qtz = 6 Hd + O_2 + 6 CO_2, into a condition of T = ca 450–500 °C and $\Delta\log fO_2$ (HM) = −9.5 to −11 (Fig. 3.1a). Nevertheless, all the reactions connect with the invariant point of Cc will affect the precipitation of magnetite.

In the retrograde stage, the precipitation of magnetite is related to the metasomatism reaction between hedenbergite and retrograde minerals of tremolite, phlogopite, clinohumite, epidote and prehnite. Therefore, all the reactions should be occurred below the reaction of 6 Cc + 2 Mt + 12 Qtz = 6 Hd + O_2 + 6 CO_2 (Fig. 3.2a). The tremolite + hedenbergite + magnetite + calcite stable assemblage is constrained by the reaction of 30 Di + 2 Mt + 12 CO_2 + 6 H_2O = 6 Tr + 6 Hd + 12 Cc + O_2 and 12 Hd + 15 Do + 2 O_2 + 3 H_2O = 21 Cc + 4 Mt + 3 Tr + 9 CO_2, with the disappearance of dolomite and diopside, respectively, under a temperature range of 370–470 °C, and a $\Delta\log fO_2$ (HM) range of −5 to −17 below QFM buffer (Fig. 3.2a).

The stable zone of clinohumite + diopside + hedenbergite + magnetite + dolomite + quartz assemblage is constrained by the reactions: 54 Di + 14 Mt +12 CO_2 + 6 H_2O = 6 Chu + 42 Hd +12 Cc + 7 O_2; and 33 Cc + 2 Mt + 3 Chu + 21 CO_2 = 6 Hd + 27 Do + O_2 + 3 H_2O. The both reactions accompany with the disappearance of calcite to be equilibrated under a temperature range of ca. 340–460 °C and a $\Delta\log fO_2$ (HM) range of −15 to −4 (Fig. 3.2b). Therefore, in the tremolite- and clinohumite-bearing system, the formation and breakdown of hedenbergite significantly affect the precipitation of magnetite (Fig. 3.2a, b). In another word, the presence of hedenbergite in the Ca–Mg–Fe–Si retrograde skarn system is not favorable for the magnetite formation because it will incorporate lots of iron from the fluid as a consequence of oxygen fugacity increase, but it will release iron as the oxygen fugacity decrease.

Phlogopite is another very important retrograde minerals for the precipitation of magnetite. It controls the formation of magnetite via the reactions: 6 Di + 2 Mt + 2 Kfs + 2 H_2O = 2 Phl + 6 Hd + O_2 and 6 Do + 6 Hd + 2 Kfs + 2 H_2O + O_2 = 12 Qtz + 2 Phl + 2 Mt + 12 Cc, accompanying a disappearance of K-feldspar. The both reactions with the Phl-in and Phl-out reactions constrain a stable zone of phlogopite + hedenbergite + diopside + quartz + calcite + magnetite assemblage at a temperature range of ca. 365–415 °C and a $\Delta \log fO_2$ (HM) range of -11 to -6 (Fig. 3.2c).

Serpentine is the most widespread retrograde mineral in the magnesian skarn zonation. It shows strongly relationship with the precipitation of magnetite through though the reactions: 24 Cc + 6 Chr +2 Mt + 12 CO_2 + 6 Hd + 18 Do + O_2 + 12 H_2O; 6 Hd + 6 Chr + 12 Cc + O_2 = 12 Di + 6 Do + 2 Mt + 12 H_2O; and 12 Cc + 18 Chr + 2 Mt = 6 Hd + 24 Fo + 6 Do + O_2 + 36 H_2O. These reactions constrain a stable zone of chrysotile + forsterite + hedenbergite + diopside + dolomite + magnetite assemblage under a temperature range of 425–455 °C and a $\Delta \log fO_2$ (HM) range of -15 to $-$ 11.5 (Fig. 3.2d).

In the calcic skarn zonation, epidote is a good indicator for the magnetite precipitation. The reactions of f and g (Table 3.3) accompany a disappearance of grossular and leave a stable zone of the epidote + andradite + hedenbergite + calcite + magnetite assemblage, which is constrained at a temperature range of ca. 410–490 °C and a $\Delta \text{low} fO_2$(HM) spectrum of ca. -3 to -14 together by the reactions of 12 Qtz + 2 Mt + 6 Cc = 6 Hd + 6 CO_2 + O_2 and 18 Qtz + 4 Mt + 18 Cc + O_2 = 6 Adr + 18 CO_2 (Fig. 3.5b). Prehnite is similar to the epidote in the calcic skarn zonation, which is a typical retrograde minerals associated with magnetite precipitation. The stable zone of prehnite + andradite + hedenbergite + magnetite + calcite assemblage is constrained by the reactions of h and i (Table 3.3), together with the reactions of 12 Qtz + 2 Mt + 6 Cc = 6 Hd + 6 CO_2 + O_2 and 18 Qtz + 4 Mt + 18 Cc + O_2 = 6 Adr + 18 CO_2 accompanying a consumption of quartz f at a temperature range of ca. 435–465 °C and a $\Delta \text{low} fO_2$(HM) spectrum of ca. -3.5 to -5.5 (Fig. 3.5a).

3.4 Conclusion

This study focused on the geochemistry of processes of the evolution and development of metasomatic zoning with respect to dolomites and other magnesian carbonate rock species. In the skarn stage, the Galinge skarn deposit developed abundant forsterite-fayalite spinel-hercynite-gahnite diopside-hedenbergite and andradite-grossular solid solution at a temperature range of 560–460 °C and $\Delta \log fO_2$ (HM) = -11 to 0. In the retrograde alteration stage, abundant magnetites coexist with tremolite, phlogopite, clinohumite, epidote and prehnite. They reflect a fluid conditions of 360–460 °C and $\Delta \log fO_2$ (HM) = −16 to −3. The paragenetic and thermodynamic results shed new light on the processes affecting skarn formation in a unique model system.

References

1. Berman RG (1988) Internally-consistent thermodynamic data for minerals in the system Na_2O-K_2O-CaO-MgO-FeO-Fe_2O_3-Al_2O_3-SiO_2-TiO_2-H_2O-CO_2. J Petrol 29(2):445–522
2. Berman RG (2007) WinTWQ (version 2.3): a software package for performing internally-consistent thermobarometric calculations. Geol Surv Can Open File 5462:1–41
3. Clark DA (2014) Magnetic effects of hydrothermal alteration in porphyry copper and iron-oxide copper–gold systems: a review. Tectonophysics 624:46–65
4. Coogan LA, Hain A, Stahl S et al (2005) Experimental determination of the diffusion coefficient for calcium in olivine between 900 °C and 1500 °C. Geochim Cosmochim Acta 69(14):3683–3694
5. Dick JM (2008) Calculation of the relative metastabilities of proteins using the CHNOSZ software package. Geochem Trans 9(1):1–17
6. Dohmen R, Becker H, Chakraborty S (2007) Fe–Mg diffusion in olivine I: experimental determination between 700 and 1200 °C as a function of composition, crystal orientation and oxygen fugacity. Phys Chem Miner 34(6):389–407
7. Dohmen R, Chakraborty S (2007) Fe–Mg diffusion in olivine II: point defect chemistry, change of diffusion mechanisms and a model for calculation of diffusion coefficients in natural olivine. Phys Chem Miner 34(6):409–430
8. Dunlop DJ, Özdemir Ö (2001) Rock magnetism: fundamentals and frontiers, vol 3. Cambridge University Press
9. Evans BW (2004) The serpentinite multisystem revisited: chrysotile is metastable. Int Geol Rev 46(6):479–506
10. Ferry JM (1996) Prograde and retrograde fluid flow during contact metamorphism of siliceous carbonate rocks from the Ballachulish aureole, Scotland. Contrib Mineral Petrol 124(3–4):235–254
11. Ferry JM (2001) Calcite inclusions in forsterite. Am Mineral 86(7–8):773–779
12. Ferry JM, Ushikubo T, Valley JW (2011) Formation of forsterite by silicification of dolomite during contact metamorphism. J Petrol 52:1619–1640
13. Frost BR, Lindsley DH (1992) Equilibria among Fe-Ti oxides, pyroxenes, olivine, and quartz: Part II. Application. Am Mineral 77:1004
14. Grigsby JD (1990) Detrital magnetite as a provenance indicator. J Sediment Res 60(6):940–951
15. Guy B, Sheppard S, Fouillac AM et al (1988) Geochemical and isotope (H, C, O, S) studies of barren and tungsten-bearing skarns of the French Pyrenees. In: Mineral deposits within the European community. Springer, pp 53–75
16. Ito M, Ganguly J (2006) Diffusion kinetics of Cr in olivine and ^{53}Mn–^{53}Cr thermochronology of early solar system objects. Geochim Cosmochim Acta 70(3):799–809
17. Lepage LD (2003) ILMAT: an Excel worksheet for ilmenite–magnetite geothermometry and geobarometry. Comput Geosci 29(5):673–678
18. Liang XJ, Qiao L (1990) Characteristics of humite group minerals formed in metasomatic experimentation and physico-chemical conditions for their formation. Acta Petrol Mineral 4:005
19. Lindsley DH, Frost BR (1992) Equilibria among Fe-Ti oxides, pyroxenes, olivine, and quartz; Part I, Theory. Am Mineral 77(9–10):987–1003
20. Luce RW, Cygan GL, Hemley JJ et al (1985) Some mineral stability relations in the system CaO·MgO·SiO_2·H_2O·HCl. Geochim Cosmochim Acta 49(2):525–538

21. Mazurov MP, Grishina SN, Istomin VE et al (2007) Metasomatism and ore formation at contacts of dolerite with saliferous rocks in the sedimentary cover of the southern Siberian platform. Geol Ore Depos 49(4):271–284
22. Meinert LD (1992) Skarns and skarn deposits. Geosci Can 19(4):145–162
23. Meinert LD (1993) Igneous petrogenesis and skarn deposits. Mineral deposit modeling 40:569–583
24. Meinert LD, Dipple GM, Nicolescu S (2005) World skarn deposits. Econ Geol 100th Anniv 299–336
25. Myers JT, Eugster HP (1983) The system Fe-Si-O: oxygen buffer calibrations to 1,500 K. Contrib Mineral Petrol 82(1):75–90
26. Nadoll P, Angerer T, Mauk JL et al (2014) The chemistry of hydrothermal magnetite: a review. Ore Geol Rev 61:1–32
27. Nadoll P, Mauk JL, Leveille RA et al (2015) Geochemistry of magnetite from porphyry Cu and skarn deposits in the southwestern United States. Mineralium Depos 50(4):493–515
28. Nielsen RL, Forsythe LM, Gallahan WE et al (1994) Major-and trace-element magnetite-melt equilibria. Chem Geol 117(1):167–191
29. Petry C, Chakraborty S, Palme H (2004) Experimental determination of Ni diffusion coefficients in olivine and their dependence on temperature, composition, oxygen fugacity, and crystallographic orientation. Geochim Cosmochim Acta 68(20):4179–4188
30. Rice JM (1980) Phase equilibria involving humite minerals in impure dolomitic limestones. Contrib Mineral Petrol 71(3):219–235
31. Sack RO, Carmichael I, Rivers ML et al (1980) Ferric-ferrous equilibria in natural silicate liquids at 1 bar. Contrib Mineral Petrol 75(4):369–376
32. Sack RO, Ghiorso MS (1991) An internally consistent model for the thermodynamic properties of Fe-Mg-titanomagnetite-aluminate spinels. Contrib Mineral Petrol 106(4):474–505
33. Spear FS (1993) Metamorphic phase equilibria and pressure-temperature-time paths. Mineralogical Society of America (Monograph), p 799
34. Suzuki K (1977) Local equilibrium during the contact metamorphism of siliceous dolomites in Kasuga-mura, Gifu-ken, Japan. Contrib Mineral Petrol 61(1):79–89
35. Van Baalen MR (1993) Titanium mobility in metamorphic systems: a review. Chem Geol 110(1):233–249
36. Vander AJ, André L (1991) Trace elements (REE) and isotopes (O, C, Sr) to characterize the metasomatic fluid sources: evidence from the skarn deposit (Fe, W, Cu) of Traversella (Ivrea, Italy). Contrib Mineral Petrol 106(3):325–339
37. Verlaguet A, Brunet F, Goffé B et al (2006) Experimental study and modeling of fluid reaction paths in the quartz–kyanite ± muscovite–water system at 0.7 GPa in the 350–550 °C range: implications for Al selective transfer during metamorphism. Geochim Cosmochim Acta 70(7):1772–1788
38. Walther JV, Helgeson HC (1980) Description and interpretation of metasomatic phase relations at high pressures and temperatures: 1. equilibrium activities of ionic species in nonideal mixtures of CO_2 and H_2O. Am J Sci 280(7):459–485
39. Yardley BW, MacKenzie WS, Guilford C (1990) Atlas of metamorphic rocks and their textures. Longman Scientific & Technical, Essex
40. Young ED, Morrison J (1992) Relations among net-transfer reaction progress, 18O–13C depletion, and fluid infiltration in a clinohumite-bearing marble. Contrib Mineral Petrol 111(3):391–408
41. Yu M, Feng CY, Liu HC et al (2016) The Iron-Titanium Oxides in the Galinge Iron Polymetallic Skarn Deposit of Qinghai Province and Their Thermodynamic Significance. Acta Geosci Sin 37(2):204–214

Chapter 4
Multistage Skarn-Related Tourmalines from the Galinge Deposit: A Significant Indicator for Varying Fluid Composition

Tourmaline, as its large range of P-T stability conditions and extremely low rates of volume diffusion has been recognized as a medium for recording geologic information [14, 60]. Tourmaline has also been recognized as an important gangue and alteration mineral in many types of intrusion-related deposits [5, 9, 43, 57]. It is relatively rare in iron skarn deposit and is almost exclusively occurred in W $\pm$ Sn skarn deposit [32]. But almost all tourmalines frequently referred to W $\pm$ Sn skarn really belongs to greisen stage, which is superimposed onto skarn. For example, Alaska Lost River Sn + W + F skarn deposit [12] and JC Yukon Sn skarn deposit [33]. In addition, Slack [53] reported tourmaline only from Mg-skarn. Therefore, the Galinge skarn-related tourmaline is the first finding of calc-skarn tourmaline.

The chemical composition of tourmaline has been proven useful as an indicator for hydrothermal ore-forming process [5, 14, 16, 23, 27, 31, 36, 44, 57]. A wide range of chemical compositions and a broad range of P-T stability field of tourmaline supergroup minerals make them as a good recorder of hydrothermal fluid evolution [60]. In this paper, we intend to summarize the compositional information of the tourmalines from the Galinge iron skarn deposits to indicate the evolution of hydrothermal fluids, and compare the obtained results with the literature data of other types of hydrothermal tourmaline-bearing deposits. Nonetheless, compilation work has been undertaken by many authors previously to see if there is a relationship between tourmaline composition and different styles of mineralisation.

M. Yu, *Metallogenic Mechanism of the Galinge Polymetallic Iron Skarn Deposit, Qiman Tagh Mountains, Qinghai Province*, Springer Theses,
https://doi.org/10.1007/978-981-10-7907-8_4

4.1 Petrography of Samples

Tourmalines from the Galinge deposit commonly occur in the V Ore Domain. They occur as idiomorphic or hypidiomorphic grains in the altered basaltic andesite rocks (Tour-I), or as anhedral crystals in metasandstones (Tour-II). Greenish blue Tour-I always intergrown with anhedral-subhedral epidote and pyrite, is replaced by chlorite (Fig. 4.2a, b). Tourmaline + epidote assemblages suggest that precipitation of epidote and tourmaline occurred at temperatures in excess of 370 °C at least during the hydrothermal evolution of the system [8]. Tour-I is generally characterized by discordant reaction zoning, which shows that the early-stage green-blue cores (Gen-1) are replaced by the late-stage yellowish blue rims (Gen-2) (Fig. 4.2c). They both developed oscillatory zonation during their crystallizing growth (Fig. 4.1). The Gen-1 tourmaline is heterogeneous in composition, but Gen-2 tourmaline is complexly zoned (combination of growth and sector zoning) (Fig. 4.2). Some of anhedral granular Tour-I crystals without clear zonation were defined other tourmalines (other) in the context (Fig. 4.2a, b). In addition, Tour-I developed thin yellowish-green overgrowth rims (Gen-3), these rims also propagate through earlier tourmaline as fracture fill (Figs. 4.1 and 4.2d). Tour-II in metasandstones is brownish green, and occurs as small veinlets and lenses with lineally deformed quartz and biotite, nevertheless Tour-II is undeformed (Fig. 4.2e, f). The amount of quartz is more than 70%.

Below the tourmalinized skarn, axinite-bearing skarn abundantly developed proximal to the granodiorite intrusion. Such non-tourmaline rocks were mainly composed

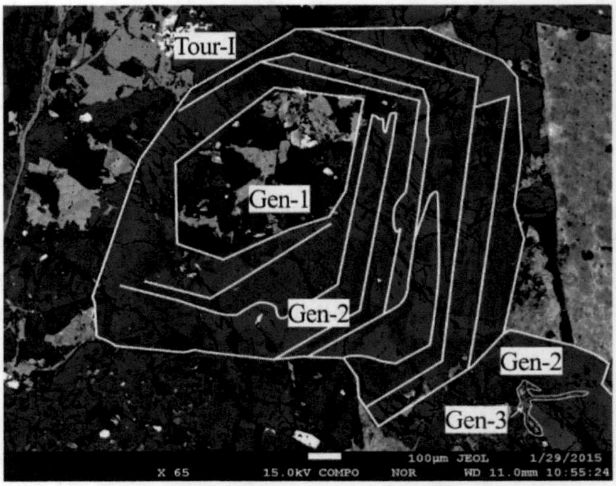

Fig. 4.1 BSE images of a single tourmaline crystal (Tour-I) from the Galinge skarn deposit. The Gen-1 tourmaline is heterogeneous in composition. Gen-2 tourmaline is complexly zoned (combination of growth and sector zoning)

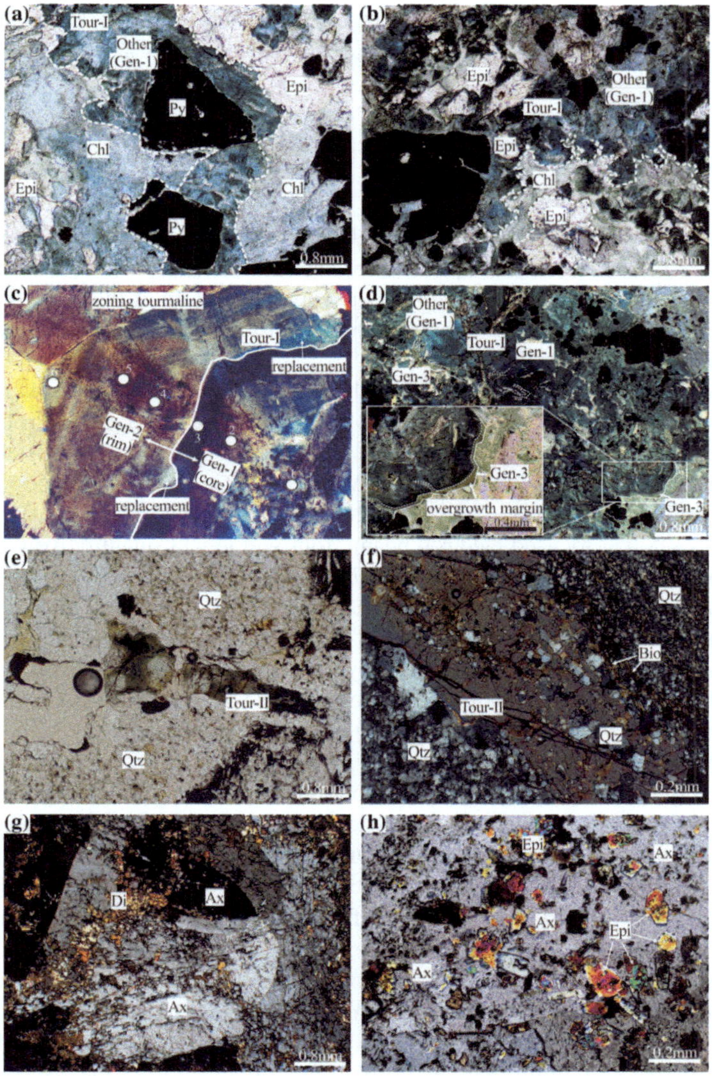

Fig. 4.2 **a** The image of Tour-I under plane light shows Tour-I coexisted with the epidote, chlorite and pyrite in the mineralization skarn. The tourmaline and epidote are formed in the same retrograde stage. The chlorite, formed in the sulfide stage with the pyrite, replaced the tourmaline and epidote. **b** The same situation with the (**a**). **c** The image of Tour-I under plane light shows a zoned tourmaline, with the rims (Gen-2) partially replaced the cores (Gen-1). **d** The image of Tour-I under plane light shows a overgrowth margin formed by the last stage fluids (Gen-3). **e** The image of Tour-II under plane light shows a brownish Green tourmaline (Tour-II) in the metasandstone, and the amount of quartz matrix is more than 70%; **f** The image of Tour-II under cross polarized light shows undeformed tourmaline and lineated quartz and biotite in metasandstone. **g** Colorless euhedral to subhedral axinite replaced by granular pyroxene. **h** Colorless euhedral to subhedral axinite with granular epidote

of axinite ± epidote ± pyroxene assemblages. The axinite is colorless and euhedral to subhedral. The epidote and pyroxene are usually enveloped by axinite (Fig. 4.2g, h).

4.2 Analytical Methods

Thin sections were made both parallel and perpendicular to the c axis of the in order to establish chemical relationships of three generation tourmalines. Optical and back-scattered electron (BSE) imaging show the complex texture of the three generation tourmaline (Fig. 4.1). These images were used for selection of areas suitable for subsequent micro-probe work.

Tourmaline electron-microprobe analyses were conducted with wavelength-dispersion spectrometry (WDS) using a JEOL JXA-823V Electron Probe Micro Analyzer (EPMA) at the Key Laboratory of Metallogeny and Mineral Assessment, Institute of Mineral Resources, Chinese Academy of Geological Sciences (Beijing) China. Analytical conditions were: Accelerating potential 15 kV, and beam current 5–10 nA, 3–5 μm focused electron beam. Tourmaline was quantitatively analyzed for K, Na, Ca, Fe, Mn, Mg, Ti, Al, Si, F, and Cl. The standards were natural and synthetic compounds, including jadeite (Si, Na, and Al), forsterite (Mg), orthoclase (K), apatite (P), wollastonite (Ca), rutile (Ti), halite (Cl), MgF_2 (F), Fe_2O_3 (Fe) and MnO (Mn). Boron couldn't be detected by the equipment. Analytical precisions for selected elements, estimated on the basis of replicate analyses of standards and samples, are <0.9% for SiO_2, <0.1% for Al_2O_3, <0.08% for FeO and MgO, and <0.03% for CaO and Na_2O.

The general formula for tourmaline can be expressed as $XY_3Z_6(T_6O_{18})(BO_3)_3V_3W$ [19]. In order to calculate the stoichiometry of the tourmalines and make estimates of the amounts of undetectable H and B, several assumptions were premised based on the general formula of tourmaline [19, 25]: (1) The amount of B_2O_3 was calculated from stoichiometry constraints on three B cations in the structural formula; (2) The structural formula is based on a normalization scheme such that the proportion of cations in the Y + Z + T site equals 15 atoms per formula unit (apfu), assuming no vacancies in these sites. If the T-site Si < 6 apfu, it could be caused by some Al incorporated into T site. If a depletion of Z-site Al is below 6 apfu, there could be different substitution mechanisms likely to keep charge balance [1]. An evaluation of likely substitution methods relevant to our tourmalines will be commented in the discussion part. Although this assumption likely not valid for tourmalines in general, is a reasonable assumption in this case because knowledge of B concentration, OH content, and oxidation state of Fe is unnecessary; (3) We assumed that Li contents are minimal in the Y site; (4) Fetot is assumed to be essentially all Fe^{2+} and does not contain Fe^{3+}; (5) Another hypothesis

Table 4.1 The microprobe analysis of the average Galinge tourmaline composition

Sample	Tour-I								Tour-II	
	Gen-1		Gen-2		Gen-3		Other			
Average (wt%)	Average (8)	R	Average (9)	R	Average (4)	R	Average (8)	R	Average (9)	R
SiO_2	35.30	0.480	35.23	0.508	35.65	0.196	34.99	0.531	34.79	0.673
TiO_2	0.36	0.201	0.36	0.148	0.32	0.064	0.10	0.083	0.77	0.236
Al_2O_3	26.20	1.021	27.51	0.932	27.09	0.716	29.16	1.260	28.22	1.079
FeO	11.84	0.670	11.26	0.587	5.08	0.559	6.91	1.250	9.06	0.715
MnO	0.02	0.025	0.01	0.011	0.03	0.017	0.03	0.014	0.03	0.031
MgO	7.29	0.368	6.85	0.267	11.30	0.464	8.83	0.833	7.20	0.342
CaO	2.90	0.282	2.32	0.205	4.19	0.273	3.15	0.491	2.53	0.470
Na_2O	1.21	0.102	1.53	0.099	0.69	0.157	1.10	0.218	1.27	0.161
K_2O	0.02	0.014	0.02	0.012	0.01	0.010	0.04	0.068	0.05	0.025
Cr_2O_3	0.02	0.011	0.02	0.011	0.02	0.008	0.01	0.005	0.21	0.316
B_2O_3	9.88	0.062	9.91	0.100	10.07	0.059	10.01	0.156	9.86	0.158
F	0.24	0.145	0.24	0.166	0.51	0.120	0.25	0.089	0.24	0.155
H_2O	3.37	0.085	3.38	0.090	3.31	0.075	3.41	0.078	3.37	0.069
Total	98.65	0.256	98.66	1.014	98.25	0.437	97.99	1.234	97.58	1.296
T(Si)	6.069	0.064	6.034	0.049	6.012	0.003	5.938	0.129	5.989	0.038
Al	0.000	0.000	0.007	0.015	0.000	0.000	0.087	0.086	0.021	0.026
B	3.000	0.000	3.000	0.000	3.000	0.000	3.000	0.000	3.000	0.000
Z(Al)	5.307	0.180	5.546	0.155	5.384	0.115	5.742	0.131	5.704	0.174

(continued)

Table 4.1 (continued)

Sample	Tour-I								Tour-II	
	Gen-1		Gen-2		Gen-3		Other			
Average (wt%)	Average (8)	R	Average (9)	R	Average (4)	R	Average (8)	R	Average (9)	R
Mg^{2+}	0.693	0.180	0.454	0.155	0.616	0.115	0.258	0.131	0.296	0.174
Y(Ti)	0.047	0.026	0.046	0.019	0.040	0.008	0.013	0.010	0.100	0.031
Fe^{2+}	1.702	0.104	1.612	0.088	0.717	0.081	0.983	0.186	1.306	0.112
Mn	0.003	0.004	0.002	0.002	0.004	0.002	0.004	0.002	0.004	0.005
Mg	1.177	0.109	1.296	0.107	2.225	0.091	1.974	0.209	1.551	0.153
Cr^{3+}	0.003	0.001	0.003	0.001	0.003	0.001	0.001	0.001	0.028	0.044
$\sum Y$	2.931	0.064	2.958	0.039	2.988	0.003	2.975	0.066	2.989	0.016
Ca	0.535	0.054	0.425	0.037	0.756	0.052	0.572	0.087	0.466	0.088
Na	0.403	0.035	0.510	0.033	0.228	0.050	0.361	0.071	0.424	0.053
K	0.005	0.003	0.004	0.003	0.003	0.002	0.009	0.016	0.011	0.005
$\sum X$	0.942	0.055	0.939	0.019	0.987	0.004	0.942	0.065	0.902	0.047
$\square^X$	0.058	0.055	0.061	0.019	0.014	0.004	0.058	0.065	0.098	0.047
$V(OH)^-$	3.000	0.000	3.000	0.000	3.000	0.000	3.000	0.000	3.000	0.000
$W(F^-)$	0.131	0.079	0.131	0.091	0.272	0.066	0.135	0.048	0.130	0.083
$(OH)^-$	0.869	0.079	0.869	0.091	0.728	0.066	0.865	0.048	0.870	0.083
$\sum W$	1.000	0.000	1.000	0.000	1.000	0.000	1.000	0.000	1.000	0.000
X_{Mg}	0.409	0.036	0.445	0.033	0.756	0.029	0.667	0.065	0.542	0.045
Group	CAG		ALG		CAG		CAG		CAG	
Name	uvite		dravite		uvite		uvite		uvite	

The ideal formula of tourmaline is $XY_3Z_6T_6O_{18}(BO_3)_3V_3W$; stoichiometric calculation based on B = 3 apfu; OH + F=4 apfu; T + Z+Y = 15. XMg = $Mg/(Fe^{2+}+Mg)$; CAG: Calcic Group; ALG: Alkalic Group is that the V site is completely occupied by OH^-. (6) Although O^{2-} substitution for OH^- preferentially happened in the W site relative to the V site, we assumed here there is no incorporation of O^{2-} into W. Whereas this assumption is probably simple, it doesn't affect the analytical uncertainties of all measured elements. (7) F is assumed to be exclusively accommodated in the W site, and the amounts of W-site OH^- and F^- equal to 4. The Galinge tourmaline compositions are listed in Table 4.1.

4.3 Chemical Composition of Tourmaline

The Galinge tourmaline is compositionally classified into calcic-group and alkali-group in terms of the Ca, Na + K and □ (Fig. 4.3b). Both Tour-I and Tour-II are intermediate member of the dravite-uvite solid solution series relatively enriched in Ca and depleted in Fe (Fig. 4.3c). A theoretically discordant growing model for three-generations of tour-I is displayed in Fig. 4.3a. The Gen-1 is mainly composed of uvite; the Gen-2 is dominated by dravite; the Gen-3 shows an uvite composition which is distinctively more enriched in Ca than Gen-1 and Gen-2 (Fig. 4.3b, c). The Other tourmalines are characterized by Ca-rich and Fe-poor to be distinguished from Gen-1 and Gen-2. The Tour-II is compositionally similar to Tour-I, which contains uvite and dravite compositions (Fig. 4.3b, c). As judged by Fig. 4.3b, c, dravite is Ca–Fe rich, whereas uvite is enriched in Na and Ca.

Replacements of Ca for Na in the X site and of Mg for Fe in the Y site are evident. Fetot and A1 show clear opposite variations, which become striking after calculation of the Fe atomic abundance. The chemical profile of the sample 21605-38 tourmalines shown in Fig. 4.4, and the image with analyzed points are given in Fig. 4.2c. It reveals that Al–Fe–Mg content is not due to a continuous evolution between two end-member compositions but to a strong oscillatory variation of the Fe content in the same crystal. The chemical composition is emphasized by moderate change between the spot 3 and 4, because the boundary between Gen-1 and Gen-2 is divided at the halfway between points 3 and 4.

Cation proportions of the Galinge tourmalines are plotted in the Al–Fe–Mg and Ca–Fe–Mg ternary diagram [24] (Fig. 4.5). The Galinge tourmalines exhibit a wide diversity of the mineral compositions. In Fig. 4.5a, almost all the Galinge tourmaline compositions plot in the Ca-rich field near the boundary between fields 3 and 4 subparallel to the Mg–Fe join. There is only a slight deviation towards more Al rich compositions with increasing Mg content. The Mg proportion is more than 40%, and Fe proportion is lower than 50%, which implies a moderate enrichment of Mg. In Fig. 4.5b, high-Al tourmalines fall into field above the schorl–dravite line. Low-Al tourmalines fall below the schorl-dravite line. Almost all the Galinge tourmaline

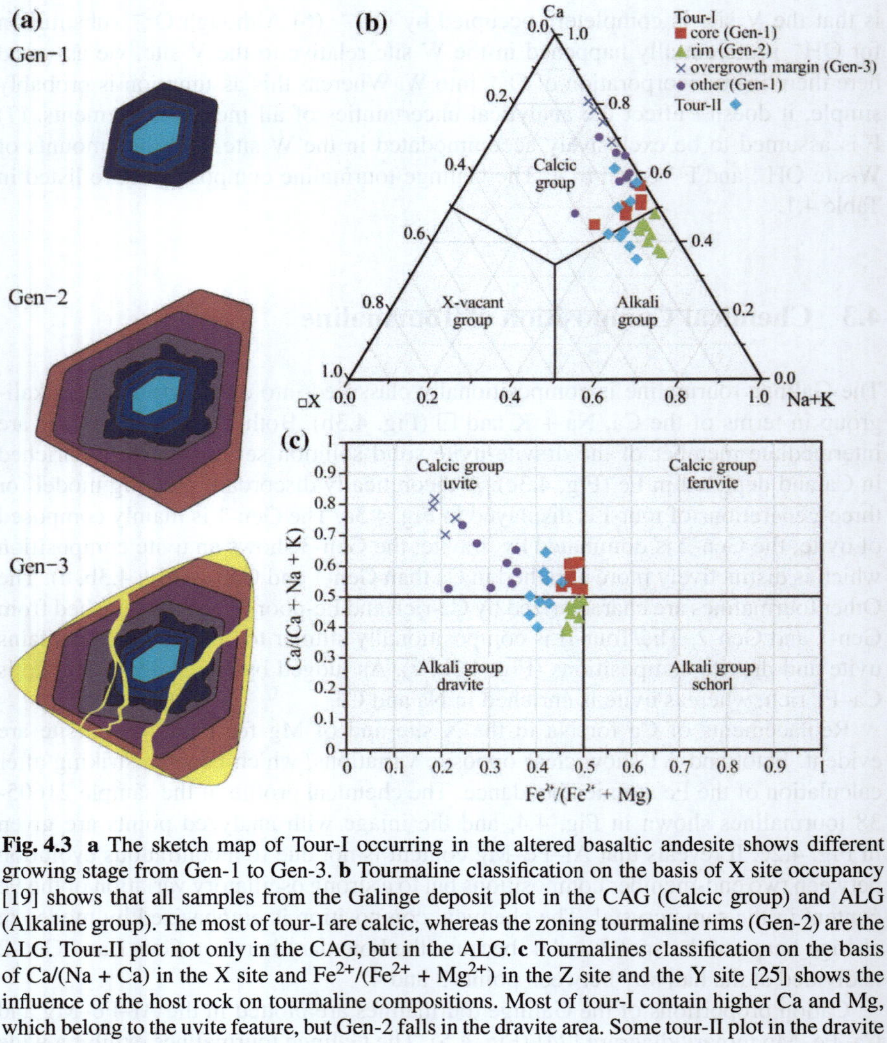

Fig. 4.3 **a** The sketch map of Tour-I occurring in the altered basaltic andesite shows different growing stage from Gen-1 to Gen-3. **b** Tourmaline classification on the basis of X site occupancy [19] shows that all samples from the Galinge deposit plot in the CAG (Calcic group) and ALG (Alkaline group). The most of tour-I are calcic, whereas the zoning tourmaline rims (Gen-2) are the ALG. Tour-II plot not only in the CAG, but in the ALG. **c** Tourmalines classification on the basis of Ca/(Na + Ca) in the X site and $Fe^{2+}/(Fe^{2+} + Mg^{2+})$ in the Z site and the Y site [25] shows the influence of the host rock on tourmaline compositions. Most of tour-I contain higher Ca and Mg, which belong to the uvite feature, but Gen-2 falls in the dravite area. Some tour-II plot in the dravite area, the rest plot in the uvite area

compositions plot in the field below the schorl-dravite line, and roughly parallel to the Mg–Fe side from the most ferrous low-Al schorl to dravite on the other side.

To evaluate the main controlling substitutions in the tourmaline, a series of binary composition diagrams are used here, and the arrays of data are compared to reference exchange-vectors that represent hypothetical substitution mechanisms. In the plot of Fe versus Mg (Fig. 4.6a), tourmaline data display a distribution roughly parallel to the $MgFe_{-1}$ exchange vector. The approximately linear data array yields the expression

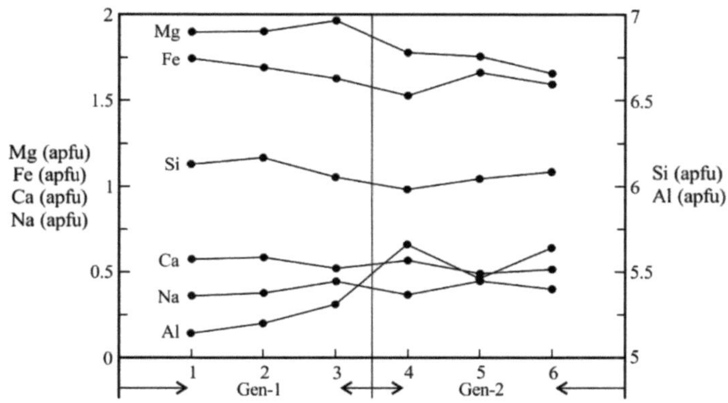

Fig. 4.4 EMP data traverse from core (Gen-1) to rim (Gen-2) of a single tourmaline crystal. These values have been normalized to 15 YZT and assume 3 B pfu. Abscissa: spot number; ordinate: calculated cation numbers. The analyzed points in this image is shown in Fig. 4.2c

Fe = −0.8903(Mg) + 2.734, with a correlation R^2 of 0.9722. Based on a slope of − 0.8903, it can be argued that most of the Mg incorporation is explained by a simple mechanism homovalent substitution of $MgFe_{-1}$. However, the minor deviation of the linear fit to the data from a slope of −1 suggests that other substitutions, such as $\square Al(NaMg)_{-1}$, $\square Al(NaFe)_{-1}$, and $FeAl_{-1}$ probably contribute to the dispersion of the data, and likely result in a weak deviation from a slope of −1. We thereby realize that the Fe^{3+} effect will be examined further. Here, we don't consider the substitution between OH^- and O^{2-} in W-site like the above assumption of formulae calculation, whereas $AlO(MgOH)_{-1}$ and $AlO(FeOH)_{-1}$ exchange vectors are parallel to $\square Al(NaMg)_{-1}$ and $\square Al(NaFe)_{-1}$ vectors on the diagram Fe versus Mg.

The Mg versus Al diagram (Fig. 4.6b) exhibits the very large range of Al and Mg content. The $\square Al(NaFe)_{-1}$, and $FeAl_{-1}$ substitution effect is excluded or neglected in Fig. 4.7b, and further as the assumption above Fe^{3+} would be excluded in the formula calculation. In addition, these data yield a relatively important indication of the $MgFe_{-1}$ and $AlMg_{-1}$ exchange vectors, and charge will be balanced by the cation substitution in the X-site. For the samples of Gen-1, Gen-2 and Tour-II, a least-squares fit to the roughly linear data produces a well-constrained line of Al = 1.0546 Mg + 4.1063, roughly paralleling to the slope of nearly 1 with a correlation coefficient R^2 equals to 0.827, may be jointly influenced by $MgFe_{-1}$ and $AlMg_{-1}$ exchange vectors, but $MgFe_{-1}$ influence shows more intensive. However, Gen-3 allows some separation of the influence of the $(CaMg)(NaAl)_{-1}$ and NaMg $(\square Al)_{-1}$ exchange vector, and The Other display a simply rough distribution consisted with the $MgFe_{-1}$ exchange vector. Therefore, $AlFe_{-1}$ exchange vector is insignificant substitution mechanism in this case.

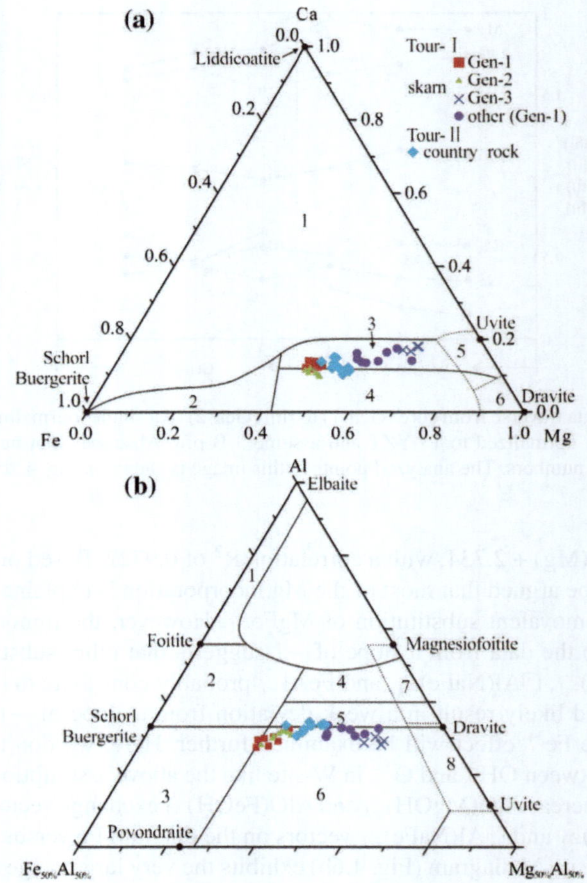

Fig. 4.5 Triangular Ca–Fe–Mg and Al–Fe–Mg plots (in total molar proportions) for the tourmalines from the Galinge deposit [24]; The fields stands for diverse rock types [24], respectively, in the (**a**): (1) Li-rich granitoid pegmatites and aplites; (2) Li-poor granitoids and their associated pegmatites and aplites; (3) Ca-rich metapelites, metapsammites and calc-silicate rocks; (4) Ca-poor metapelites, metapsammites and quartz-tourmaline rocks; (5) Metacarbonates; (6) Metaultramafics. In the (**b**): (1) Li-rich granitoid pegmatites and aplites; (2) Li-poor granitoids and their associated pegmatites and aplites; (3) Fe^{3+}-rich quartz-tourmaline rocks (hydrothermally altered granites); (4) Metapelites and metapsammites coexisting with an Al-saturating phase; (5) Metapelites and metapsammites not coexisting with an Al-saturating phase; (6) Fe^{3+}-rich quartz-tourmaline rocks, calc-silicate rocks, and metapelites; (7) Low-Ca metaultramafics and Cr, V-rich metasediments; (8) Metacarbonates and meta-pyroxenites

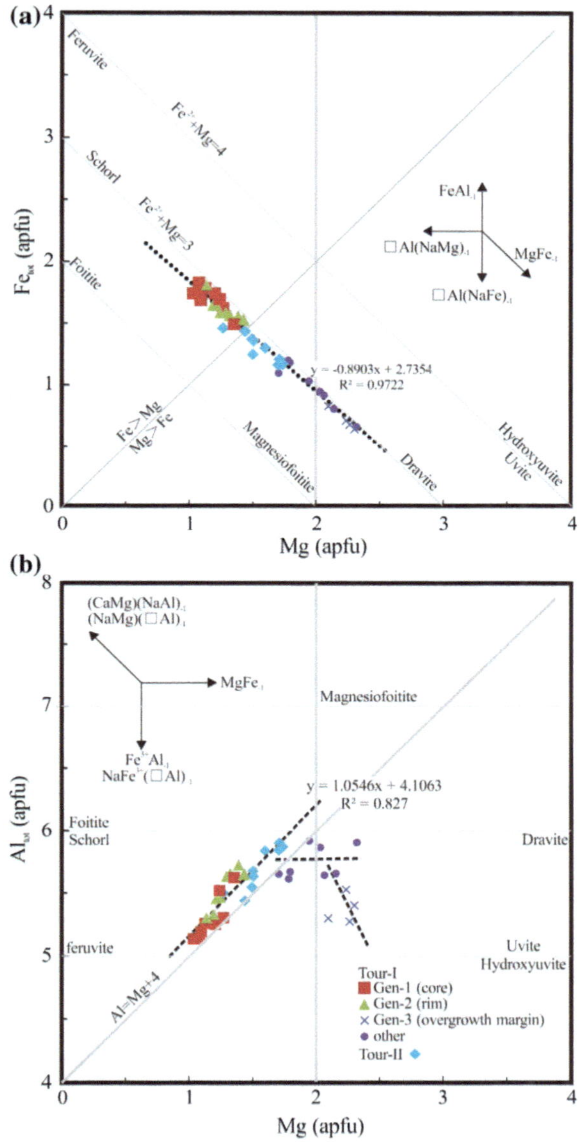

Fig. 4.6 Relationship between Mg and Fe or Al for Galinge tourmalines including Tour-I and Tour-II. A few of Tour-I without clear zonation were defined other in the legend. Arrows labeled with the exchange vectors represent directions of the vector but not the magnitude. **a** Mg versus Fe in the referring to the octahedral YZ sites for the Galinge tourmalines. The Mg and Fe contents of the different type tourmalines have defined arrangements; Reference line has a slope of -1, consistent with the exchange vector $MgFe_{-1}$. **b** Mg versus Al. Reference line has a slope of +1 for the samples of Gen-1, Gen-2 and Tour-II, consistent with the exchange vector of $(\square Mg_2Al)(NaFe^{2+}{}_{-2}Mg_{-1})_{-1}$ with the progression from Gen-1 to Gen-2, but Gen-3 shows possible exchange vectors as following $(CaMg)(NaAl)_{-1}$, $NaMg(\square Al)_{-1}$ and $Mg(OH)(AlO)_{-1}$

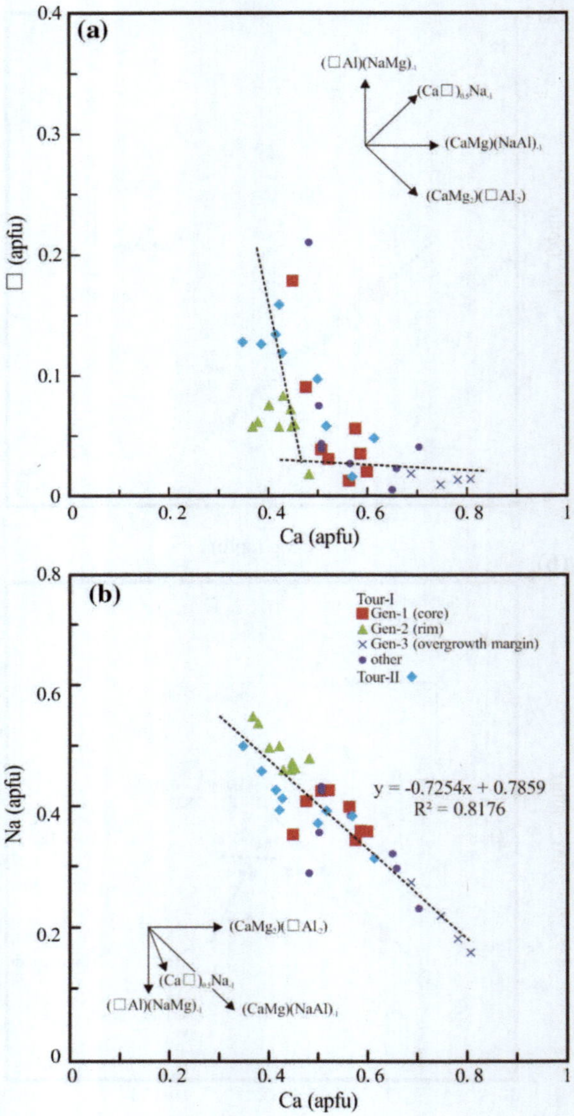

Fig. 4.7 **a** Correlation of X-site vacancies with Ca contents. The two dashed lines represent two linear squares fit to the data. The exchange vectors is referenced on [21]. **b** Correlation of Na contents with Ca contents

In the plot of Ca versus $\square_X$ (Fig. 4.7a), there is an apparent correlation in agreement with ($\square$Al)(NaMg)$_{-1}$ scheme for Gen-1, Gen-2 and Tour-II, and (CaMg)(NaAl)$_{-1}$ schemes for Gen-3 and Other. In respect of the correlation between Na and Ca in X-site (Fig. 4.7b), it is predominately controlled by the (CaMg)(NaAl)$_{-1}$ substitution schemes.

4.4 Controls on Compositional Variations in Tourmaline

The Galinge tourmaline could be divided into two types in respect of their host rocks: Tour-I is are hosted in the mafic intermediate-basic volcanic rocks; Tour-II is developed in metasandstones. Their chemical composition belongs to uvite-dravite solid solution series, and is featured by high Ca and low Al content and wide range of Fe and Mg concentration. The chemical stability of common tourmaline compositions can now be explained in terms of the activity or concentration products of the constituent chemical components for each structural site [34]. The extensive variations of Mg and Fe contents implies that MgFe$_{-1}$ and MgAl$_{-1}$ substitution mechanism is dominant, and charge balanced by the cations in the X-site, as followed (CaMg)(NaAl)$_{-1}$ and NaMg ($\square$Al)$_{-1}$. They were believed to be formed as a consequence of reaction between B-bearing magma-derived hydrothermal fluids and host rocks. In many hydrothermal-related deposits, the tourmaline chemistry is dominated by the nature of the host rock, but in other cases the tourmaline chemistry develops a "mixed" chemical signature influenced by the chemistry of the fluids. When hydrothermal fluid becomes dominant, the B-bearing fluids can lead to extensive tourmalinization. Tourmalinization is relatively selective in that B-bearing fluids will preferentially tourmalinize rocks with the requisite Mg–Fe–Ca–Al constituents, and exhibit uvite-dravite solid solutions. More typically, tourmalines found in metamorphic rocks are more magnesian with variable amounts of Al and Ca [38]. The ratios Fe/(Fe + Mg) of the Gen-1 and Gen-2 tourmalines are relatively invariant, but much higher than the Gen-3 tourmalines (including the other tourmaline), which indicates that tourmalines in the early stage (Gen-1 and Gen-2) are much more equilibrated with the mafic host rocks, but the growth of Gen-3 are compositionally controlled by the fluid.

Henry and Dutrow [21] describe compositionally-zoned magnesian tourmaline, which chiefly involves cation changes in X-site with Mg and Al, and is best described by the exchange vector (CaMg)(NaAl)$_{-1}$ and $\square$Al(NaMg)$_{-1}$. Such compositional growth-zoning tourmalines commonly records insightful information on the varying conditions of its evolving host and hydrothermal fluid signature [8]. The alkaline content of Na + K is slightly increased from the Gen-1 to the Gen-2 in zoned tourmalines, but extensively decreased to the lowest content in Gen-3. The alkaline content of tourmalines is generally associated with the salinities of the forming-fluid, and van Hinsberg et al. [60] thinks that povondraite component may be preferentially formed in highly saline fluids. Therefore, the extensively decreased alkaline content from Gen-2 to Gen-3 indicates a strong decrease in the salinities of the hydrothermal fluid. Highly saline fluids are critical in forming rich deposits because the high salinities permit greater concentrations of metal-chloride complexes to be transported by the fluids. In addition, the Fe content in the Gen-1 and Gen-2 tourmalines are much higher than the Gen-3, which suggests that large amounts of Fe are precipitated to mineralize with decreasing concentration of metal-chloride complexes during Gen-3; Therefore, the most of tour-I are formed in the early stage (involving Gen-1 and Gen-2) and predominately equilibrate with the host rock composition, displaying a narrow compositional range (Fe/(Fe + Mg)) due to the buffer of mafic rocks. The tourmaline composition in late stage (Gen-3) is distinguished by higher Ca/(Ca + Na + K) ratios, which is due to the continual fluid influx and related to the main ore-forming event [41]. Moreover, tourmaline is considerably more stable in the acidic solution than the alkali solution [23, 48], so we postulate the hydrothermal fluid in the early stage is probably more met-acidic than late stage fluid, and the metal precipitation may be related with the varying pH values of the fluid.

4.5 Implications for Skarn Deposit

Tourmaline rarely occurred in skarn deposits, almost exclusively in Sn ± W skarns near the contacts of granitoid plutons [32]. However, in many deposits tourmaline frequently referred to skarn really belongs to greisen, which is superimposed onto skarn. For example, the tourmalines from the Lost River Sn–W–F deposit are a part of greisen mineralization. In addition, Slack [53] reported tourmaline only from Mg-skarn. The Galinge skarn-related tourmaline is the first finding of calc-skarn tourmaline, and thus discussion on relations between skarn process and tourmaline formation become considerably necessary.

In the Galinge skarn deposit, tourmaline are occurred with epidote and pyrite, a typical skarn-type assemblage, which are believed to have formed from granite-derived magmatic-hydrothermal fluids, based on the close special association with plutons. Axinite + epidote + pyroxene assemblages abundantly occurred beneath the tourmalinized skarns indicates that the reactions of "Tour + Ep + Hd = Ax" act as barriers to constraint B into axinite with a consequence of tourmaline disappearance. The pyroxene shows lightly hedenbergite-rich composition. Axinite formation is not only broadly associated with the involvement of host rock components [40], but also reflects metasomatic introduction of B, Fe and Mn [13, 42]. In most cases, axinite is a retrograde mineral in skarn, which is typically formed by replacing earlier epidote, tourmaline, amphibole, garnet, clinopyroxene and feldspar [42] under relatively low temperatures (~350 °C) and CO_2 mol fractions [17]. Isolated occurrences of axinite formed at ≥500 °C [18]. The common association of axinite and epidote probably suggests a moderately oxidizing environment proximal to the intrusion. In this case, the presence of the typical tourmaline-epidote-pyrite and axinite-epidote-pyroxene hydrothermal assemblages indicates a temperature range between 350 and 500 °C for the early hydrothermal assemblages at the Galinge skarn deposit.

Chemical compositions of hydrothermal ore-related tourmalines have been compiled for this study from different hydrothermal type deposit, mainly including (Table 4.2): (1) Cu ± Mo ± Au breccia pipe and porphyry deposit; (2) granitoid

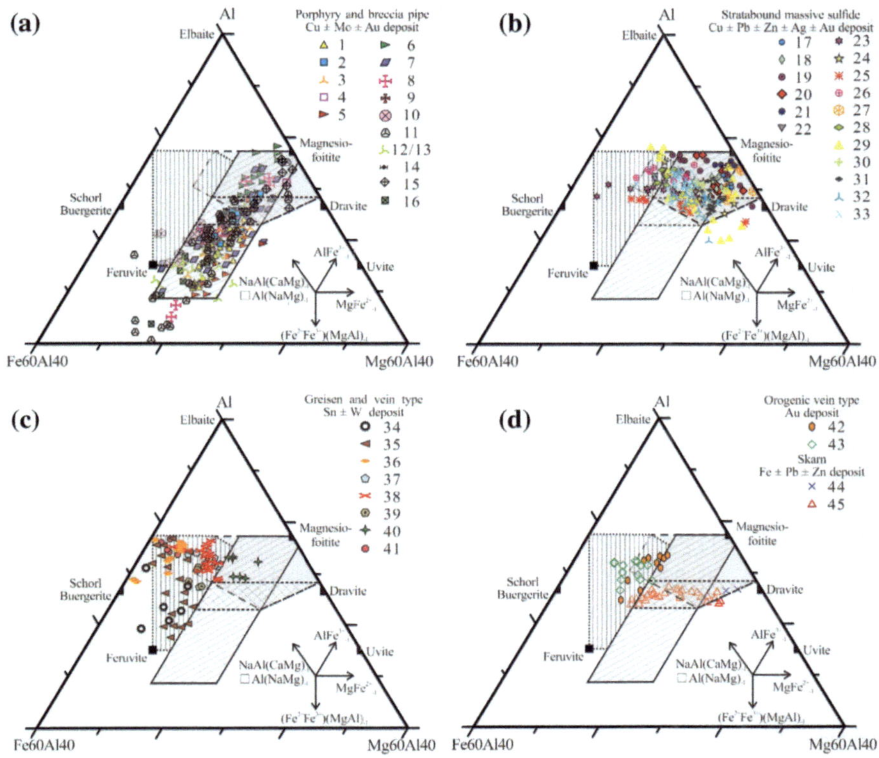

Fig. 4.8 Ternary cation plots of microprobe analyses of tourmalines from various hydrothermal type deposits, including: **a** copper ± molybdenum ± gold breccia pipe and porphyry deposit; **b** granitoid related Tin ± tungsten greisens and veins deposit; **c** various types of stratabound ores, in both metasedimentary and metavolcanic terranes; **d** gold-quartz vein type deposit within greenschist- or amphibolite-grade metavolcanic and metasedimentary rocks. Different shadow area defined different hydrothermal type deposit, and encompassed most of tourmaline plots from each type

related Sn ± W greisen and vein type deposit; (3) various types of stratabound ores, in both metasedimentary and metavolcanic terranes; (4) Au-quartz vein type deposit within greenschist- or amphibolite-grade metavolcanic and metasedimentary rocks. Analyses were considered suitable if the tourmaline samples were clearly related to the ore deposits under consideration. Cation proportions of the tourmalines are plotted on the Fe40Al60–Al–Mg40Al60 ternary diagram, which was used as an identified diagram in respect of tourmaline forming environment [24].

Their composition in the Al–Fe–Mg plots (Fig. 4.8) approximately parallel to the Mg–Fe margin, overwhelmingly controlled by the $Fe^{2+}Mg_{-1}$ exchange vectors, which is significantly different with those occurred in the other hydrothermal deposit, especially vein and greisen type deposit that is mainly controlled by the $(Fe^{2+}Fe^{3+})(MgAl)_{-1}$ exchange vectors.

Table 4.2 Selected tourmaline-bearing hydrothermal ore deposits

Name	Geology	Metals	Location	References
Coxheath	porphyry	Cu–Mo–Au	Nova Scotia, Canada	[35]
Salikvan	porphyry	Cu–Mo	Turkey	[61]
Bajan-Ula	porphyry	Mo–Cu	Mongolia	[31]
Vetka	porphyry	Cu	Chukchi Peninsula, Russia	[3]
Ol'khovka	porphyry	Cu	Chukchi Peninsula, Russia	[5]
Peschanka	porphyry	Cu	Chukchi Peninsula, Russia	[5]
Satrv-Khudag	porphyry	Cu	Mongolia	[5]
Kafan	porphyry	Cu	Armenia	[5]
Aktogai	porphyry	Cu	Kazakhsatn	[5]
Darasun ore district	porphyry	Au	Chita, Russia	[4]
Kljuchevsky	porphyry	Au	Russia	[4]
Kariisky	porphyry	Au	Russia	[4]
Donoso	breccia pipe	Cu	Chili	[50]
Big Bell	stratabound massive sulfide	Au	Murchison, Australia	[29]
Mount Gibson	stratabound massive sulfide	Au	Murchison, Australia	[29]
Black Hawk	stratabound massive sulfide	Cu–Zn–Pb	Maine	[59]
Elizabeth	stratabound massive sulfide	Cu–Zn–Ag	Vermont	[59]
Bleikvassli	stratabound massive sulfide	Pb–Zn–Ag	Norway	[59]
Guangxi Sn ore district	stratabound massive sulfide	Sn–Cu–Pb–Zn	China	[36]
Kidd Creek	stratabound massive sulfide	Cu–Zn–Pb	Timmins, Ontario, Canada	[54]
Dome Rock	stratabound massive sulfide	Cu–Co–Au	Australia	[26]

(continued)

Table 4.2 (continued)

Name	Geology	Metals	Location	References
Boken Hill ore district	stratabound massive sulfide	Pb–Zn–Ag	Australia	[55]
Rum Jungle	stratabound massive sulfide	Cu–Pb–Zn	Australia	[7]
Bieluwutu ore district	stratabound massive sulfide	Cu–Pb–Zn–Ag	Inner Mongolia, China	[39]
Zhongtiaoshan	stratabound massive sulfide	Cu	Shanxi, China	[58]
Golden Dyke Dome	stratabound massive sulfide	Au	Northern Australia	[45]
Deri	stratabound massive sulfide	Zn–Pb–Cu	Rajasthan, India	[11]
Yindongzi-Tongmugou	stratabound massive sulfide	Pb–Zn	Qinling, China	[27]
Cuyun ore district	stratabound massive sulfide	Fe	Minnesota, America	[10]
Mramorny ore district	vein and greisen	Sn	Chukchi Peninsula, Russia	[2]
Guangxi ore district	vein and greisen	Sn	China	[37]
Tygerberg ore district	vein	Sn–Zn–(W)	South Africa	[47]
Kaneuchi	vein	W–Sn	Japan	[49]
Baiganhu ore district	vein and greisen	W	China	[63]
Kirwans Hill	vein and greisen	Sn–W–Au	South Island, New Zealand	[44]
Doctor Hill	greisen	Sn–W	South Island, New Zealand	[44]
Yunlong	Vein type	Sn	Yunnan, China	[62]
Variscan-Karkonosze	greisen	Sn–W	Poland	[51]
Timmins-Porcupine ore district	vein type	Au	Ontario, Canada	[30]
Bin Yauri ore district	vein type	Au	Nigeria	[16]

Tourmalines from porphyry and breccia pipe Cu ± Mo ± Au deposits are apparently contain significant ferric iron, and is predominately controlled by the $Fe^{3+}Al_{-1}$ substitution mechanism. Almost all plotted positions fall well below the dravite-feruvite join and approximately parallel to the Al–Fe margin (Fig. 4.8a). Baksheev et al. [5] found that the common feature of tourmalines from all type of porphyry deposits is starting crystallization of Fe-rich tourmalines at high fO_2 causing the high Fe^{3+}/Fetot value and subsequent precipitation of Fe-poorer tourmalines with a lower Fe^{3+}/Fetot value caused by sulphide deposition at increasing H_2S in mineralizing fluid. More likely that oxidized fluids are better at transported metals (in most base metal systems), and then it reflect progressive reduction of the system during mineralisation to stabilize sulphides.

Tourmalines from granitoid-related Sn ± W vein and greisen type deposits contain relatively less Mg and Fe than tourmalines from porphyry and breccia pipe deposits (Fig. 4.8b). They are effected by the $(Fe^{2+}Fe^{3+})(MgAl)_{-1}$ substitution mechanism, and trends towards feruvite compositions (Fig. 4.8b). These data, Mg content and Fe^{3+}/Fetot value increasing from tin to copper deposits, suggest that tourmalines in the porphyry and breccia pipe deposits formed under relatively oxidizing hydrothermal conditions, in opposition to most the tourmalines from the Sn ± W deposits that formed mainly under reducing conditions [53].

Stratabound massive sulfide Cu ± Pb ± Zn ± Ag ± Au deposits contain compositionally diverse tourmalines from porphyry Cu ± Mo ± Au and granitoid-related Sn ± W vein and greisen deposits (Fig. 4.8c). They are relative approximate to Mg-rich and Fe-poor dravite compositions, and in most cases approaching magnesio-foitite which is commonly deficiency in alkali component (Fig. 4.8c). Fe is generally low in most massive-sulfide tourmalines, as exhibited in Fig. 4.8c. The most of data mainly plot above the schorl-dravite join, which are controlled by the $(\square Al)(NaMg)_{-1}$ and $(NaAl)(CaMg)_{-1}$ exchange vectors. The compositional variations in massive-sulfide tourmalines may reflect a mixing of high temperature, Fe-rich endmember hydrothermal fluid with cold, Mg-rich entrained seawater [54]. Moreover, tourmaline compositions were controlled either by the chemistry of the hydrothermal fluid in settings proximal to submarine systems with relatively high water/rock ratios [28, 56], or by bulk compositions of host rocks in settings distal from submarine systems with relatively low water/rock ratios [55]. In terms of the formation of Mg-rich tourmaline in submarine-hydrothermal system, there are two completely different points: (1) sulfide-silicate reactions during regional and/or metamorphism, which remove Fe from tourmaline to form additional S-poor sulfides such as pyrrhotite, and produce new Mg-rich tourmaline [22, 54, 59]; (2) mainly influenced by sea-water dominated Mg-rich hydrothermal fluids during mineralization, and predated regional metamorphism that overprints the deposit [15, 27, 28]. The presence Fe-rich tourmalines (Fe > Mg) in massive sulfide deposits is believed to reflect proximity to submarine volcanic centers in which the tourmaline compositions were controlled largely by the chemistry of the surrounding clastic sedimentary and felsic metavolcanic rocks [6, 54].

Tourmalines from the Bin Yauri and Timmins-Porcupine Au-quartz vein type deposit have intermediate schorl-dravite compositions (Fig. 4.8d), with little or no buergerite, ferridravite, or uvite components [16, 30]. They show a similar exchange vectors with those from porphyry deposit, controlling by the $Fe^{3+}Al_{-1}$ substitution mechanism. Tourmaline is a typically minor gangue mineral within the veins and/or wall rocks, whereas their temporal relations with gold mineralization are ambiguities.

Tourmalines from various types of hydrothermal deposits have a distinguished composition field in terms of Al, Mg and Fe cation proportions and substitution mechanisms. The Galinge skarn-related tourmalines is overwhelmingly controlled by the $MgFe_{-1}$ substitution mechanisms. This is different with tourmalines occurred in porphyry, VMS and vein-greisen type deposit, which respectively controlled by $Fe^{3+}Al_{-1}$, $(CaMg)(NaAl)_{-1}$ and $(NaMg)(\square Al)_{-1}$, and $(Fe^{2+}Fe^{3+})(MgAl)_{-1}$ substitution mechanisms. On the basis of the general substitution mechanism and chemical variations of tourmaline, they were considered as a good guidance for exploration by using their extensively varying composition [5, 15, 20, 44, 46, 52].

References

1. Bačík P, Uher P, Sýkora M, Lipka J (2008) Low-Al tourmalines of the schorl—dravite–povondraite series in redeposited tourmalinites from the Western Carpathians. Slovakia. Can Mineral 46(5):1117–1129
2. Baksheev IA, Tikhomirov PL, Yapaskurt VO, Vigasina MF, Prokof Ev VY, Ustinov VI (2009) Tourmaline of the Mramorny Tin Cluster, Chukotka Peninsula. Russia. Can Mineral 47(5):1177–1194
3. Baksheev IA, Chitalin AF, Yapaskurt VO, Vigasina MF, Bryzgalov IA, Ustinov VI (2010) Tourmaline in the Vetka porphyry copper-molybdenum deposit of the Chukchi Peninsula of Russia. Mosc Univ Geol Bull 65(1):27–38
4. Baksheev IA, Prokof Ev VY, Yapaskurt VO, Vigasina MF, Zorina LD, Solov Ev VN (2011) Ferric-iron-rich tourmaline from the Darasun gold deposit, Transbaikalia. Russia. Can Mineral 49(1):263–276
5. Baksheev IA, Prokof'Ev VY, Zaraisky GP et al (2012) Tourmaline as a prospecting guide for the porphyry-style deposits. Eur J Mineral 24(6):957–979
6. Bandyopadhyay BK, Slack JF, Palmer MR et al (1993) Tourmalinites associated with stratabound massive sulphide deposits in the Proterozoic Sakoli Group, Nagpur district, central India. In: Proceedings of Eight Quad IAGOD Symposium, E. Schweizerbart'sche Verlagsbuchhandlung, Stuttgart, pp 867–885
7. Bone Y (1988) The geological setting of tourmalinite at Rum Jungle, NT, Australia—genetic and economic implications. Min Depos 23(1):34–41
8. Cavarretta G, Puxeddu M (1990) Schorl-dravite-ferridravite tourmalines deposited by hydrothermal magmatic fluids during early evolution of the Larderello geothermal field, Italy. Econ Geol 85(6):1236–1251
9. Clarke DB, Readon NC, Chatterjee AK, Gregoire DC (1989) Tourmaline composition as a guide tomineral exploration; a reconnaissance study from NovaScotia using discriminant function analysis. Econ Geol 84:1921–1935
10. Cleland JM, Morey GB, McSwiggen PL (1996) Significance of tourmaline-rich rocks in the North Range Group of the Cuyuna Iron Range, east-central Minnesota. Econ Geol 91(7):1282–1291
11. Deb M, Tiwary A, Palmer MR (1997) Tourmaline in Proterozoic massive sulfide deposits from Rajasthan. India. Min Depos 32(1):94–99
12. Dobson DC (1982) Geology and alteration of the Lost River tin-tungsten-fluorine deposit, Alaska. Econ Geol 77(4):1033–1052

13. Dube B, Guha J (1993) Factors controlling the occurrence of ferro-axinite within Archean gold-copper-rich quartz veins; Cooke Mine, Chibougamau area, Abitibi greenstone belt. Can Mineral 31(4):905–916
14. Dutrow BL, Henry DJ (2011) Tourmaline: a geologic DVD. Elements 7:301–306
15. Ethier VG, Campbell FA (1977) Tourmaline concentrations in Proterozoic sediments of the southern Cordillera of Canada and their economic significance. Can J Earth Sci 14(10):2348–2363
16. Garba I (1996) Tourmalinization related to Late Proterozoic-Early Palaeozoic lode gold mineralization in the Bin Yauri area, Nigeria. Miner Depos 31(3):201–209
17. Grew ES (1996) Borosilicates (exclusive of tourmaline) and boron in rock-forming minerals in metamorphic environments. Rev Mineral Geochem 33(1):387–502
18. Grew ES, Anovitz LM (1996) Boron: mineralogy, petrology and geochemistry. Mineral Soc Am, Washington DC
19. Hawthorne FC, Henry DJ (1999) Classification of the minerals of the tourmaline group. Eur J Mineral 11(2):201–215
20. Hellingwerf RH, Gatedal K, Gallagher V, Baker JH (1994) Tourmaline in the central Swedish ore district. Min Depos 29(2):189–205
21. Henry DJ, Dutrow BL (1990) Ca substitution in Li poor aluminous tourmaline. Can Mineral 28:111–124
22. Henry DJ, Dutrow BL (1992) Tourmaline in a low grade clastic metasedimentary rock: an example of the petrogenetic potential of tourmaline. Contrib Mineral Petrol 112(2–3):203–218
23. Henry DJ, Dutrow BL (1996) Metamorphic tourmaline and its petrologic applications. Rev Mineral Geochem 33(1):503–557
24. Henry DJ, Guidotti CV (1985) Tourmaline as a petrogenetic indicator mineral- An example from the staurolite-grade metapelites of NW Maine. Am Mineral 70(1–2):1–15
25. Henry DJ, Novák M, Hawthorne FC et al (2011) Nomenclature of the tourmaline-supergroup minerals. Am Mineral 96(5–6):895–913
26. Ito O, Plimer IR (1987) The significance of tourmaline in the stratiform Dome Rock deposit. Australia. Min Geol 37(206):403–418
27. Jiang SY, Palmer MR, Li YH et al (1995) Chemical compositions of tourmaline in the Yindongzi-Tongmugou Pb-Zn deposits, Qinling, China: implications for hydrothermal ore-forming processes. Miner Depos 30(3–4):225–234
28. Jiang SY, Palmer MR, Slack JF et al (1998) Paragenesis and chemistry of multistage tourmaline formation in the Sullivan Pb-Zn-Ag deposit, British Columbia. Econ Geol 93(1):47–67
29. Jiang SY, Palmer MR, Yeats CJ (2002) Chemical and boron isotopic compositions of tourmaline from the Archean Big Bell and Mount Gibson gold deposits, Murchison Province, Yilgarn Craton. Western Australia. Chem Geol 188(3):229–247
30. King RW, Kerrich R (1989) Chromian dravite associated with ultramafic-rock-hosted Archean lode gold deposits, Timmins-Porcupine District, Ontario. Can Mineral 27:419–426
31. Koval PV, Zorina LD, Kitajev NA, Spiridonov AM, Ariunbileg S (1991) The use of tourmaline in geochemical prospecting for gold and copper mineralization. J Geochem Explor 40:349–360
32. Kwak TA (2012) W-Sn skarn deposits: and related metamorphic skarns and granitoids. Elsevier, Netherland, pp 1–449
33. Layne GD, Spooner E (1991) The JC tin skarn deposit, southern Yukon Territory; I, Geology, paragenesis, and fluid inclusion microthermometry. Econ Geol 86(1):29–47
34. London D (2011) Experimental synthesis and stability of tourmaline: a historical overview. Can Mineral 49(1):117–136
35. Lynch G, Ortega J (1997) Hydrothermal alteration and tourmaline-albite equilibria at the Coxheath porphyry Cu-Mo-Au deposit. Nova Scotia. Can Mineral 35(1):79–94
36. Mao J (1995) Tourmalinite from northern Guangxi, China. Miner Deposita 30(3–4):235–245
37. Mao JW, Wang PA, Wang DH, Bi CS (1993) The trace of tourmaline for rock-forming and metallogenic environments and its applied conditions. Geol Rev 39(6):497–507
38. Morgan GB, London D (1989) Experimental reactions of amphibolite with boron-bearing aqueous fluids at 200 MPa: implications for tourmaline stability and partial melting in mafic rocks. Contrib Miner Petr 102(3):281–297

39. Nie FJ, Zhang HT, Sun H, Fun JT (1990) Discovery of tourmalinites in the Bieluwuto copper metallogenic district, Nei Mongol, and their geological significance. Geol Rev 36(5):467–472
40. Ozaki M (1972) Chemical composition and occurrence of axinite. Kumamoto J Sci Geol 9(2):1–34
41. Pal DC, Trumbull RB, Wiedenbeck M (2010) Chemical and boron isotope compositions of tourmaline from the Jaduguda U (–Cu–Fe) deposit, Singhbhum shear zone, India: Implications for the sources and evolution of mineralizing fluids. Chem Geol 277(3):245–260
42. Pertsev NN (1971) Parageneses of boron minerals in magnesian skarns. Nauka, Moscow
43. Pirajno F (2008) Hydrothermal processes and mineralsystems. Springer, Netherlands
44. Pirajno F, Smithies RH (1992) The FeO/(FeOt+MgO) ratio of tourmaline: a useful indicator of spatial variations in granite-related hydrothermal mineral deposits. J Geochem Explor 42:371–381
45. Plimer IR (1986) Tourmalinites from the Golden Dyke dome, northern Australia. Min Depos 21(4):263–270
46. Robinson GD (1989) Stream sediment tourmaline geochemistry in massive sulfide exploration: an example from Virginia. USA. J Geochem Explor 34(2):173–188
47. Rozendaal A, Bruwer L (1995) Tourmaline nodules: indicators of hydrothermal alteration and Sn-Zn-(W) mineralization in the Cape Granite Suite. South Africa. J Afr Earth Sci 21(1):141–155
48. Sanyal A, Nugent M, Reeder RJ et al (2000) Seawater pH control on the boron isotopic composition of calcite: evidence from inorganic calcite precipitation experiments. Geochim Cosmochim Acta 64(9):1551–1555
49. Shibue Y (1984) Chemical compositions of tourmaline in the vein-type tungsten deposits of the Kaneuchi mine. Japan. Min Depos 19(4):298–303
50. Skewes MA, Holmgren C, Stern CR (2003) The Donoso copper-rich, tourmaline-bearing breccia pipe in central Chile: petrologic, fluid inclusion and stable isotope evidence for an origin from magmatic fluids. Min Depos 38(1):2–21
51. Słaby E, Kozłowski A (2005) Composition of tourmalines from tin-tungsten-bearing country rock of the Variscan Karkonosze granitoid-a record of the wall rock and hydrothermal fluid interaction. Neues Jahrbuch für Mineralogie-Abhandlungen. J Mineral Geochem 181(3):245–263
52. Slack JF (1982) Tourmaline in Appalachian-Caledonian massive sulphide deposits and its exploration significance. Trans Inst Min Metall, Sect B: Appl Earth Sci 91(May):81–89
53. Slack JF (1996) Tourmaline associations with hydrothermal ore deposits. Rev Mineral Geochem 33(1):559–643
54. Slack JF, Coad PR (1989) Multiple hydrothermal and metamorphic events in the Kidd Creek volcanogenic massive sulphide deposit, Timmins, Ontario: evidence from tourmalines and chlorites. Can J Earth Sci 26(4):694–715
55. Slack JF, Palmer MR, Stevens BP et al (1993) Origin and significance of tourmaline-rich rocks in the Broken Hill district, Australia. Econ Geol 88(3):505–541
56. Slack JF, Shaw DR, Leitch C et al (2000) Tourmalinites and coticules from the Sullivan Pb-Zn-Ag deposit and vicinity, British Columbia: Geology, geochemistry, and genesis. Geol Environ Sullivan Depos Br C: Geol Assoc Can Miner Depos Div Spec Publ 1:736–767
57. Slack JF, Trumbull RB (2011) Tourmaline as a recorder of ore-forming processes. Elements 7:321–326
58. Sun HT, Ge CH (1989) Discovery of Banded Tourmalinite and Mineralized Tourmaline-Rich Chemical Sedimentary-Rocks from Stratabound and Stratiform Copper-Deposits in Zhongtiaoshan District. Shanxi Province. Chin Sci Bull 34(10):846–851
59. Taylor BE, Slack JF (1984) Tourmalines from Appalachian-Caledonian massive sulfide deposits; textural, chemical, and isotopic relationships. Econ Geol 79(7):1703–1726
60. van Hinsberg VJ, Henry DJ, Marschall HR (2011) Tourmaline: an ideal indicator of its host environment. Can Mineral 49:1–16
61. Yauvz F, Ali I, Jiang SY (1999) Tourmaline compositions from the Salikvan porphyry Cu-Mo deposit and vicinity, northeastern Turkey. Can Mineral 37:1007–1023

62. Yu JM, Jiang SY (2003) Chemical composition of tourmaline from the Yunlong tin deposit, Yunnan, China: implications for ore genesis and mineral exploration. Min Pet 77(1–2):67–84
63. Zheng Z, Chen YJ, Deng XH, Yue SW, Chen HJ (2016) Muscovite 40Ar/39Ar dating of the Baiganhu W-Sn orefield, Qimantag, East Kunlun Mountains, and its geological implications. Geol China 43(4):1341–1352

Chapter 5
Multistage Amphiboles from the Galinge Skarn Deposit: Evidence of Igneous Rocks Replacement

Hydrothermal fluids can intensively alter the mineralogy, texture and chemistry of the host rocks. As chemical components are transferred from one geological environment into another, it may promote fluid-mineral reactions as the system moves back toward equilibrium. Evidence of these interactions is common in the residual bulk composition, mineralogy and textures [1, 4]. If metasomatism is moderate, clear mineralogical evidence is preserved in the form of skarns or metasomatic minerals. Nevertheless, if alteration effects are very intensive, as a consequence, only subtle textures and mineralogical features are preserved [3]. As such, mineralogy and textures of multistage amphiboles may provide important clues of the controlling process of fluid-rock interaction in the Galinge deposit.

This chapter reports on an interesting textural and mineralogical feature: Cl-rich amphiboles replacing edenite in the first stage and actinolite replacing amphiboles in the second stage, as observed in the exocontact zone of granodiorite intrusions in the Galinge skarn deposit. Cl-rich amphiboles have been reported in many skarn deposit, e.g., the highest Cl concentrations (7.24 wt%) were reported from the Dashkesan skarn Fe deposit in Azerbaijan [11]; Jacobson [10] reported Cl-rich calcic amphiboles containing 5.34 and 6.51 wt% Cl, respectively, from skarn in Dashkesan and mylonite in St. Paul's Rock, and Dick and Robinson [2] reported Cl-rich calcic amphibole containing 3.09 wt% Cl from a sphalerite skarn in southern Yukon. The amphibole group is characterized by extensive chemical substitutions and compositionally diverse members. This chemical complexity makes amphibole a good recorder for the compositional transformation of fluids. Amphibole is also useful for tracking the compositional evolution of igneous and metamorphic rocks, and for fluid infiltration and leaching fingerprinting. Especially, Cl-rich amphiboles shed light on the volatile contributions to mineralization. This study aims to relate the amphibole compositional variation to the chemical/physical variation in the magma-derived hydrothermal fluid stages, and determine the link between amphibole formation and igneous rock replacement.

M. Yu, *Metallogenic Mechanism of the Galinge Polymetallic Iron Skarn Deposit, Qiman Tagh Mountains, Qinghai Province*, Springer Theses,
https://doi.org/10.1007/978-981-10-7907-8_5

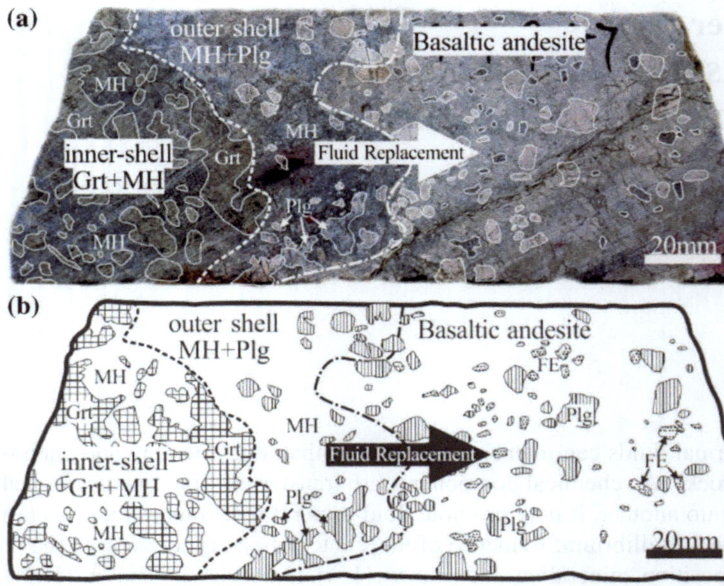

Fig. 5.1 **a, b** Photograph and sketch showing the reaction front in a hand specimen. The inner shell contains an assemblage of garnet and amphibole, and the outer shell develops only amphibole. The arrow shows the fluid replacement direction. Grt-garnet; MH-magnesian hastingsite; FE-ferro edenite

5.1 Petrography

The Galinge multistage amphiboles are predominantly associated with altered basaltic andesite (Fig. 5.1). Specimens studied in detail shows a clear and variable degree of replacement between the infiltrating fluids and the basaltic andesite along the replacement direction (Fig. 5.1a). In the less altered parts, phenocrysts of plagioclase and amphibole are well preserved; in the moderately altered parts (outer shell about 3 cm wide) (Fig. 5.1b), only plagioclase phenocrysts are preserved, whilst the amphibole phenocrysts and groundmass are completely altered to elongated secondary amphibole aggregates. In the intensively altered parts (inner shell) (Fig. 5.1b), the basaltic andesite rock was completely altered to a rock composed of greenish yellow garnets and dark green amphiboles, a few garnet grains from pseudomorphs have replacing elongated feldspars.

Amphiboles in the altered basaltic andesite can be petrographically and compositionally divided into three types, i.e., Amp-I, Amp-II and Amp-III (Figs. 5.2 and 5.3). Under the microscope, Amp-II (Gen-2: Generation 2) formed after Amp-I (Gen-1: Generation 1) (Fig. 5.2a–c), as supported by the distinctive replacement of Amp-I by Amp-II. Amp-II, in turn, is partially replaced by Amp-III (Gen-3:

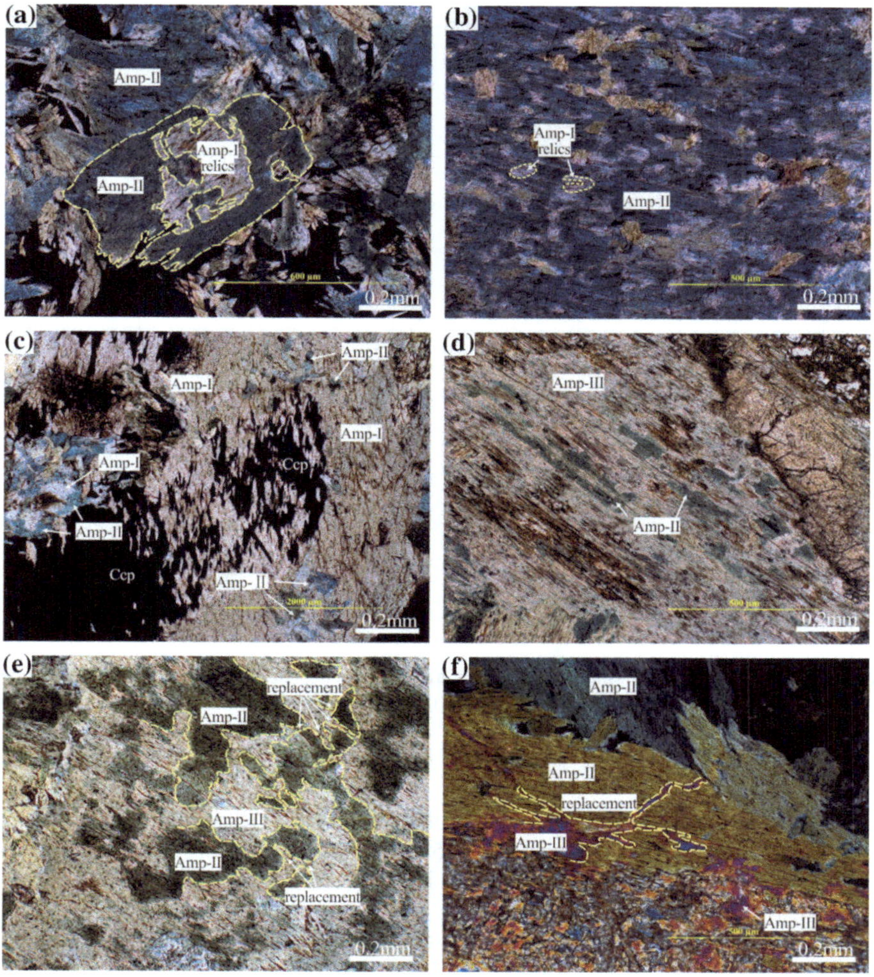

Fig. 5.2 Photomicrographs of the three amphibole generations. **a** Zoned grains of amphibole with relics of brown cores (Amp-I) and bluish-green rims (Amp-II), plane polarized light. **b** Fibrous amphibole displaying the texture of the host brown Amp-I and the bluish-green Amp-II growth in the fibers as observed in sections cut parallel to the fiber length, plane polarized light. **c** Amp-II (and ore minerals) replacing Amp-I along the rhombic cleavages, plane polarized light. **d** Host Amp-II being replaced by Amp-III, plane polarized light. **e** Host Amp-II and Amp-III fiber growth, plane polarized light. **f** Amp-III in a dendritic aggregate vein cutting Amp-II grains, cross-polarized light

Generation 3) (Fig. 5.2d–f). Optical and back-scattered electron (BSE) imaging show similar complex textures of the three amphibole generations (Fig. 5.3).

The earliest Amp-I may have preserved the primary igneous edenitic composition of the basaltic andesite, and it coexists with plagioclase phenocrysts in the least altered rocks. Subsequently, it was replaced by Amp-II and occurring as pseudomorphs.

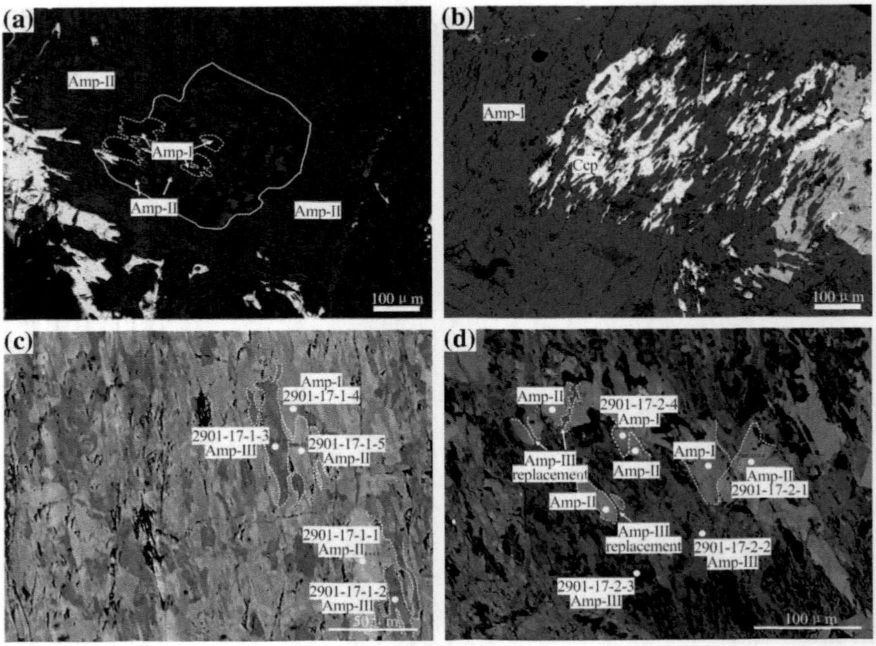

Fig. 5.3 BSE images of amphiboles from the Galinge skarn deposit. The figures **a–d** clearly show the replacement relationships, and Amp-I to Amp-III show grey scales. The numbered points show the electron microprobe analyses spots

Cores of brown amphibole (Amp-I) are rimmed and partly to nearly totally replaced by bluish green amphibole (Amp-II) (Fig. 5.2a). In some places, ore minerals (±Amp-II) replaced Amp-I along the rhombic cleavages (Fig. 5.2c).

Amp-II is the major constituent of the ore-bearing skarn from the Ca-SK zones, and was mainly associated with moderately to extensively altered basaltic andesite. In moderately altered rocks, Amp-II coexists with anhedral plagioclase phenocryst, which occur as fine grained aggregates in the groundmass. In addition, Amp-II always occurs as prismatic bluish green crystals up to 1 cm long in interstices of very coarse-grained and strongly birefringent grossular-andradite in extensively altered basaltic andesite (Fig. 5.1). Some Amp-II crystals locally show concentric oscillatory zonation. Amp-II generally coexists with garnets, magnetite, chalcopyrite, pyrrhotite, pyrite, and some plagioclase with in larger garnet pseudomorphs. Amp-II also occurs in garnet-free skarn rocks; in this case the coexisting accessory minerals include apatite, sphene, actinolite, magnetite, pyrite, pyrrhotite and chalcopyrite.

The light green Amp-III (actinolite), is common in the Galinge skarn deposit. Bluish-green Amp-II is clearly crosscut and replaced by the Amp-III (Fig. 5.2d, e). Besides, Amp-III in dendritic veins crosscut Amp-II grains. In some places, Amp-III aggregates occur together with the Amp-II (Fig. 5.2f).

5.2 Analytical Method

Electron-microprobe analyses of amphibole were conducted with wavelength-dispersion spectrometry (WSD) using a JOEL JXA-823V Electron Probe Micro Analyzer (EPMA) at the MLR Key Laboratory of Metallogeny and Mineral Assessment, Institute of Mineral Resources, Chinese Academy of Geological Sciences (Beijing) China. Analytical conditions were: accelerating potential 15 kV, beam current 5–10 nA, 3–5 μm focused electron beam. The analyses were calibrated using jadeite (Si, Na, and Al), forsterite (Mg), orthoclase (K), apatite (P), wollastonite (Ca), rutile (Ti), halite (Cl), and synthetic oxide (Cr, Mn, Fe, Ni) as standards.

In order to calculate the stoichiometry of the amphiboles and estimate the amounts of H_2O, several assumptions were made. The general formula of amphiboles can be expressed as $W_{0-1}X_2Y_5Z_8O_{22}(OH,O,F)_2$, where W represents cations occupying the A-site; X and Y are octahedrally coordinated cations occupying the M4 and M1-M3 positions, respectively. Z represents tetrahedrally coordinated small cations in the T-site. The calculation of amphibole formulae was based on a total cation charge of 46 (i.e., $\sum O = 22$; $\sum OH, Cl = 2$), and was determined on the assumption that Si + Al + Ti + Fe + Mg + Ca = 15 (15eNK: Si + Al^{IV} = 8; Al^{VI} + Ti + Fe + Mg + Ca = 7); Concentration of H_2O was calculated assuming $\sum O3 = 2.0$ (OH, F, Cl) apfu.

5.3 Mineral Composition

The chemical composition of amphibole group minerals is very complex, and thus the problems inherent in the classification of such a group of minerals are obvious. Currently, two classification schemes are largely used: the first one is based on the cations A, M4 and T sites (Fig. 5.4a) [12–14]; the second one is based on the cations A, M4 and M1-M3 sites of the amphibole structure (Fig. 5.4b) [7]. Both classification schemes both highlight the compositional changes from Amp-I to Amp-III.

The Galinge amphiboles show a considerable variation with respect to $Mg^{2+}/(Fe^{2+} + Mg)$ and Si (apfu) (Fig. 5.4a) [14]. The igneous Amp-I shows ferro-edenitic hornblende (FE) composition. Amp-II is magnesian-hastingsite (MH); Amp-III is ferro-actinolite (FA). The Si content ranges widely from the MH to the FA. The MH has the lowest Si content (5.635–6.278 apfu), followed by FE (6.505–6.817 apfu), and FA having the highest Si content (7.383–7.851 apfu) (Table 5.1). However, almost all X_{Mg} ($X_{Mg} = Mg/(Fe^{2+} + Mg)$) values are below 0.5 (Table 5.1, Fig. 5.4a). In Fig. 5.4b, Na_A (apfu) versus $(Al + Fe^{3+} + 2Ti)_C$ (apfu) [8], the Na content increases with $(Al + Fe^{3+} + 2Ti)_C$. The Amp-II has the highest Na content (0.333–0.490 apfu), followed by FE (0.267–0.35 apfu), whereas the FA has the lowest Na content (0.03–0.162 apfu) (Table 5.1, Fig. 5.4b).

In the Ca–Fe^{2+}–Mg ternary diagram (Fig. 5.5), the distinct compositions can be distinguished for amphiboles from different generations. The Fe and Mg contents vary largely among the different amphibole generations. In addition, the Ca contents

Table 5.1 Representative microprobe analysis of Amp-I, Amp-II and Amp-III from the Galinge deposit

	Amp-I			Amp-II						Amp-III					
Sample	2901-17-2-4	Average (7)	R	2901-11	Average (6)	R	21605-87	Average (15)	R	2901-12-1-2	Average (8)	R	4204-15-1.3	Average (4)	R
Name	FE	FE		MH	MH		MH	MH		FA	FA		FA	FA	
Ore Domain	II	II		II	II		V	V		II	II		IV	IV	
SiO_2	42.69	42.19	0.66	38.35	38.88	0.61	36.05	36.81	1.22	50.66	50.22	1.68	49.13	49.68	1.15
TiO_2	0.00	0.11	0.06	0.10	0.11	0.05	0.33	0.33	0.16	0.02	0.04	0.04	0.00	0.03	0.02
Al_2O_3	8.01	8.72	0.77	12.15	11.59	0.67	14.56	12.87	1.08	1.19	2.56	1.20	3.36	2.10	1.02
FeO	28.34	27.97	0.54	24.20	24.78	2.48	23.13	23.09	2.45	21.37	23.09	3.01	28.72	30.31	0.95
MnO	0.51	0.45	0.08	0.37	0.39	0.08	0.30	0.25	0.05	0.40	0.89	0.59	0.23	0.37	0.11
MgO	4.28	4.42	0.42	8.67	6.39	1.49	5.68	6.21	1.31	9.79	8.16	2.28	5.58	4.37	0.85
CaO	11.56	11.57	0.14	11.46	11.81	0.29	12.13	11.83	0.35	11.73	11.97	0.34	11.84	11.86	0.37
Na_2O	0.88	0.99	0.09	1.12	1.28	0.15	1.31	1.33	0.13	0.36	0.25	0.13	0.37	0.26	0.07
K_2O	1.17	1.28	0.12	2.24	2.13	0.23	3.13	2.73	0.29	0.38	0.20	0.17	0.50	0.33	0.11
F	0.23	0.10	0.14	0.00	0.01	0.02	0.04	0.11	0.08	0.00	0.09	0.11	0.11	0.15	0.03
Cl	0.42	0.39	0.15	1.08	1.20	0.23	1.90	1.91	0.62	0.03	0.02	0.01	0.04	0.04	0.01
H_2O	1.66	1.75	0.07	1.68	1.59	0.08	1.38	1.32	0.16	1.94	1.91	0.08	1.89	1.85	0.03
Total	98.09	98.18	0.47	101.41	98.86	1.76	99.94	98.80	1.07	95.92	97.50	0.92	99.88	99.50	1.01
Si	6.763	6.664	0.111	5.894	6.120	0.132	5.749	5.922	0.144	7.798	7.665	0.146	7.507	7.683	0.152
$^{[4]}Al$	1.237	1.336	0.111	2.106	1.880	0.132	2.251	2.078	0.144	0.202	0.335	0.146	0.493	0.317	0.152
$\sum T$	8.000	8.000	0.000	8.000	8.000	0.000	8.000	8.000	0.000	8.000	8.000	0.000	8.000	8.000	0.000
$^{[6]}Al$	0.258	0.287	0.044	0.095	0.270	0.109	0.481	0.359	0.133	0.014	0.128	0.081	0.112	0.066	0.034

(continued)

Table 5.1 (continued)

	Amp-I			Amp-II						Amp-III					
Sample	2901-17-2-4	Average (7)	R	2901-11	Average (6)	R	21605-87	Average (15)	R	2901-12-1-2	Average (8)	R	4204-15-1.3	Average (4)	R
Name	FE	FE		MH	MH		MH	MH		FA	FA		FA	FA	
Ore Domain	II	II		II	II		V	V		II	II		IV	IV	
Ti	0.000	0.013	0.007	0.012	0.013	0.005	0.040	0.040	0.019	0.002	0.004	0.005	0.000	0.004	0.003
Fe^{3+}	0.472	0.461	0.067	1.215	0.766	0.216	0.649	0.663	0.187	0.001	0.084	0.072	0.176	0.101	0.093
Fe^{2+}	3.228	3.180	0.128	1.757	2.409	0.427	2.359	2.368	0.308	2.749	2.860	0.390	3.474	3.809	0.224
Mn	0.069	0.061	0.011	0.048	0.053	0.011	0.040	0.034	0.008	0.052	0.115	0.077	0.030	0.049	0.014
Mg	1.011	1.041	0.096	1.987	1.497	0.334	1.359	1.494	0.300	2.247	1.851	0.494	1.271	1.005	0.183
Ca	1.962	1.957	0.021	1.886	1.992	0.055	2.073	2.041	0.062	1.934	1.958	0.042	1.938	1.966	0.076
$\sum M1,2,3$	7.000	7.000	0.000	7.000	7.000	0.000	7.000	7.000	0.000	7.000	7.000	0.000	7.000	7.000	0.000
$^{[A]}Na$	0.270	0.304	0.028	0.333	0.390	0.048	0.405	0.413	0.037	0.108	0.075	0.038	0.108	0.079	0.019
K	0.237	0.259	0.024	0.439	0.427	0.046	0.636	0.562	0.066	0.074	0.039	0.034	0.097	0.064	0.021
$\sum A$	0.507	0.563	0.050	0.772	0.817	0.086	1.041	0.975	0.075	0.182	0.115	0.070	0.206	0.143	0.038
F	0.129	0.054	0.077	0.000	0.005	0.010	0.020	0.061	0.046	0.000	0.051	0.061	0.060	0.084	0.017
Cl	0.113	0.105	0.039	0.281	0.321	0.063	0.511	0.521	0.178	0.007	0.006	0.002	0.010	0.009	0.002
OH	1.758	1.842	0.070	1.719	1.674	0.071	1.469	1.418	0.153	1.993	1.943	0.062	1.930	1.906	0.016
$\sum O3$	2.000	2.000	0.000	2.000	2.000	0.000	2.000	2.000	0.000	2.000	2.000	0.000	2.000	2.000	0.000

(continued)

Table 5.1 (continued)

	Amp-I			Amp-II						Amp-III					
Sample	2901-17-2-4	Average (7)	R	2901-11	Average (6)	R	21605-87	Average (15)	R	2901-12-1-2	Average (8)	R	4204-15-1.3	Average (4)	R
Name	FE	FE		MH	MH		MH	MH		FA	FA		FA	FA	
Ore Domain	II	II		II	II		V	V		II	II		IV	IV	
X_{Mg}	0.238	0.247	0.024	0.531	0.386	0.095	0.366	0.387	0.078	0.450	0.391	0.096	0.268	0.209	0.038
X_{Cl}	0.060	0.054	0.019	0.141	0.161	0.032	0.258	0.267	0.087	0.004	0.003	0.001	0.005	0.005	0.001
X_K	0.468	0.459	0.012	0.569	0.523	0.025	0.611	0.575	0.036	0.408	0.307	0.093	0.473	0.443	0.042
X_{Al}	0.155	0.167	0.014	0.263	0.235	0.017	0.281	0.260	0.018	0.025	0.042	0.018	0.062	0.040	0.019

FE: ferro-edenite; MH: magnesian-hastingsite; FA: ferro-actinolite

The general formulae of amphiboles: $W_{0\text{–}1}X_2Y_5Z_8O_{22}$ (OH, O, F)$_2$

The structural formulae calculation based on 15eNK: Si + $^{[4]}$Al = 8; $^{[6]}$Al + Ti + Fe + Mg + Ca = 7)

X_{Mg} = Mg/(Mg + Fe^{2+}); X_{Cl} = Cl/(Cl + OH^-); X_K = K/(K + Na); X_{Al} = Al/(Al + Si)

The formulae of MH: $(Na, K)_{0.930}(Ca, Mg, Mn, Fe^{2+})_{5.941}(Fe^{3+}, Al, Ti)_{1.059}(Al, Si)_8O_{22}(F, Cl, OH)_2$

The formulae of FE: $(Na, K)_{0.563}(Ca, Mg, Mn, Fe^{2+})_{6.239}(Fe^{3+}, Al, Ti)_{0.761}(Al, Si)_8O_{22}(F, Cl, OH)_2$

The formulae of FA: $(Na, K)_{0.123}(Ca, Mg, Mn, Fe^{2+})_{6.798}(Fe^{3+}, Al, Ti)_{0.202}(Al, Si)_8O_{22}(F, Cl, OH)_2$

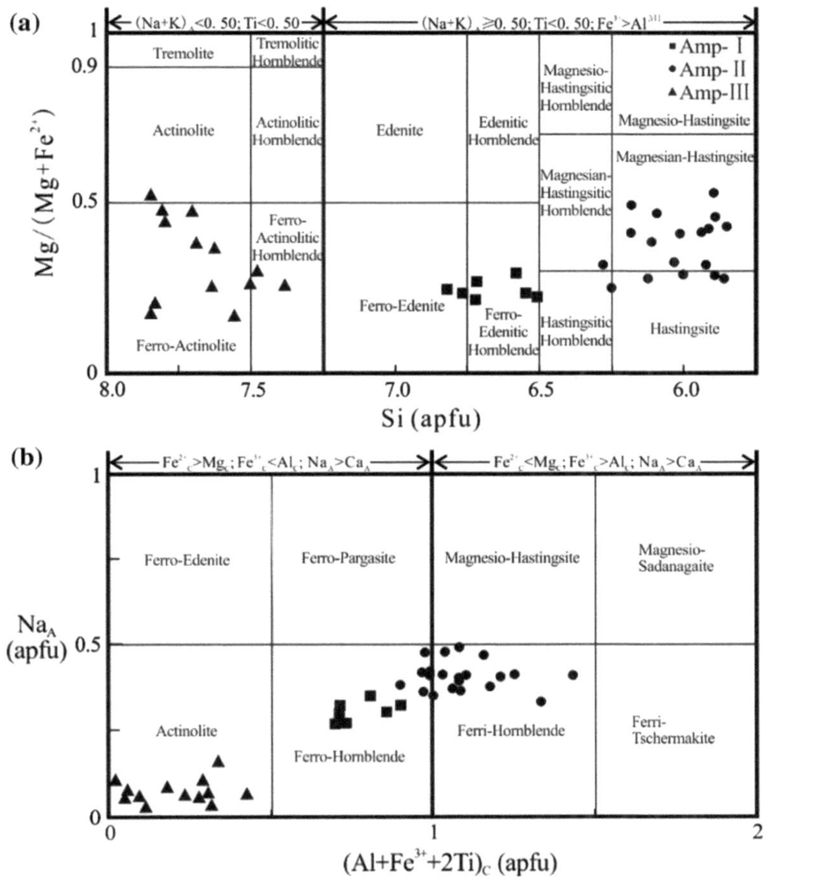

Fig. 5.4 **a** Amphibole classification diagram [12]. Amphiboles from the Galinge skarn deposit vary in composition from Amp-I to Amp-III, i.e., ferro-edenite (FE) → magnesian hastingsite (MH) → ferro-actinolite (FA). **b** Compositional variations from Amp-I to Amp-III based on Na_A (apfu) content and $(Al + Fe^{3+} + 2Ti)_C$ (apfu) content [8]

are clearly distinguish Amp-I, Amp-II and Amp-III, e.g., the Ca content increases from Amp-I to Amp-II, and then decreases from Amp-II to Amp-III.

In Fig. 5.6a, (Na + K) values (apfu) in the A-site are plotted against Si in the tetrahedral site, showing a strongly negative correlation for all amphibole generations. The (Na + K) values are markedly higher in Amp-II MH (0.721–7.851 apfu), when compared to the Amp-I FE (0.487–0.649 apfu), but much lower in the Amp-III FA (0.034–0.281 apfu).

The three amphibole types are: Amp-I, Amp-II and Amp-III contain 0.582–0.176, 3.161–0.894 and 0.039–0.017 wt% Cl, respectively (Fig. 5.7a, b). In Fig. 5.7a, the mole fraction of K in the A-site ($X_K = K/(K + Na)$) versus X_{Cl} ($X_{Cl} = Cl/(Cl + OH)$)

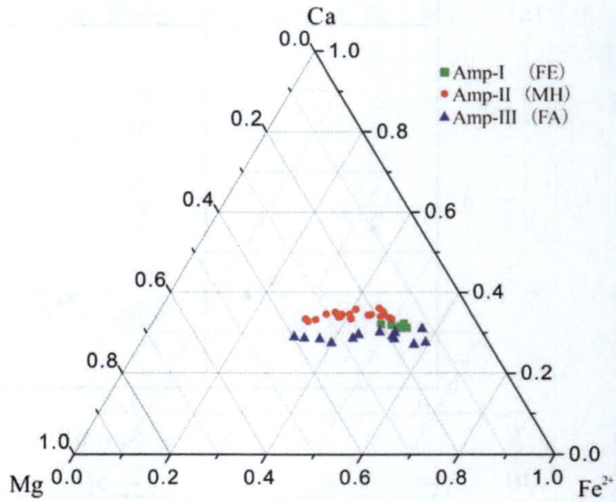

Fig. 5.5 Triangular Ca–Fe–Mg plot (in total molar proportions) for Amp-I, Amp-II and Amp-III from the Galinge skarn deposit

shows extensive positive correlation for Amp-II and Amp-III. Amp-II has elevated X_K (0.491–0.629) and X_{Cl} (0.118–0.439); whereas, Amp-III has low X_K (0.129–0.481) and X_{Cl} (0.002–0.006). In contrast, Amp-I shows no obvious correlation between X_{Cl} and X_K. Moreover, the mole fraction of Al versus X_{Cl} shows similar correlations (Fig. 5.7b).

5.4 Replacement of Mafic Igneous Rocks

Mafic igneous rocks replaced by magmatic-hydrothermal fluids are reported in many skarn deposits [5, 16]. There is a close link between igneous rock composition and skarn mineralogy. In general, skarn formed in volcanic rocks initially reflects the bulk composition of the protolith, but is later altered by the influx of metasomatic fluids.

Amphibole formation requires excessive amount of Si, Al, Fe, Mg, Ca, Na, K and Ti from the fluids and host rocks. Calcium, Na and K are likely to be the main constituents of magma derived hydrothermal fluid, as these elements are easily transported in saline solutions. Alteration of igneous protoliths to skarn assemblages generally results in a high concentration and strong mobility of elements such as Ti, V, and Al in calc-silicate minerals [16]. For example, the breakdown of igneous feldspar can release Ca, Al, Na and K into the hydrothermal fluids. Therefore, Ca, Al, Na and K contents show an increasing trend from Amp-I to Amp-II. However, the common formation of magnesian-hastingsite, consuming substantial Al, K and Na, and later magma evolved fluid is characterized by low saline with relatively low K, Na and Ca content. Such factors may result in the depletion of Al, K, Na and Ca from Amp-II to Amp-III (Figs. 5.6 and 5.7). In addition, the mole fraction of K in

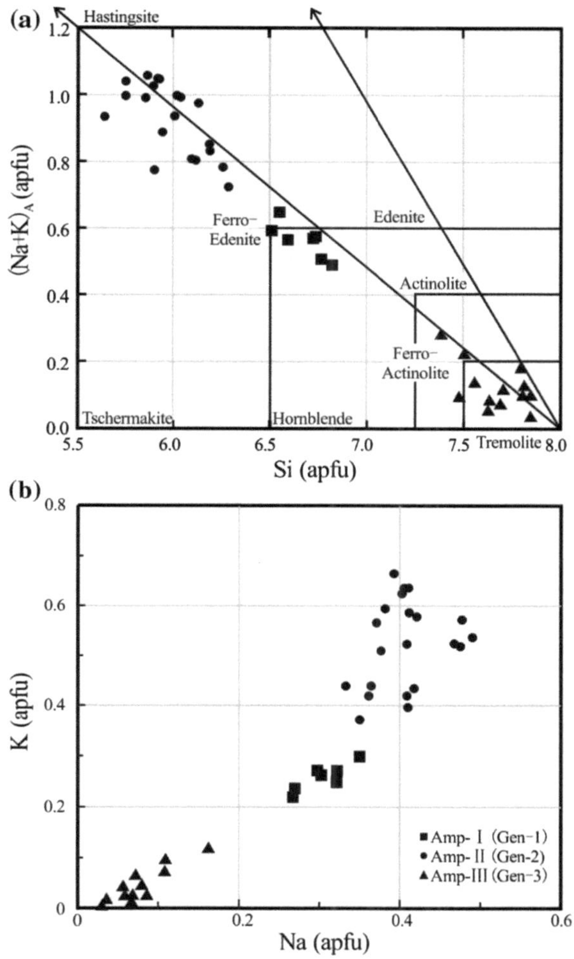

Fig. 5.6 **a** Plot of (Na + K) versus Si of the amphiboles from Galinge skarn deposit. **b** Plot of K versus Na of the amphiboles from Galinge skarn deposit

the A-site shows strong positive correlation with X_{Cl} for the Amp-II and Amp-III but very different slopes (Fig. 5.7), which indicates that they were probably formed under two distinct fluid environments. Amp-II may be associated with high saline fluid, but Amp-III may be yielded by low saline fluid. Nonetheless, Amp-I shows no obvious X_{Cl} and X_K correlation, implying that Amp-I was not generated by the replacement of hydrothermal fluids but most probably crystallized from the magmatic melt. The X_{Al} values from Amp-I to Amp-III shows a positive correlation with X_{Cl} values, indicating that the Al concentration of the octahedral site could influence the Cl incorporation.

Large amount of mafic Amp-II has extensively replaced the basaltic andesite rocks, showing that Fe and Mg were efficiently mobilized with the hydrothermal fluids. Compared with the Galinge dacite and rhyolite, the Galinge basaltic andesite

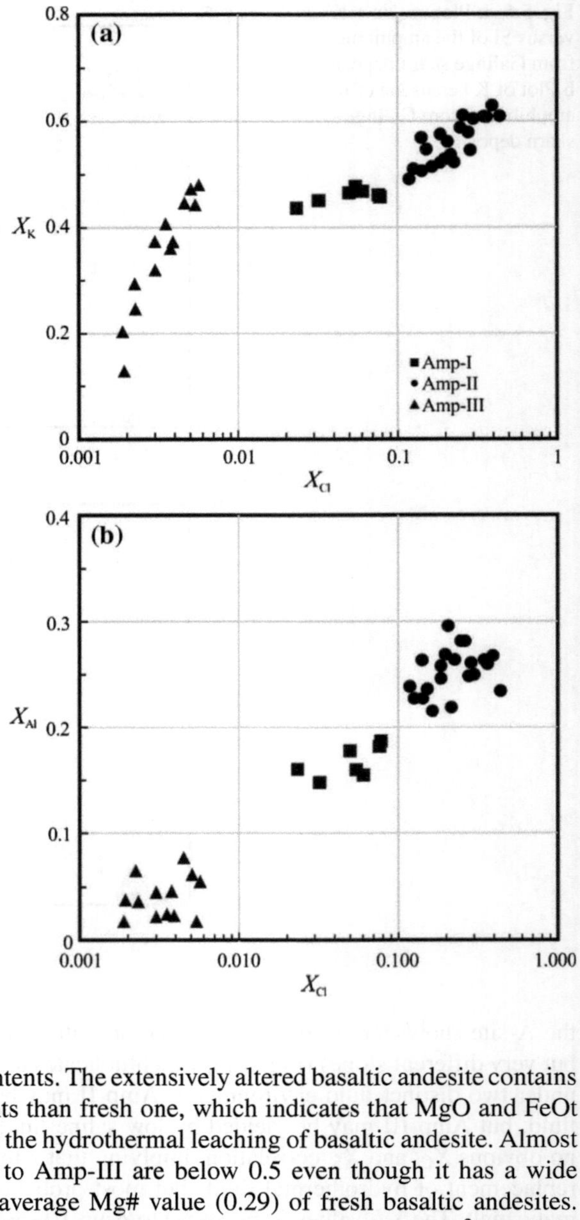

Fig. 5.7 **a** Plot of X_K versus X_{Cl} of Amp-1 to Amp-3 amphiboles. The X_{Cl} values show strong linear correlations with the X_K values. **b** Plot of X_{Al} versus X_{Cl} of Amp-I to Amp-III amphiboles. The X_{Cl} values show linear correlations with the X_K values

has higher MgO and FeOt contents. The extensively altered basaltic andesite contains lower MgO and FeOt contents than fresh one, which indicates that MgO and FeOt was likely to be mobilized by the hydrothermal leaching of basaltic andesite. Almost all X_{Mg} values from Amp-I to Amp-III are below 0.5 even though it has a wide range, but it is close to the average Mg# value (0.29) of fresh basaltic andesites. Furthermore, different amphibole generations show a wide range of Fe^{2+} and Mg concentrations likely reflecting the compositional variability of hydrothermal fluids and host rocks (Fig. 5.5).

The complete breakdown of primary plagioclase and the formation of Cl-rich amphibole indicate that a fluid phase was introduced along intergranular boundaries and microfractures throughout the fracture zones. Nevertheless, the large compositional variations of amphibole on the cm-scale suggest that the fluid flux was moderately low. This is supported by the incomplete breakdown of plagioclase along the ca. 3 cm wide metasomatism front (Fig. 5.1). Formation of magnesian hastingsite may have been associated with the decomposition of Amp-I and groundmass, since Mg and Fe content was preferentially incorporated into magnesian hastingsite. As there was no garnet formation at the metasomatic front, Amp-I and the groundmass of basaltic rocks may have supplied sufficient Ca for the Amp-II formation. Decomposition of feldspar can supply Ca and Si for the garnet formation in addition to Ca and Si in the fluid. A schematic reaction could be written as:

$$\begin{aligned} &CaAl_2Si_2O_8\,\text{(anorthite)} + SiO_2\,\text{(quartz)}\\ &\quad + 2CaCO_3\,\text{(calcite)} = Ca_3Al_2Si_3O_{12}\,\text{(grossular)} + 2CO_2 \end{aligned}$$

The narrow range of Ca for the three amphibole generations possibly reflects the restricted plagioclase composition (Fig. 5.5). The continuous introduction of fluid phase may have resulted in a system far diverted from the thermodynamic equilibrium.

Titanium from the Galinge fresh basaltic andesite rocks is dominantly hosted in sphene and intergrowths of ilmenite and titaniferous magnetite. When the ilmenite and titaniferous magnetite were dissolved later by volatile-rich hydrothermal fluids, titanium would be released into the hydrosaline fluids. The dissolved Ti would then be incorporated into sphene. The sphene grains in the skarn system developed a core-rim texture, which suggests that the growth of sphene may have occurred in two stages. The early stage cores might be replacement relicts of the basaltic andesite during hydrothermal leaching process, and the second stage rims were probably formed during the skarn stage. The varying composition of the two stage sphenes reflects their hydrothermal fluid evolution.

5.5 Incorporation of Cl

In previous studies on Cl-rich amphibole, some factors for the substitution of OH by Cl in the amphibole structure have been introduced [6, 19]. Enami et al. [6] argued that the Cl content of amphibole increases with increasing Cl concentration of the coexisting fluid in Cl-poor environments. The high Cl content in the MH (up to 3.16 wt%) reflect the early hydrothermal fluid was enriched in volatile phase. Morrison [19, 23] found that the Cl content difference between the Cl-rich and Cl-poor amphiboles may reflect equilibration at significantly different temperatures and varying fluid salinity. Therefore, the Cl-rich MH was formed by the replacement of original igneous FE through the early high-temperature and high-saline hydrothermal fluid, however, the formation of Cl-poor FA is associated with the later low-temperature and low-saline fluid alteration of the Cl-rich MH. Oberti et al. [21] argued that the crystal-structure refinement of amphiboles shows strong compositional correlation with Cl incorporation. Furthermore, many studies of Cl-bearing amphiboles emphasized the positive correlations of X_{Al} and X_K with X_{Cl} (Fig. 5.7), which suggests that the Cl content is interrelated with on A-site occupancy. [19, 24, 25]. Sato et al. [23] pointed out that the incorporation of Cl in amphibole is governed by the crystal chemical constraints. Substitution of Cl for OH expands the sizes of octahedral sites [15], and Al substitution for Si could cause the deformation of double chains to fit the expanded octahedral strips. The effects of Fe substitution into octahedral sites on the Cl incorporation may be a chemical one [25].

Munoz and Swenson [20] and Morrison [19] have confirmed that the Cl–OH exchange reaction between amphibole and a coexisting fluid or silicate melt is modeled as follows:

$$OH_{amp} + Cl_{fluid\ or\ melt} = Cl_{amp} + OH_{fluid\ or\ melt} \tag{5.1}$$

The equilibrium constants (K) for these reactions can be given as:

$$\log K = \log(Cl/OH)_{amp} - \log(Cl/OH)_{fluid\ or\ melt} \tag{5.2}$$

The Cl distribution may be ideally expressed by:

$$\ln(Cl/OH)_{amp} = \ln(Cl/OH)_{fluid} - \Delta G^0/RT \tag{5.3}$$

where ΔG^0 is the free energy changes of the end-members in the reaction (5.1), R and T are ideal gas constant and temperature, respectively. Obviously, the ΔG^0 varies with the cation contents in amphibole. Since $^{[4]}Al$ and Fe^{2+} are the principal compositional control on Cl incorporation, the ΔG^0 may be related with the product of the $^{[4]}Al$ and Fe^{2+} content. Sato et al. [23] suggested that the temperature and composition was related with ΔG^0. Therefore, Eq. (5.3) may be written as [23]:

$$\ln(Cl/OH)_{amp} = \ln(Cl/OH)_{fluid} + A^{[4]}Al \cdot Fe^{2+}/RT + B/RT \tag{5.4}$$

If T and $\ln(Cl/OH)_{fluid}$ values are given, amphiboles with variable $\ln(Cl/OH)_{amp}$ should be as a linear function of $^{[4]}Al{\cdot}Fe^{2+}$. Thus, three distinct trends in the $^{[4]}Al{\cdot}Fe^{2+}$ versus ln(Cl/OH) plots (Fig. 5.8) for the three generations amphiboles (FE, MH and FA) probably indicate the replacement and growth of amphiboles underwent variable physical-chemical conditions of the coexisting hydrothermal fluids [23]. They also reflect the $\ln(Cl/OH)_{fluid}$ values of the MH-forming fluids were higher than the ones of the FA-forming fluid. Therefore, the Cl concentrations of MH-forming fluids are higher than ones of the FA-forming fluid. As a consequence, the formation of MH might be associated with the early high-temperature and high-saline liquid, and the formation of FA would be yielded from later low-temperature and low-saline fluids.

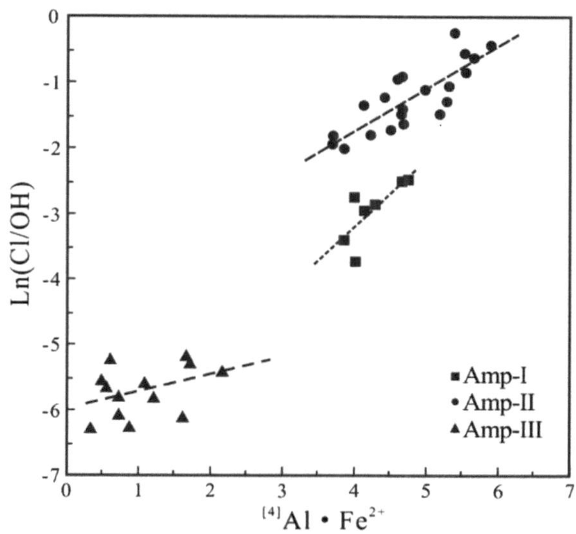

Fig. 5.8 $^{[4]}Al \cdot Fe^{2+}$ versus ln(Cl/OH) diagram for Amp-I, Amp-II and Amp-III from the Galinge deposit

5.6 Implications of Fluid Inclusions

Skarn systems studies have revealed that the magmatic fluid evolution is a dominant factor for the formation of diverse skarn minerals. The Galinge hydrothermal events are divided into two major stages as most skarn deposits comprised [18, 22]. Early high-temperature and high-saline fluids exsolved from semi-solidified magma carrying large amounts of dissolved salts, followed by a later stage of low-temperature and low-saline fluids generated probably through meteoric water dilution. The halite-bearing inclusions coexisting with the vapor-rich inclusions are considered to be the consequence of phase separation at largely constant pressure and temperature, which can be taken as an evidence for boiling in the fluid evolution [9, 29]. The results of boiling, especially emanation of volatiles and potential shift in pH and oxidation state, have become well appreciated with regard to the ore-forming processes.

An early prograde stage with anhydrous minerals, such as garnet and pyroxene, yields from relatively high-temperature and high-saline liquid; and a later retrograde stage with hydrous minerals, such as epidote, amphibole, and chlorite plus sulfide ore minerals, forms from low-temperature and low-saline fluids. However, exceptions occur, e.g., the Cl-rich MH documented in this study was formed coeval with the garnet. In terms of the paragenetic sequence, the Cl-rich MH would be formed by the replacement of original igneous FE through the early high-temperature and high-saline hydrothermal fluid; the formation of Cl-poor FA would be associated with the later low-temperature and low-saline fluid alteration. The differences in Cl content among the amphiboles may reflect equilibration at significantly different temperatures and broadly varying fluid salinity [19, 23]. We suggest that, during an early hydrothermal event, a relatively high-temperature and high-saline fluid altered the original FE to the Cl-rich MH at the crystal margins adjacent to circulating fluids. Late-stage hydrothermal alteration by a low Cl content and low-temperature fluid produced FA at the rims of grains.

The segregation and release of an aqueous phase from the crystallizing melt are of particular importance to skarn formation [17]. The H_2O solubility in silicate melts saturated in aqueous vapor and/or hydrosaline liquid varies inversely with Cl^- content. The presence of magmatic volatile phase buffers the H_2O and Cl^- concentrations of the silicate melt. Hydrosaline liquids are stable with higher Cl contents in the melt, and H_2O solubility is highly sensitive to Cl content. Thus, Cl is much more likely to affect the exsolution of hydrosaline liquid vapor from highly-evolved, H_2O-poor, silicic melts because these melts saturate in these phases if they contain as little as several thousand ppm Cl [28]. Since the solubility of Cl^- in silicic melts is much lower than that in mafic melts [26], large amounts of Cl^- would be exsolved from magmas along with fractional crystallization [27].

5.7 Concluding Remarks

The multistage amphiboles in skarn Fe deposit are rarely reported. Amphiboles from the Galinge skarn deposit are compositionally and paragenetically divided into three generations: Amp-I: Ferro-edenitic hornblende (FE); Amp-II: Deep bluish-green magnesian-hastingsite (MH); Amp-III: Light greenish-beige ferro-actinolite (FA). Both Mg and Fe contents of the MH and FA vary widely, which may have been associated with the interactions between magmatic-derived hydrothermal fluids and the basaltic andesite.

Further mineralogical studies on multistage amphiboles provided insights into the formation mechanisms that control Cl incorporation into OH^--site [21]. Our work supports that the varying Cl content in amphiboles is associated with different temperature and broadly different salinity of the hydrothermal fluids [19]. In prograde alteration stage, high-temperature and high-saline fluids exsolved from the granodioritic magma may have replaced the original FE in the mafic igneous rocks to form the Cl-rich MH, and succeeded by the Cl-poor FA in retrograde alteration stage (with low-temperature and low-saline fluids). The present study suggest that crystal-structure refinements of amphiboles are also remarkably effective on Cl incorporation.

The Cl-rich fluids could result in increased concentrations of some elements such as Fe, Al, Cu, Mg, Ca and Ti in solutions, and transport them as metal chloride complexes. The mafic content were predominately held by MH and FA, Ca and Al incorporated into garnet, and Ti and Fe into sphene.

References

1. Baker J, Spiegelman M (1995) Modelling an infiltration-driven geochemical front. Earth Planet Sci Lett 136:87–96
2. Dick LA, Robinson GW (1979) Chlorine-bearing potassian hastingsite from a sphalerite skarn in southern Yukon. Can Mineral 17:25–26
3. Duke EF (1995) Contrasting scales of element mobility in metamorphic rocks near Harney Peak Granite, Black Hills, South Dakota. Geol Soc Am Bull 107:274–285
4. Dutrow BL, Foster CT, Henry DJ (1999) Tourmaline-rich pseudomorphs in sillimanite zone metapetites: demarcation of an infiltration front. Am Mineral 84:794–805
5. Einaudi MT, Burt DM (1982) Introduction; terminology, classification, and composition of skarn deposits. Econ Geol 77:745–754
6. Enami M, Liou JG, Bird DK (1992) Cl-bearing amphibole in the Salton Sea geothermal system, California. Can Mineral 30:1077–1092
7. Hawthorne FC, Oberti R (2006) On the classification of amphiboles. Can Mineral 44:1–21
8. Hawthorne FC, Oberti R, Harlow GE, Maresch WV, Martin RF, Schumacher JC, Welch MD (2012) Nomenclature of the amphibole supergroup. Am Mineral 97:2031–2048
9. Hemley JJ, Cygan GL, Fein JB, Robinson GR, D'Angelo WM (1992) Hydrothermal ore-forming processes in the light of studies in rock-buffered systems; I, Iron-copper-zinc-lead sulfide solubility relations. Econ Geol 87:1–22
10. Jacobson SS (1975) Dashkesanite: high-chlorine amphibole from St. Paul's rocks, Equatorial Atlantic and Transcaucasia, USSR. Smithson Contrib Earth Sci 14:17–20
11. Krutov GA (1936) Dashkesanite, a new chlorine amphibole of the hastingsite group. Mineral Abstr 6:438
12. Leake BE (1978) Nomenclature of amphiboles. Can Mineral 16:501–520
13. Leake BE, Woolley AR, Arps C, Birch WD, Gilbert MC, Grice JD, Hawthorne E, Kato A, Kisch HJ, Krivovichev VG (1997) Nomenclature of amphiboles report of the subcommittee on amphiboles of the International mineralogical association commission on new minerals and mineral names. Eur J Mineral 61:295–361
14. Leake BE, Woolley AR, Birch WD, Burke EA, Ferraris G, Grice JD, Hawthorne FC, Kisch HJ, Krivovichev VG, Schumacher JC (2004) Nomenclature of amphiboles: additions and revisions to the International Mineralogical Association's amphibole nomenclature. Am Mineral 89:883–887
15. Makino K, Tomita K, Suwa K (1993) Effect of chlorine on the crystal structure of a chlorine-rich hastingsite. Mineral Mag 57:677–686
16. Meinert LD (1984) Mineralogy and petrology of iron skarns in western British Columbia, Canada. Econ Geol 79:869–882
17. Meinert LD (1993) Igneous petrogenesis and skarn deposits. Miner Depos Model 40:569–583
18. Meinert LD, Hedenquist JW, Satoh H, Matsuhisa Y (2003) Formation of anhydrous and hydrous skarn in Cu-Au ore deposits by magmatic fluids. Econ Geol 98:147–156
19. Morrison J (1991) Compositional constraints on the incorporation of Cl into amphiboles. Am Mineral 76:1920–1930
20. Munoz JL, Swenson A (1981) Chloride-hydroxyl exchange in biotite and estimation of relative HCl/HF activities in hydrothermal fluids. Econ Geol 76:2212–2221
21. Oberti R, Ungaretti L, Cannillo E, Hawthorne FC (1993) The mechanism of Cl incorporation in amphibole. Am Mineral 78:746–752
22. Öztürk YY, Helvacı C, Satır M (2008) The influence of meteoric water on skarn formation and late-stage hydrothermal alteration at the Evciler skarn occurrences, Kazdağ, NW Turkey. Ore Geol Rev 34:271–284
23. Sato H, Yamaguchi Y, Makino K (1997) Cl incorporation into successively zoned amphiboles from the Ramnes cauldron, Norway. Am Mineral 82:316–324
24. Suwa K, Enami M, Horiuchi T (1987) Chlorine-rich potassium hastingsite from West Ongul Island, Liitzow-Holm bay, East Antarctica. Mineral Mag 51:709–714

25. Vanko DA (1986) High-chlorine amphiboles from oceanic rocks: product of highly saline hydrothermal fluids. Am Mineral 71:51–59
26. Webster JD (1997) Exsolution of magmatic volatile phases from Cl-enriched mineralizing granitic magmas and implications for ore metal transport. Geochim Cosmochim Acta 61:1017–1029
27. Webster JD, Kinzler RJ, Mathez EA (1999) Chloride and water solubility in basalt and andesite melts and implications for magmatic degassing. Geochim Cosmochim Acta 63:729–738
28. Webster JD, Rebbert CR (1998) Experimental investigation of H_2O and Cl^- solubilities in F-enriched silicate liquids; implications for volatile saturation of topaz rhyolite magmas. Contrib Mineral Petrol 132:198–207
29. Wilkinson JJ (2001) Fluid inclusions in hydrothermal ore deposits. Lithos 55:229–272

Chapter 6
Formation and Breakdown of Ilvaite in the Large Galinge Skarn Fe Deposit, Western China: A Record of Multistage Retrograde Alteration

Ilvaite, a mixed-valence Fe sorosilicate [17, 18], has long been recognized as an accessory gangue mineral in Fe, Zn and Pb–Zn skarn deposits [2, 9, 14, 38]. The ilvaite group comprises the Mn-dominant endmember manganilvaite, ideally ($CaFe^{2+}Fe^{3+}(Mn, Fe^{2+})(Si_2O_7)O(OH)$) and the Fe-dominant ilvaite, ideally ($CaFe_2^{2+}Fe^{3+}(Si_2O_7)O(OH)$) [7, 43]. The Mn-rich species occurs exclusively in marble-hosted distal Zn–(Pb) skarn deposits, and is closely associated with Mn-rich clinopyroxene (hedenbergite–johannsenite) ± rhodonite exoskarn minerals [26, 38]. In contrast, the Fe-rich species occurs in skarn Fe, Au, Sn and W skarn deposits, and is associated with exoskarn minerals [16, 25, 30]. Ilvaite is rarely occurs in the Cu skarn deposit, but it was found in Igarapé Bahia Cu-Au deposits which is a typical Fe-oxide Cu-Au-(U-REE) deposits in the Carajás region [41]. Graser and Markl [21] reported that the ilvaite, epidote and garnet occur as an endoskarn assemblage along fracture zones. Ilvaite is commonly considered as a retrograde alteration mineral in various skarn deposits [15, 32–34], but Franchini et al. [16] suggested that ilvaite crystallizes on both prograde and retrograde paths. Nonetheless, since ilvaite is not present in large quantities in hydrothermal systems its potential for recording hydrothermal ore-forming processes has long been underestimated.

This study documents the occurrence of ilvaite and the associated Ca-rich aluminosilicates from the Galinge Fe skarn deposit, and elucidates the physicochemical conditions of their formation and evolution. Mineral textures and stability relationships provide valuable information on their origin and relative timing of formation. In this study, substantial amounts of ilvaite geochemical data were compiled from various Fe, Zn ± Pb, Au, and Sn ± W skarn deposits for comparing the ilvaite composition with the type of mineralization.

6.1 Petrography of the Ilvaite-Bearing Association

Ilvaite is a common accessory mineral that is typically formed in the retrograde stage in the Galinge exoskarn, and is mainly concentrated in the area between the

M. Yu, *Metallogenic Mechanism of the Galinge Polymetallic Iron Skarn Deposit, Qiman Tagh Mountains, Qinghai Province*, Springer Theses,
https://doi.org/10.1007/978-981-10-7907-8_6

Ore Domain II and IV. Ilvaite coexists with magnesio-hastingsite and magnetite (Fig. 6.1a, b), and some andradite relics are locally kept in the system (Fig. 6.1e). On the microprobe image of Fig. 6.1c, d, ilvaite occurs as spongy resinous masses or aggregates coexisting with ferro-actinolite, quartz and magnetite. Many grains of quartz, magnetite and ferro-actinolite fill the fissures of ilvaite, which explains the formation of magnetite and ferroactinolite after ilvaite. Furthermore, ilvaite also occurs as euhedral prismatic crystals in calcite veins, and the ilvaite crystals are locally crosscut by later calcite (Fig. 6.1e, f). The petrographic analysis gives a total of three different ilvaite-bearing replacement textures: (1) Andradite and magnetite replaced by ilvaite; (2) ilvaite with quartz breakdown to ferro-actinolite and magnetite; (3) ilvaite breakdown to calcite, ferroactinolite and pyrite. The formation of the three associations is completely documented in the retrograde alteration stage, suggesting that they record a continuous variation in temperature and oxidation state, although each texture shows strong replacement relationship.

Under the microscope, magnesio-hastingsite is bluish-green to dark green, and forms anhedral to subhedral lamellae crystals. Replacement by ilvaite and ferro-actinolite is very common (Fig. 6.1a, b). Pleochroic yellowish-green or bluish green hastingsite is found intergrown with epidote or replacing garnet.

6.2 Analytical Methods

Optical imaging was conducted to investigate the retrograde skarn mineral relations in ilvaite-bearing associations. These images were used to locate points for microanalysis of ilvaite and associated silicates. Electron microprobe analyses were conducted by wavelength-dispersive spectrometry (WDS) in a JEOL JXA-823V electron probe microanalyser (EPMA) at the MLR Key Laboratory of Metallogeny and Mineral Assessment, Institute of Mineral Resources, Chinese Academy of Geological Sciences (Beijing) China. The accelerating potential was 15 kV, probe current 5–10 nA, and the beam was focused to a spot size of 3–5 μm. The analyses were calibrated using jadeite (Si, Na, and Al), forsterite (Mg), orthoclase (K), apatite (P), wollastonite (Ca), rutile (Ti), halite (Cl), and synthetic oxide (Cr, Mn, Fe, Ni) as standards.

6.3 Chemical Composition of the Ilvaite-Bearing Association

The chemical compositions of representative ilvaites from Stages I to III are very similar, and show a particularly Fe-rich character, close to the pure Fe-end-member (Fig. 6.2). Mineral chemistry shows that the FeO_{tot} content ranges from 46.03 to 49.78 wt% (Table 6.1). The MnO content ranges from 0.21 to 1.47 wt%. Mn^{2+}–Fe^{2+}

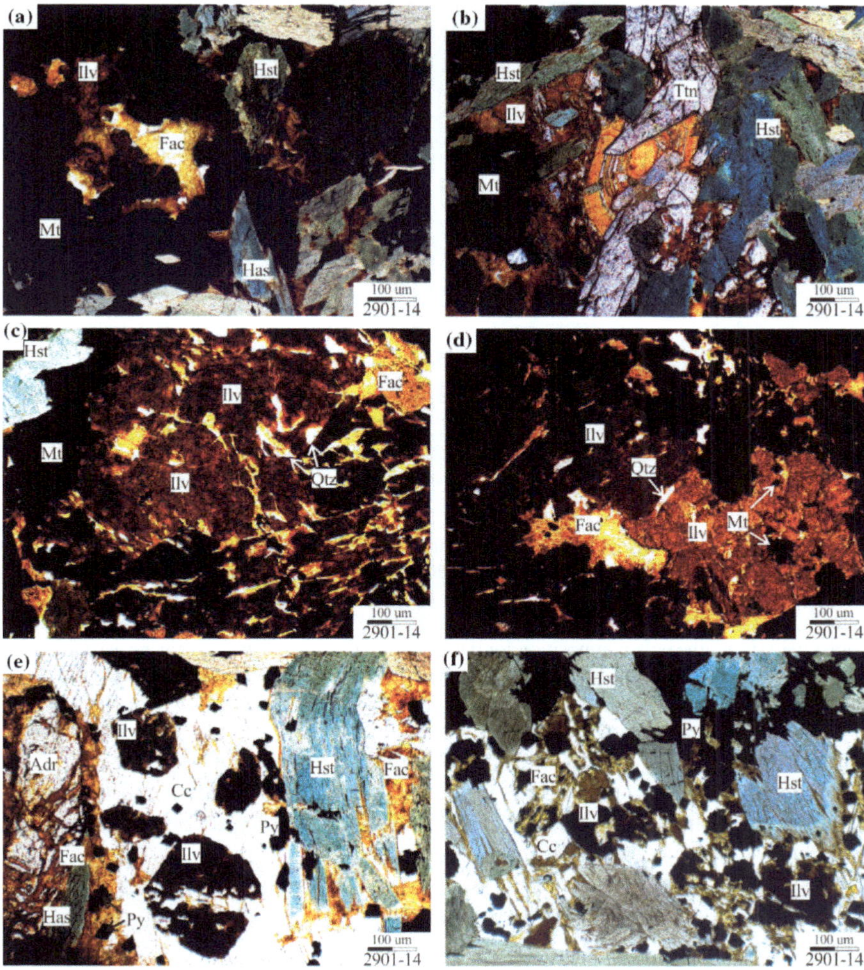

Fig. 6.1 **a**, **b** Subhedral to anhedral scarlet ilvaite associated with hastingsite, andradite, ferro-actinolite, titanite and magnetite, and locally enveloped by magnetite. **c**, **d** Anhedral ilvaite aggregates with magnetite and quartz inclusions, replaced by ferro-actinolite around the rims and fissures. **e**, **f** Dark brown subhedral to euhedral ilvaite in calcite-quartz-sulfide veins, crosscut and filled by calcite along fissures, and replaced by ferro-actinolite around the rims. Abbreviations are as in Table 6.1

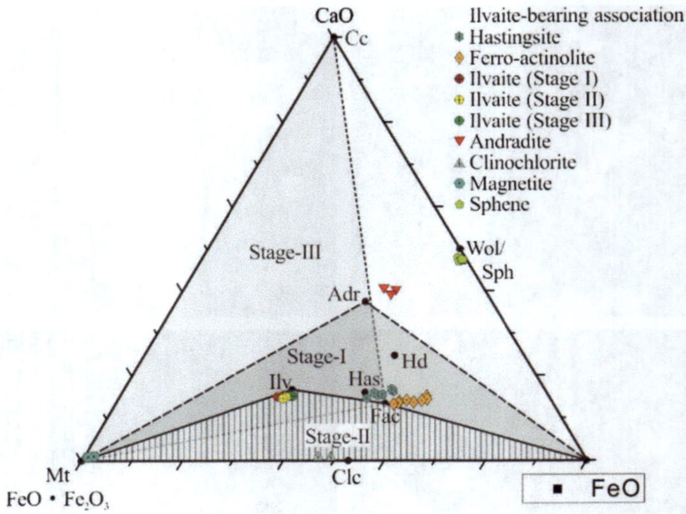

Fig. 6.2 The compositions of ilvaite and associated minerals (andradite + hastingsite + ilvaite + ferroactinolite + magnetite) are plotted in the SiO_2–CaO–FeO_{tot} ternary diagram. The Ca–Fe–Si ternary plot shows the composition of crystalline phases considered for stability calculations in the system Ca–Fe–Si–C–O–H, assuming Al_2O_3, Na_2O, H_2O and O_2 are saturated in the system. The shaded field shows the bulk composition for the stable relationship of ilvaite-bearing associations in the different stages. The solid black circle corresponds to the theoretical value of different minerals occurring in the ilvaite-bearing system

replacement unambiguously occurs at the M(2) site [7, 10]. Only small amounts of Al_2O_3 are present, ranging from 0.21 to 0.98 wt%, and the CaO content is between 13.62 and 14.06 wt%.

The garnet coexisting with ilvaite is a grossular-andradite solid solution, containing moderate Al_2O_3 (8.65–9.76 wt%) and FeO_{tot} (15.29–17.43 wt%) contents; the average composition is $Adr_{55.2}Gro_{42.9}Pyr_{1.9}$.

Previous microprobe work distinguished two compositionally different but texturally identical secondary amphibole types, lying within the compositional fields for hastingsite and ferro-actinolite (Fig. 6.2) [46]. The associated hastingsite shows chlorine-rich character (Cl: 0.89–1.58 wt%), a low Si content (5.89–6.27 a.p.f.u.) and X_{Mg} from 0.25 to 0.53. The ferro-actinolite replacing ilvaite shows a higher Si content than hastingsite, from 7.383 to 7.808 a.p.f.u. (Fig. 6.2), and a wide X_{Mg} range of 0.25 to 0.48. The Cl content of ferro-actinolite is close to the detection limit.

In the ilvaite-bearing samples, the FeO_{tot} of magnetite ranges from 89.39 to 93.38 wt% (Fe_2O_3: 62.89–68.23 wt%; FeO: 31.44–32.79 wt%) (Table 6.1). The chlorite is Fe-rich chamosite ($X_{Fe} = 0.66$–0.67), and the TiO_2 of titanite ranges from 27.93 to 32.56 wt% (mean = 29.91 wt%; Table 6.1).

Table 6.1 Electron microprobe analyses of minerals from the ilvaite-bearing association

Mineral	Hst	Hst	Hst	Hst	Fac	Fac	Fac	Fac	Fac	Fac		Ilv (Stage-I)	Ilv (Stage-I)	Ilv (Stage-II)	Ilv (Stage-II)	Ilv (Stage-III)	Ilv (Stage-III)
Sample (wt%)	2901-11	2901-17-1-3	2901-12-1-5	2901-12-2-1	2901-17-1-5	2901-17-2-2	2901-17-2-3	2901-17-3-1	2901-12-1-3	2901-12-2-2		2901-14-3-1	2901-14-3-2	2901-14-6-1	2901-14-7-1	2901-14-1-1	2901-14-2-2
SiO_2	38.35	39.25	38.16	39.96	47.70	50.45	48.95	49.90	52.37	50.76		29.29	29.98	30.65	30.77	30.75	31.27
TiO_2	0.10	0.14	0.07	0.20	0.08	0.14	0.00	0.01	0.05	0.00		0.16	0.00	0.03	0.12	0.18	0.08
Al_2O_3	12.15	10.42	11.97	10.90	4.96	2.18	2.94	2.74	1.27	2.41		0.89	0.58	0.98	0.64	0.33	0.21
FeO	24.20	28.76	25.78	26.26	27.36	19.85	26.48	24.61	20.77	23.46		49.78	48.27	49.23	47.98	47.78	46.03
MnO	0.37	0.34	0.43	0.29	0.46	0.31	1.79	0.41	0.34	1.71		0.21	0.31	0.93	1.24	1.32	1.47
MgO	8.67	4.27	5.49	5.43	5.26	10.03	5.02	7.52	10.61	7.95		0.74	0.52	0.67	0.92	0.47	0.38
CaO	11.46	11.58	12.12	11.85	11.41	12.44	11.81	12.12	12.39	11.72		13.83	14.05	13.62	13.81	14.06	13.74
Na_2O	1.12	1.17	1.55	1.15	0.54	0.30	0.20	0.12	0.22	0.22		0.00	0.00	0.03	0.02	0.05	0.00
K_2O	2.24	2.06	2.57	1.85	0.61	0.15	0.14	0.10	0.20	0.05		0.00	0.00	0.2	0.01	0.02	0.00
F	0.00	0.05	0.00	0.00	0.26	0.19	0.12	0.27	0.00	0.00		–	–	–	–	–	–
Cl	1.08	1.58	1.39	1.23	0.03	0.02	0.02	0.03	0.03	0.02		–	–	–	–	–	–
H_2O	1.68	1.46	1.54	1.59	1.79	1.85	1.85	1.81	2.00	1.97		2.20	2.20	2.20	2.20	2.23	2.21
Total	101.4	99.62	99.52	99.12	98.66	96.05	97.47	97.82	98.24	98.30		97.10	95.91	98.54	97.71	97.20	95.39
	Based on 15 cations and 23 oxygens											Based on 8.5 oxygens and 6 cations					
Si	5.894	6.249	6.031	6.278	7.383	7.705	7.637	7.629	7.808	7.692	Si	2.109	2.171	2.161	2.177	2.193	2.253
$Al^{[IV]}$	2.106	1.751	1.969	1.722	0.617	0.295	0.363	0.371	0.192	0.308	Ti	0.009	0.000	0.002	0.006	0.010	0.004
$\sum T$	8.000	8.000	8.000	8.000	8.000	8.000	8.000	8.000	8.000	8.000	Al	0.076	0.050	0.081	0.053	0.028	0.018
$Al^{[VI]}$	0.095	0.204	0.261	0.296	0.287	0.097	0.178	0.122	0.031	0.123	Fe^{3+}	0.922	0.819	0.829	0.788	0.776	0.638

(continued)

Table 6.1 (continued)

Mineral	Hst	Hst	Hst	Hst	Fac	Fac	Fac	Fac	Fac	Fac		Ilv (Stage-I)	Ilv (Stage-I)	Ilv (Stage-II)	Ilv (Stage-II)	Ilv (Stage-III)	Ilv (Stage-III)
Sample (wt%)	2901-11	2901-17-1-3	2901-12-1-5	2901-12-2-1	2901-17-1-5	2901-17-2-2	2901-17-2-3	2901-17-3-1	2901-12-1-3	2901-12-2-2		2901-14-3-1	2901-14-3-2	2901-14-6-1	2901-14-7-1	2901-14-1-1	2901-14-2-2
Ti	0.012	0.017	0.008	0.023	0.009	0.016	0.000	0.001	0.006	0.000	Fe^{2+}	2.069	2.098	2.068	2.045	2.068	2.130
Fe^{3+}	1.215	0.733	0.698	0.659	0.030	0.050	0.099	0.190	0.050	0.110	Fe	2.991	2.917	2.896	2.833	2.844	2.768
Fe^{2+}	1.757	3.013	2.631	2.716	3.508	2.479	3.345	2.934	2.533	2.850	Mn	0.013	0.019	0.056	0.074	0.080	0.089
Mn	0.048	0.046	0.058	0.039	0.060	0.040	0.237	0.053	0.042	0.220	Mg	0.079	0.056	0.070	0.097	0.050	0.041
Mg	1.987	1.012	1.293	1.273	1.214	2.283	1.168	1.714	2.358	1.795	$\sum M$	3.168	3.042	3.105	3.064	3.012	2.920
Ca	1.886	1.975	2.052	1.994	1.892	2.035	1.974	1.985	1.980	1.902	Ca	1.067	1.090	1.029	1.047	1.075	1.060
$\sum M$	7.000	7.000	7.000	7.000	7.000	7.000	7.000	7.000	7.000	7.000	Na	0.000	0.000	0.004	0.003	0.006	0.000
[A]Na	0.333	0.361	0.475	0.350	0.162	0.088	0.059	0.036	0.062	0.065	K	0.000	0.000	0.009	0.000	0.001	0.000
K	0.439	0.419	0.518	0.371	0.120	0.029	0.028	0.020	0.037	0.010	$\sum A$	1.067	1.090	1.041	1.050	1.082	1.060
$\sum A$	0.772	0.780	0.993	0.721	0.281	0.117	0.087	0.056	0.099	0.075	OH^-	1.000	1.000	1.000	1.000	1.000	1.000
F	0.000	0.028	0.000	0.000	0.144	0.105	0.068	0.147	0.000	0.000	Fe^{2+}/Fe^{3+}	2.243	2.561	2.496	2.597	2.664	3.339
Cl	0.281	0.426	0.371	0.328	0.008	0.004	0.006	0.007	0.008	0.004	Xilv	92.06	93.59	89.45	85.92	89.13	89.63
OH^-	1.719	1.546	1.629	1.672	1.848	1.891	1.926	1.846	1.992	1.996							
$\sum O_3$	2.000	2.000	2.000	2.000	2.000	2.000	2.000	2.000	2.000	2.000							
X_{Mg}	0.531	0.251	0.329	0.319	0.257	0.479	0.259	0.369	0.482	0.386							

(continued)

Table 6.1 (continued)

Mineral	Ttn	Ttn	Ttn	Ttn	Ttn		Adr	Adr	Adr		Cha	Cha		Mt	Mt	Mt
Sample (wt%)	2901-14-7-3	2901-14-7-4	2901-14-5-2	2901-14-6-1	2901-14-6-2		2901-11-1	2901-14-1-3	2901-14-2-1		2901-17-1-2	2901-17-1-6		2901-17-5-1	2901-14-2-1	2901-14-3-1
SiO_2	31.50	30.83	31.48	31.00	30.94		37.23	36.93	35.48		30.91	28.39	SiO_2	0.60	0.95	2.01
TiO_2	29.29	30.87	32.03	32.56	28.40		0.29	0.33	0.42		0.37	0.07	TiO_2	0.00	0.03	0.00
Al_2O_3	4.63	2.72	3.58	3.13	2.97		9.76	8.74	8.65		15.29	16.85	Al_2O_3	0.08	0.12	0.11
FeO	0.98	0.82	0.43	0.60	0.95		15.29	16.69	17.43		31.65	32.17	FeO	93.38	89.89	89.39
MnO	0.03	0.04	0.02	0.02	0.02		0.58	0.62	0.48		0.70	0.88	MnO	0.11	0.12	0.06
MgO	0.03	0.01	0.01	0.00	0.00		0.19	0.18	0.10		8.96	8.89	MgO	0.02	0.03	0.00
CaO	28.63	28.27	28.94	28.96	28.24		35.48	34.67	36.34		0.09	0.59	CaO	0.00	0.00	0.17
Na_2O	0.00	0.00	0.00	0.03	0.00		–	–	–		0.01	0.06	Cr_2O_3	0.02	0.00	0.02
K_2O	0.00	0.00	0.00	0.00	0.00		–	–	–		3.51	0.19	ZnO	0.08	0.01	0.12
F	0.90	0.44	0.88	0.49	0.85		–	–	–		–	–	V_2O_3	0.00	0.00	0.00
Cl	–	–	–	–	–		–	–	–		–	–				
H_2O	–	–	–	–	–		–	–	–		10.734	10.683				
Total	95.99	94.00	97.37	96.78	92.37		98.81	98.17	98.89		102.23	98.77	Total	94.29	91.15	91.87
	Based on 3 cations									Based on 8 cations						

(continued)

Table 6.1 (continued)

Mineral	Ttn	Ttn	Ttn	Ttn	Ttn		Adr	Adr	Adr		Cha	Cha		Mt	Mt	Mt
Sample (wt%)	2901-14-7-3	2901-14-7-4	2901-14-5-2	2901-14-6-1	2901-14-6-2		2901-11-1	2901-14-1-3	2901-14-2-1		2901-17-1-2	2901-17-1-6		2901-17-5-1	2901-14-2-1	2901-14-3-1
Si	1.043	1.048	1.036	1.025	1.069	Si	2.963	2.973	2.838	Si	3.450	3.185	Si	0.040	0.063	0.134
Ti	0.730	0.789	0.793	0.809	0.738	$Al^{[IV]}$	0.037	0.027	0.162	$Al^{[IV]}$	0.550	0.815	Ti	0.000	0.001	0.000
$Al^{[IV]}$	0.181	0.109	0.139	0.122	0.121	$\sum$	3.000	3.000	3.000	$\sum$	4.000	4.000	Al	0.006	0.009	0.008
Fe^{2+}	0.027	0.023	0.012	0.017	0.027	$Al^{[VI]}$	0.878	0.802	0.653	$Al^{[VI]}$	0.792	0.670	Fe^{3+}	3.418	3.254	3.151
Mn	0.001	0.001	0.001	0.001	0.001	Ti	0.009	0.010	0.013	Ti	0.015	0.003	Fe^{2+}	1.781	1.751	1.826
Mg	0.002	0.000	0.000	0.000	0.000	Fe^{3+}	1.016	1.121	1.163	Fe^{2+}	2.948	3.012	Mn	0.006	0.006	0.003
Ca	1.016	1.029	1.020	1.026	1.045	$\sum$	1.903	1.933	1.829	Mn	0.066	0.084	Mg	0.002	0.003	0.000
$\sum M$	3.000	3.000	3.000	2.998	3.000	Fe^{2+}	0.000	0.000	0.000	Mg	1.492	1.487	Ca	0.000	0.000	0.012
Na	0.000	0.000	0.000	0.002	0.000	Mn	0.039	0.042	0.032	X	6.686	6.745	Cr	0.001	0.000	0.001
K	0.000	0.000	0.000	0.000	0.000	Mg	0.022	0.022	0.011	$\sum M$	12.000	12.000	Zn	0.004	0.001	0.006
F	0.094	0.048	0.092	0.052	0.093	Ca	3.026	2.990	3.114	Ca	0.011	0.071	V	0.000	0.000	0.000
						$\sum$	3.087	3.055	3.158	Na	0.005	0.025	$\sum$	5.259	5.105	5.154
						Adr%	52.48	57.02	63.05	K	1.326	0.073	Fe_2O_3	68.23	64.96	62.89
						Gro%	45.39	40.79	35.38	$\sum A$	1.343	0.168	FeO	31.99	31.44	32.79
						Pyr%	2.12	2.19	1.58	OH^-	8.000	8.000	Total	101.13	97.78	98.27
										X_{Fe}	66.397	66.951				
										Cha%	65.42	65.73				
										Clc%	33.11	32.45				
										Pen%	1.47	1.83				

6.4 Thermodynamic Relations of Ilvaite-Bearing Associations

Based on petrographic analysis, ilvaite occurs in three different stages based on the irreversible replacement of preexisting minerals follow by the breakdown of ilvaite (Fig. 6.1). In Stage I, the garnet is altered to bluish-green amphibole (hastingsite), which is further altered into ilvaite. In Stage II, the ilvaite with SiO_2 breaks down into ferro-actinolite and magnetite, these product minerals can coexist in partial equilibrium, although ilvaite remains unstable through the reaction. In Stage III, the remaining ilvaite also replaced by ferro-actinolite and magnetite, but in this case calcite is also part of the secondary product. After the initial precipitation of magnetite it is further altered to pyrite which indicates that there was a source of sulfide in the fluid.

Ilvaite is stable at medium to low temperatures (<450 °C) and high fluid pressures and occurs in a wide fO_2 range because the Fe in ilvaite is not fully reduced [22]. Delgado and Soler [11] considered that the replacement relationships of ilvaite requires that ilvaites unlikely form under pressures below 2 kbar and show that this mineral can only develop from fluids with low or very low CO_2 content. However, we found that the activity of reacted minerals and H_2O in the system also affect the stability of ilvaite, and the suitable geochemical conditions can yield ilvaite under low pressues of <2 kbar. A lower P_{H_2O}/P_{total} also decrease the thermal stability range of ilvaite [22]. In addition, relatively low pH favors the formation of ilvaite [47]. However, the major temperature and oxidation state effects on ilvaite stability can be assessed by considering reactions between neutral species.

Figure 6.3 depicts the mineral stabilities calculated using thermodynamic data from Berman [4] supplemented with data for Adr, Fac, Hd, and Ilv [11]. The diagram is constructed for unit activity of mineral phases except for ilvaite (0.7), ferro-actinolite (0.8), and andradite (0.7). These estimates of activity are calculated with the ideal mixing approximation from representative mole fractions of the endmembers obtained from the microprobe analysis. The activity of H_2O was set to 1. Values slightly lower than this that are caused by mixing with CO_2 and ionic species do not significantly affect the thermodynamic model. Grunerite was not included in these calculations because it was not identified in the Galinge deposit. Stable reactions in the system were identified using winTWQ [5], then were normalized to the hematite-magnetite buffer and plotted using CHNOSZ [12]. In the model, the pressure range based on the calculation of amphibole geobarometer is 57–26 Mpa. Therefore, we set a pressure of 50 Mpa for the replacement model. In the model, although all the replacement reactions involve oxygen, and the coexisting minerals present in each of reactions serve to define oxygen fugacity, it is too little to result in detectable growth or breakdown of ilvaite. Therefore, The growth or breakdown of ilvaite either requires the presence of a much more abundant oxidising agent in place of dissolved oxygen, most likely sulphate, or is an open system accompanied by addition or removal of Fe, Ca or Si, or both.

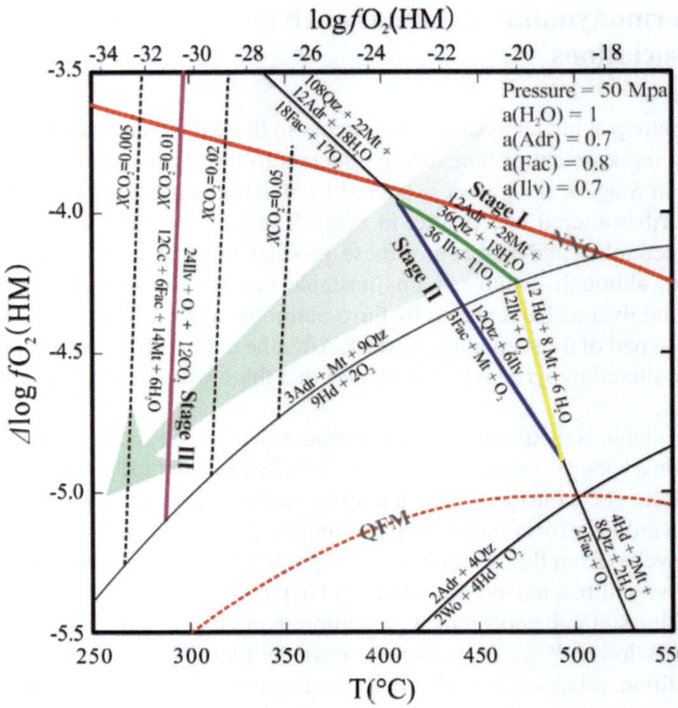

Fig. 6.3 Temperature-log fO_2 diagram for the Ca–Fe–Si–O–H–C system computed at $P = 50$ Mpa and unit activity of mineral phases except for ilvaite (0.7), andradite (0.7) and ferro-actinolite (0.8). The line for the reaction in which CO_2 is a reactant is drawn for three values of X_{CO_2}. The outer top x-axis shows the values of oxygen fugacity in the hematite-magnetite (HM) buffer; the fO_2 scale on the y-axis is normalized with respect to the HM buffer [11]. Reactions involving Ilv are shown in bold; dotted red lines correspond to the indicated mineral buffers (QFM: quartz–fayalite–magnetite; NNO: nickel–nickel oxide). The arrow indicates the postulated trajectory of fO_2 during skarn development

At high temperatures, and fO_2 values between the HM buffer and hedenbergite stability field, formation of ilvaite are occurs through the replacement of andradite and magnetite, and accompanied by the disappearance of quartz. Stage I ilvaite is formed by the replacement reaction of $12Ca_3Fe_2{}^{3+}Si_3O_{12(andradite)} + 28Fe^{2+}Fe_2{}^{3+}O_{4(magnetite)} + 36SiO_{2(quartz)} + 18H_2O \rightarrow 36CaFe^{2+}(Fe^{2+}Fe^{3+})(Si_2O_7)O(OH)_{(ilvaite)} + 11O_2$ at 400–470 °C at an $\Delta \log fO_2$(HM) range of ca. −4 to −4.2 (Fig. 6.3). This reaction defines the maximum temperature for the formation of ilvaite in Stage I. Another reaction that could possibly form ilvaite is the breakdown of hedenbergite according to $6CaFe^{2+}(Fe^{2+}Fe^{3+})(Si_2O_7)O(OH)_{(ilvaite)} + 1/2O_2 \rightarrow 6CaFe^{2+}Si_2O_{6(hedenbergite)} + 4Fe^{2+}Fe_2{}^{3+}O_{4(magnetite)} + 3H_2O$. However, we did not find any evidence for this reaction in the petrographic analyses. Therefore, the formation of ilvaite must be constrained to higher oxygen fugacities than that of the reaction of 3Adr + Mt + 9Qtz → 9Hd + $2O_2$.

In order to interpret the metasomatic relationship we have hypothesize that the associations formed by the breakdown of ilvaite are in partial equilibrium with fluid. The reaction corresponding to the breakdown of ilvaite to give ferro-actinolite and magnetite defines the low-*T* stability of ilvaite in Stage II (Fig. 6.3). In Stage II, the replaced relationship corresponds to the reaction $6CaFe^{2+}(Fe^{2+}Fe^{3+})(Si_2O_7)O(OH)_{(ilvaite)}$ + $12SiO_{2(quartz)}$ → $3Ca_2Fe_5{}^{2+}[Si_8O_{22}](OH)_{2(ferro\text{-}actinolite)}$ + $Fe^{2+}Fe_2{}^{3+}O_{4(magnetite)}$ + O_2, which takes place at 400–440 °C, and an $\Delta \log fO_2$(HM) range of −4 to −4.4 (Fig. 6.3).

Any ilvaite that remains after Stage II, is available for lower temperature of breakdown in Stage III. The breakdown of ilvaite at Stage III is tentatively related to the reaction $24CaFe^{2+}(Fe^{2+}Fe^{3+})(Si_2O_7)O(OH)_{(ilvaite)}$ + $12CO_2$ + O_2 → $14Fe^{2+}Fe_2{}^{3+}O_{4(magnetite)}$ + $12CaCO_{3(calcite)}$ + $6Ca_2Fe_5{}^{2+}[Si_8O_{22}](OH)_{2(ferro\text{-}actinolite)}$ + $6H_2O$ (Fig. 6.3) at ca. 270–350 °C under relatively low fO_2 (up to −4.5 log units below HM). However, unlike the reaction in the preceding stages, this reaction is intensively affected by the activity of CO_2 of the fluid (Fig. 6.3). The temperature where this occurs will increase from 270 to 350 °C when the $X(CO_2)$ values are elevated from 0.005 to 0.05 (Fig. 6.3), and the fO_2 is up to −5.2 log units below HM.

6.5 Comparison with Other Ilvaite-Bearing Skarn Deposit Types

Ilvaite is a common accessory mineral in various skarn- and vein-type deposits occurring mostly in Fe and Zn–Pb skarns, and rarely in Au and Sn–W skarns (Table 6.2). However, it rarely occurs in Cu skarn deposits, and it was only reported in a Cu–Au IOCG deposit and associated with skarns [41]. Ilvaite occurs typically as disseminated crystals, veins or aggregates in carbonates (mainly limestone) [34] and to a lesser degree in intermediate-mafic rocks [3, 45]. The formation of ilvaite has been attributed to either pneumatolytic alteration or hydrothermal replacement [21, 37]. In most cases, ilvaite has been regarded as a retrograde mineral. However, ilvaite from the Nikolaevsky Pb–Zn–Ag skarn deposit is exclusively formed in the prograde stage, and mainly occurs in a garnet–olivine–wollastonite–ilvaite–hedenbergite assemblage [39]. Graser and Markl [21] suggest that the formation conditions of ilvaite in the Ilimaussaq complex, bearing the epidote + albite + andradite ± prehnite assemblage, are ca. 500 °C and at a range of oxygen fugacities that includes the HM buffer. The model presented here also indicates that ilvaite may have crystallized in a wide temperature range (ca. 400–500 °C), but at an oxygen fugacity below the HM buffer and above NNO buffer (Fig. 6.3) in stage I.

The variable composition of ilvaite makes it potentially useful for indicating the geochemical environment of formation [16]. The chemistry of ilvaite is deeply influenced by the host rocks and magmatic-hydrothermal fluids, together with mineral

Table 6.2 Selected ilvaite-bearing skarn and vein type deposits, and a IOCG deposit

Deposit	Type	Location	Ore minerals	Mineral associations	References
Mt. Lindsay	Skarn	Tasmania, Australia	Sn–W–F–Be	Sn-Amp + Mt + Po + Sn-Sph + Sn-Ilv ± Ccp ± Py ± Apy ± Fl	[25]
Sarrahera	Skarn	Central Pyrenees, Spain	Au–As	Adr + Hd + Fac + Ilv + Qtz + Cc + Sch ± Stil ± Chl ± Mt ± Apy ± Po ± Ccp ± Py ± Au ± Bi	[11]
Fortitude	Skarn	Nevada, America	Au	Hd + Fac + Cc + Qtz + Ilv + Mt + Po	[16]
Arties-Escunhau	Skarn	Central Pyrenees, Spain	Au	Wol + Hd + Qtz + Cc + Fac + Ilv + Mt + Epi + Ax ± Stil ± Chl ± Po ± Ccp ± Py	[11]
Chillagoe district	Skarn	north-east Queensland	Au	Wol + Adr + Ilv + Px + Cc + Qtz ± Act ± Ser ± Chl ± Ccp ± Hem	[27]
Cavnic	Skarn	East Carpathians, Romania	Au	Sph + Gn + Mt + Hem + Py + Ccp + Au + Bi	[23]
Igarapé Bahia	IOCG	Carajás, Brazil	Cu–Au	Cc + Ccp + Mt + Ilv + Qtz + Fl ± Tour Mt + Stil + Chl + Amp + Tour + Qtz + Cc + Ser ± Sch ± Bi ± Apa ± Mnz ± Fl ± Epi ± Ilv	[41]
Ilímaussaq	Skarn	South Greenland	U–Zr–Nb–Ta–REE	Epi + Ab + Gt + Ilv ± Preh	[21]
Ossikovo	Skarn	Madan ore district, Bulgaria	Pb–Zn–(Mn)	Px +Ilv ± Rdn	[7]
Mogilata	Skarn	Madan ore district, Bulgaria	Pb–Zn–(Mn)	Px + Ilv ± Rdn	[7]
Govedarnika	Skarn	Laki district, Bulgaria	Pb–Zn–(Mn)	Px + Rdn + Ilv	[7]
Groundhog	Skarn	New Mexico	Pb–Zn	Amp + Ilv + Qtz + Cc + Clay + Hem + Sph + Chl + Py + Thd + Gn + Ccp.	[31]
Broken Hill	Skarn	Australia	Pb–Zn	Rdn + Grs + Cc + Sph + Gn + Ccp + Ilm	[38]
Ban Ban	Skarn	Australia	Pb–Zn	Sph + Gt + Cc + Qtz + Py + Mt + Tre	[1]

(continued)

Table 6.2 (continued)

Deposit	Type	Location	Ore minerals	Mineral associations	References
Perda Niedda	Skarn	Sardinia, Italy	Pb–Zn	–	[6]
Oridda	Skarn	Sardinia, Italy	Pb–Zn	–	[10]
Skjerpemyr	Skarn	Oslo Graben, Norway	Pb–Zn	Qtz + Hd + Adr + Ilv + Sph + Gn	[26]
Konnerud	Skarn	Drammen, Norway	Pb–Zn–Cu	Gt + Ilv + Qtz + Cc + Sph + Gn + Ccp	[26]
Motukokako	Skarn	Cape Brett, New Zealand	Zn–Pb	Ilv + Bat	[8]
Gualilan	Skarn	western Argentina	Zn–Pb–(Ag–Au)	Px + Gt + Py + Sph + Qtz + Ilv + Cc	[28]
Makeng	Skarn	Fujian, China	Fe–Pb–Zn	Jo + Hd + Rdn + Adr + Ilv + Mt + Sph + Gn	[47]
Yangshan	Skarn	Fujian, China	Fe–Pb–Zn	Jo + Hd + Rdn + Adr + Ilv + Mt + Sph + Gn	[47]
Elstrom-Laronningen	Vein	Oslo Graben, Norway	Fe–Zn	Qtz + Mt + Hd + Fl + Ilv + Cc +Epi + Py + Sph ± Po ± Ccp	[26]
Galinge	Skarn	Qinghai, China	Fe	Adr + Hst + Ilv + Fac + Sph + Chl + Mt Qtz + Ilv + Fac + Mt Cc + Ilv + Fac + Py	This study
Black Perry Mountain	Skarn	Australia	Fe	Hd + Adr + Ilv + Qtz + Cc + Mt + Act + Epi + Bat + Stil	[19, 20]
Iron Hill	Skarn	Western British Columbia, Canada	Fe	Px + Ilv + Amp + Mt + Qtz + Epi	[30]
Cihai	Skarn	Xinjiang, China	Fe	Hd + Px + Mt + Ilv + Py	[42]
Cinco Villas	Skarn	Western Pyrenees, Spain	Fe	Ilv + Hd + Gt + Epi + Qtz + Cc + Act + idocrase + Mt	[36]
Seriphos	Skarn	Greece	Fe	–	[10, 40]

(continued)

Table 6.2 (continued)

Deposit	Type	Location	Ore minerals	Mineral associations	References
San Leon	Skarn	Sardini, Italy	Fe	Hd + Gt + Epi + Qtz + Ilv + Fl + Cc ± Sch ± Bat + Mt	[44]
Forsum	Vein	Oslo Graben, Norway	Fe	Qtz + Mt ± Fl ± Hd ± Adr ± Ilv ± Cc	[26]
Capo Calamita	Skarn	East Elba, Italy	Fe	Fst + Adr + Mt + Ilv + Hem + Qtz Po + Fac + Ilv + Mt + Epi + Cc + Qtz	[13]
Rio Marina	Skarn	East Elba, Italy	Fe	Hd + Ilv + Epi + Qtz + Cc + Po + Py	[13]
Elba	Skarn	Italy	Fe	–	[40]
Tyrnyauz	Skarn	North Caucasus	Mo–W–F	Fl + Adr + Ves + Wol + Epi + Act + Xo + Ilv + Gn + Cc + Qtz + Zeo + Apt	[24]

Note Mineral abbreviations listed in Table 6.1. Stil: stilpnomelane; Rdn: rhodonite; Bat: babingtonite; Sch: scheelite; Ves: vesuvianite; Xo: xonotlite; Ser: sericite; Ab: albite; Fst: ferrosilite; Zeo: zeolite; Apt: apophyllite; Tre: tremolite; Mnz: monazite; Thd: tetrahedrite composition of the prograde stage skarn [42]. Therefore, it is necessary to compare the studied ilvaite with that of the other skarn deposits. The ilvaite compositions from various skarn deposits are plotted in Fig. 6.4, and associated references for all presented data are listed in Table 6.2. The ilvaite composition diagram for various important skarn deposits in the world shows that the data points fall into distinctive fields (Fig. 6.4).

Ilvaite from Fe skarn deposits is compositionally different from the ilvaite from Au, Fe–Pb–Zn, Pb–Zn and Sn skarn deposits. It has higher Mg than those from Au deposits, and lower Mn than the Pb–Zn– and Fe–Pb–Zn skarns (Fig. 6.4a). Ilvaite from Sn skarns has the lowest Fe^{3+} content of all skarn types, with total Fe contents intermediate between that of Zn and Au–Fe skarn deposits (Fig. 6.4b). We suppose the varying Fe^{3+}/Fe^{2+} ratios from Sn to Au skarn deposits is probably due to an increase in the fO_2 of fluids. Most ilvaite from Fe-skarn deposits occurs in the skarn replacement of limestone, and grows simultaneously with amphibole and magnetite, as a replacement of Fe-rich pyroxene. Generally, ilvaite compositions in Fe skarn deposits are more Fe-rich than those of Zn–Pb deposits, which are Mn-rich (Fig. 6.4a). The varying Fe–Mn content of ilvaite from Fe to Pb–Zn skarns may be as a consequence of a decrease in temperature (Fig. 6.4a). In addition, most MgO in ilvaite was suggested to be derived from the alteration of the prograde-stage Mg-bearing silicates [42], which usually takes place at ca. 450 °C under low pressure [44]. This temperature is near the stability limit for ilvaite (Fig. 6.3).

Mn-rich members of the ilvaite-manganilvaite solid solution occur in the distal parts of Pb–Zn skarn deposits [7, 47], and appear as: (1) radiating crystals or nets among the interstices of pyroxene aggregates, (2) aggregates or nets in wollastonite along the skarn–marble front, and (3) along fine quartz-calcite veins enveloped by pyroxene or rhodonite aggregates. Generally, the retrograde alteration started with the occurrence of ilvaite. In some cases, the clinopyroxene and andradite skarn is crosscut by calcite-quartz-ilvaite veins [27]. Logan [28] reported quartz fluid inclusion data from the ilvaite–quartz–calcite retrograde assemblage in the Gualilan Zn–Pb–(Ag–Au) skarn deposit, in which temperatures and salinities range from 237 to 251 °C and 1.4–2 wt% NaCl eq., respectively. Based on these observations, ilvaite has a broad stability field from 440 to 230 °C. In addition, Bonev et al. [7] argued that the presence of manganilvaite in some deposits reflects the evolution of the hydrothermal environment from a reduced to a more oxidized and hydrated conditions. However, a trend toward relatively reducing conditions at lower temperature is consistent with the petrographic and compositional observations in this study (Fig. 6.3).

Very rarely, ilvaite also occurs in hydrothermal veins as an accessory mineral [26]. For instance, at the Fossum Fe ore field (in the Oslo Rift near Skien), ilvaite occurs in hydrothermal veins (quartz + magnetite ± fluorite ± hedenbergite ± andradite ± ilvaite ± calcite) that intruded the Proterozoic granitic gneisses [26]. The ilvaite from the hydrothermal veins contains generally much higher Mn than those from massive Au and Sn skarn deposits, also similar to some ilvaite components from Fe skarn [23].

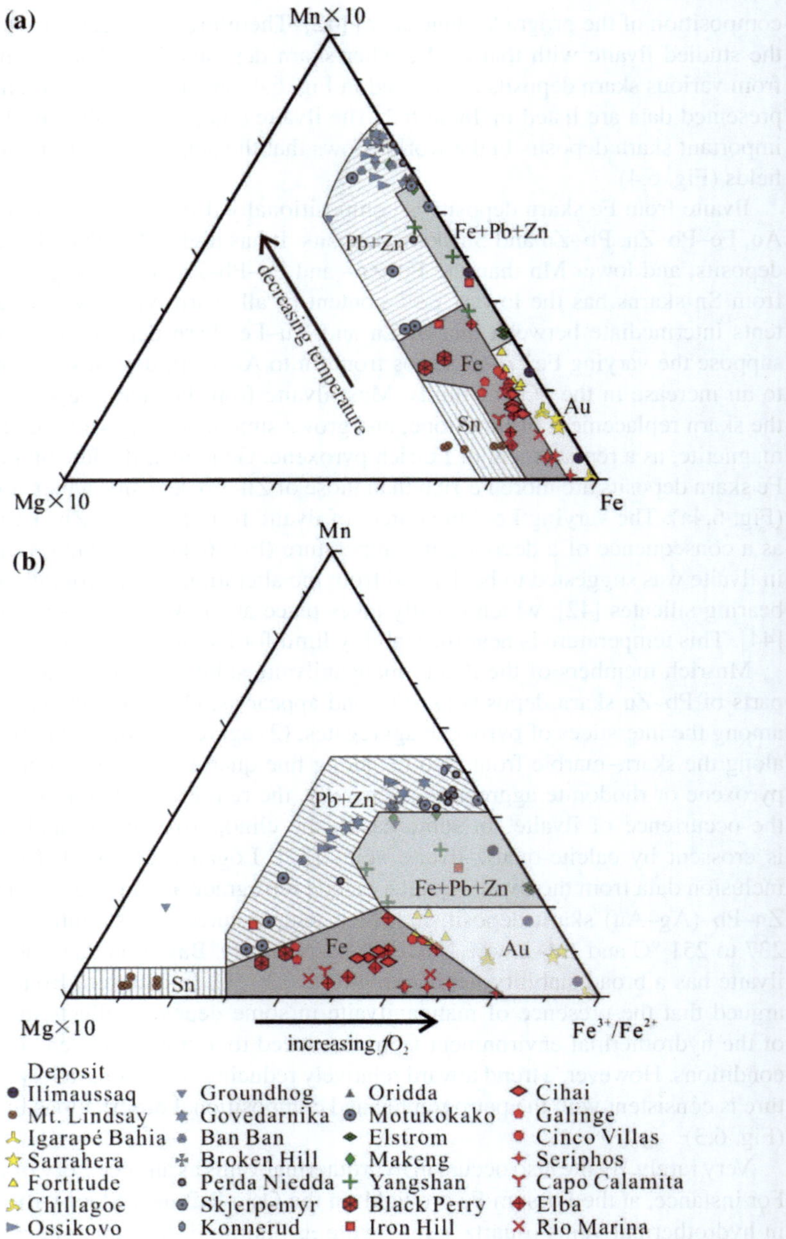

Fig. 6.4 Ternary plot of microprobe analyses of ilvaite from various skarn deposits. **a** The Fe_{tot}-10 × Mn-10 × Mg ternary plot. **b** The Fe^{2+}/Fe^{3+}-Mn-10 × Mg ternary plot. Different shadow fields define different skarn deposits

Comparisons of electron microprobe analyses demonstrate that the ilvaite from Au-skarn deposits have higher Fe and lower Mn than Pb–Zn skarns, which is compositionally similar to ilvaite from Fe-skarn deposits [16] (Fig. 6.4). However, the Au-skarn ilvaite contains lower MgO than its Fe-skarn counterpart (Fig. 6.4). Ilvaite in Au-skarn deposits generally occurs in both breccia groundmass and veins as subhedral crystals [16]. Observed textures in the Fortitude Au skarn suggest that ilvaite formed during the breakdown of iron-rich pyroxene in magnetite-rich exoskarn, similar to the ferroactinolite, or replaced calcite grains at the marble front [16], probably suggesting an origin from the reaction: $Fe^{2+}Fe_2{}^{3+}O_{4(magnetite)} + CaCO_{3(calcite)} + 2SiO_{2(quartz)} + 1/2H_2O = CaFe^{2+}(Fe^{2+}Fe^{3+})(Si_2O_7)O(OH)_{(ilvaite)} + CO_2 + 1/4O_2$. The ilvaite from the Cavnic deposit was found enveloping native gold, and native gold forming a little veins on the fine crack in the ilvaite, which reflects that the ilvaite crystallization took place at ca 400 °C before gold formation [23]. The presence of coeval rhodonite further constrains the temperature to be higher than 400 °C, because synthesis experiments show rhodonite to be stable in the presence of water only at ca. 425–450 °C at 3 kbar [29, 35].

In Arties-Escunhau Au skarn deposits, Ilvaite is formed before ferro-actinolite, according to the reaction $10CaFe^{2+}(Fe^{2+}Fe^{3+})(Si_2O_7)O(OH)_{(ilvaite)} + 4CO_2 + 4SiO_{2(quartz)} = 4CaCO_{3(calcite)} + 3Ca_2Fe_5{}^{2+}[Si_8O_{22}](OH)_{2(ferro\text{-}actinolite)} + 5Fe^{2+}Fe_2{}^{3+}O_{4(magnetite)} + 2H_2O$ [11]. However, Dünkel [13] suggested that ilvaite may have formed by replacing ferro-actinolite in the Capo Calamita skarn deposit. This reaction, which corresponds to Stage III in this study, allows the formation of ilvaite to proceed at temperatures as low as 270–350 °C, depending on the X_{CO_2} (Fig. 6.3).

6.6 Conclusion

The Galinge skarn followed a decreasing temperature path that comprises a prograde- and retrograde alteration stages, and quartz-calcite-sulfide stage. The prograde stage shows a high temperature (>500 °C) and a more oxidizing environment (above the NNO oxygen buffer). Previous studies concerning the stability of ilvaite allow us to determine the physicochemical conditions of the retrograde alteration. At Galinge, ilvaite is mainly concentrated in the area between the Ore Domain II and IV. In terms of the formation and breakdown of ilvaite showing intensive replacement textures, it occurs a wide temperature range of 270–500 °C and oxygen fugacity that is approximately 3.5–4.0 log units below hematite-magnetite. Some low-temperature (<250 °C) minerals may have formed after the retrograde alteration. The garnet + ilvaite + magnetite association was formed at a relatively high temperature range of ca. 400–470 °C, and under an fO_2 of approximately −4 to −4.2 log units below hematite-magnetite, corresponding to the Stage I conditions. The ferro-actinolite + ilvaite + quartz + magnetite association was formed in Stage II at a medium temperature (400–440 °C) and under an fO_2 of ca. −4 to −4.4 log units blow HM. Moreover, the ilvaite stability field extends to more reducing conditions (up to −5.2

log units blow HM), at a lower temperature range (ca. 275–375 °C). In this stage, the activity of CO_2 significantly affects the stability of ilvaite.

Although there are some similarities in minerals and hydrothermal fluid conditions of various skarn deposit types, the major differences in the compositions of ilvaite could arise from subtle changes of fO_2, temperature, fluid components and prograde silicates composition.

References

1. Ashley PM (1980) Geology of the Ban Ban zinc deposit, a sulfide-bearing skarn, southeast Queensland, Australia. Econ Geol 75(1):15–29
2. Bartholomé P, Dimanche F (1968) On the paragenesis of ilvaite in Italian skarns. Publication/Laboratoires de geologie de la Faculte des sciences appliquees. Universite de Liege, p 51
3. Barton M, van Bergen MJ (1984) Secondary ilvaite in a dolerite dyke from Rogaland, SW Norway. Mineral Mag 48:449–456
4. Berman RG (1988) Internally-consistent thermodynamic data for minerals in the system Na_2O-K_2O-CaO-MgO-FeO-Fe_2O_3-Al_2O_3-SiO_2-TiO_2-H_2O-CO_2. J Petrol 29(2):445–522
5. Berman RG (2007) WinTWQ (version 2.3): a software package for performing internally-consistent thermobarometric calculations. Geol Surv Can Open File 5462:41
6. Bonazzi P, Bindi L (2002) Structural properties and heat-induced oxidation-dehydrogenation of manganoan ilvaite from Perda Niedda mine, Sardinia, Italy. Am Mineral 87(7):845–852
7. Bonev IK, Vassileva RD, Zotov N, Kouzmanov K (2005) Manganilvaite, CaFe2 + Fe3 + (Mn, Fe2 +)(Si2O7)O(OH), a new mineral of the ilvaite group from Pb-Zn skarn deposits in the Rhodope Moutntains, Bulgaria. Can Mineral 43(3):1027–1042
8. Brathwaite RL, Isaac MJ, Challis GA, Brook FJ (1990) Tertiary limestone and Zn-Pb mineralised skarn at Motukokako, Cape Brett, northern New Zealand. J R Soc N Z 20(4):427–438
9. Burt DM (1971) Multisystems analysis of the relative stabilities of babingtonite and ilvaite. Carnegie Inst Wash Year Book 70:189–197
10. Carrozzini B (1994) Crystal structure refinements of ilvaite: new relationships between chemical composition and crystallographic parameters. Eur J Mineral 6(4):465–479
11. Delgado J, Soler A (2010) Ilvaite stability in skarns from the northern contact of the Maladeta batholith, Central Pyrenees (Spain). Eur J Mineral 22(3):363–380
12. Dick JM (2008) Calculation of the relative metastabilities of proteins using the CHNOSZ software package. Geochem Trans 9(1):1–17
13. Dünkel I (2002) The genesis of East Elba iron ore deposits and their interrelation with Messinian tectonics. Universität Tübingen
14. Einaudi MT, Burt DM (1982) A special issue devoted to skarn deposits (Introduction-terminology, classification, and composition of skarn deposits). Econ Geol 77(4):745–754
15. Einaudi MT, Meinert LD, Newberry RJ (1981) Skarn deposits. Econ Geol 75:317–391
16. Franchini MB, Meinert LD, Vallés JM (2002) First occurrence of ilvaite in a gold skarn deposit. Econ Geol 97(5):1119–1126
17. Ghose S, Sen Gupta PK, Schlemper EO (1985) Electron ordering in ilvaite, a mixed-valence iron silicate: crystal structure refinement at 138 K. Am Miner 70:1248–1252
18. Ghose S, Tsukimura K, Hatch DM (1989) Phase transitions in ilvaite, a mixed-valence iron silicate. Phys Chem Miner 16(5):483–496
19. Gole MJ (1972) Iron Calc-silicate Rocks at Black Perry Mountain, Talbingo, Southern New South Wales. Macquarie University, Sydney
20. Gole MJ (1981) Ca-Fe-Si skarns containing babingtonite: first known occurrence in Australia. Can Mineral 19:269–277

21. Graser G, Markl G (2008) Ca-rich ilvaite–epidote–hydrogarnet endoskarns: a record of late-magmatic fluid influx into the persodic Ilímaussaq Complex, South Greenland. J Petrol 49(2):239–265
22. Gustafson WI (1974) The stability of andradite, hedenbergite, and related minerals in the system Ca–Fe–Si–O–H. J Petrol 15(3):455–496
23. Hirtopanu P, Andersen JC, Hartopanu I, Udubasa SS (2012) Ilvaite from the Cavnic deposit, Romania. Rom J Engl Stud 62–65
24. Kulikov IV, Boyarskaya RV (1989) Hydrosaline melts in inclusions in fluorite from the Tyrnyauz Deposit. Int Geol Rev 31(10):1039–1054
25. Kwak TA (1983) The geology and geochemistry of the zoned, Sn-W-F-Be skarns at Mt. Lindsay, Tasmania, Australia. Econ Geol 78(7):1440–1465
26. Larsen AO, Dahlgren S (2002) Ilvaite from the Oslo Graben, Norway. Neues Jahrbuch für Mineralogie-Monatshefte 2002(4):169–181
27. Lehrmann B (2012) Polymetallic mineralisation in the Chillagoe district of north-east Queensland: insights into base metal rich intrusion-related gold systems. James Cook University
28. Logan MAV (2000) Mineralogy and geochemistry of the Gualilán skarn deposit in the Precordillera of western Argentina. Ore Geol Rev 17(1):113–138
29. Maresch WV, Mottana A (1976) The pyroxmangite-rhodonite transformation for the $MnSiO_3$ composition. Contrib Miner Petrol 55(1):69–79
30. Meinert LD (1984) Mineralogy and petrology of iron skarns in western British Columbia, Canada. Econ Geol 79(5):869–882
31. Meinert LD (1987) Skarn zonation and fluid evolution in the Groundhog mine, Central mining district, New Mexico. Econ Geol 82(3):523–545
32. Meinert LD (1992) Skarns and skarn deposits. Geosci Can 19(4):145–162
33. Meinert LD (1993) Igneous petrogenesis and skarn deposits. Miner Depos Model 40:569–583
34. Misra KC (2000) Understanding mineral deposits. In: Understanding mineral deposits. Springer, pp 414–449
35. Momoi H (1974) Hydrothermal crystallization of $MnSiO_3$ polymorphs. Mineral J 7(4):359–373
36. Pesquera A, Velasco F (1986) An occurrence of ilvaite layers in the Cinco Villas metasomatic rocks, Western Pyrenees (Spain). Mineral Mag 50:653–656
37. Petersen OV, Micheelsen HI, Leonardsen ES (1995) Bavenite, $Ca_4Be_3Al[Si_9O_{25}(OH)_3]$, from the Ilímaussaq alkaline complex, South Greenland. Neues Jahrbuch Für Mineralogie, Monatsheft
38. Plimer IR, Ashley PM (1978) Manganoan ilvaite from Broken Hill, NSW and Ban Ban, Queensland, Australia. Mineral Mag 42(321):85–88
39. Rogulina LI, Sveshnikova OL (2008) The Nikolaevsky base-metal skarn deposit, Primorye, Russia. Geol Ore Depos 50(1):60–74
40. Schiener A (1933) Lievrit von Seriphos. Z Für Krist-Cryst Mater 85(1):89–118
41. Tallarico FHB (2002) Occurrence of ilvaite in the Igarapé Bahia Cu-Au deposit, Carajás Province, Brazil. Rev Bras Geociências 32(1):149–152
42. Tang PZ, Wang YW, Wang JB, Long LL, Zhang HQ, Liao Z (2011) Finding and significance of ilvaite in the Cihai Iron Deposit, Xinjiang Autonomic Region, China. Acta Mineral Sin 1:003
43. Vassileva RD, Bonev IK, Zotov N (2001) High-Mn ilvaites from the skarn Pb-Zn deposits in the Central Rhodopes. In: Mineral deposits at the beginning of the 21st century. Balkema Publ., Lisse, pp 925–928
44. Verkaeren J, Bartholome P (1979) Petrology of the San Leone magnetite skarn deposit (SW Sardinia). Econ Geol 74(1):53–66
45. Wang YS (1994) Analysis on special case of ilvaite enrichment in a certain iron deposit. Qinghai Geol 2:19–20
46. Yu M, Feng CY, Zhu YF, Mao JW, Zhao YM, Li DX (2016) Multistage amphiboles from the Galinge iron skarn deposit in Qiman Tagh, western China: evidence of igneous rocks replacement. Mineral Petrol 1–17
47. Zhao YM, Tan HJ, Xu ZN, Yuan RG, Bi C, Zheng RL, Li DX, Sun JH (1983) Makeng type calcic skarn iron deposit in the Southwest of Fujian province. Bulletin of the institute of mineral deposit Chinese academy of geological sciences (Album)

Chapter 7
Fluid Evolution and Stable Isotope Characters

Fresh and altered granodiorite samples which are in contact with the ore body were used for fluid inclusion thermometry. The original and secondary quartzes in samples were used for testing. Different minerals reflect different primary environment, the original quartz reflect the property of the early magmatic fluid, the secondary quartzes reflect the interaction between the residual fluid and plutons, and the inclusions in the skarn minerals determine the lowest crystal temperature.

In skarn mineral phases, the composition is consistently correlated with fluid inclusion salinity and temperature. Fluid inclusions are particularly useful in documenting the temporal and spatial evolution of skarn-forming fluids, and provide evidence for the temperature and salinity shifts between skarn- and retrograde alteration stage. For example, fluid inclusions of prograde skarn minerals have homogenization temperatures above 300 °C, and salinities up to 50 wt% NaCl equivalent; whereas, inclusions in the retrograde stage have homogenization temperatures up to 250 °C, and salinities up to 25 wt% NaCl equivalent [8]. Early hydrothermal fluids carrying large amount of dissolved salts exsolved from magma, and then a low-temperature and low-saline fluid was yielded probably through mixing with meteoric water [9].

7.1 Classification of Fluid Inclusions

The Galinge calc-alkaline granodiorite was closely related to the skarn mineralization. The common occurrence of halite-bearing fluid inclusions in anhedral quartz from the Galinge granodiorite suggests that high-saline and high-temperature hydrosaline fluids in the prograde stage were exsolved from the semi-solidified granodioritic

M. Yu, *Metallogenic Mechanism of the Galinge Polymetallic Iron Skarn Deposit, Qiman Tagh Mountains, Qinghai Province*, Springer Theses,
https://doi.org/10.1007/978-981-10-7907-8_7

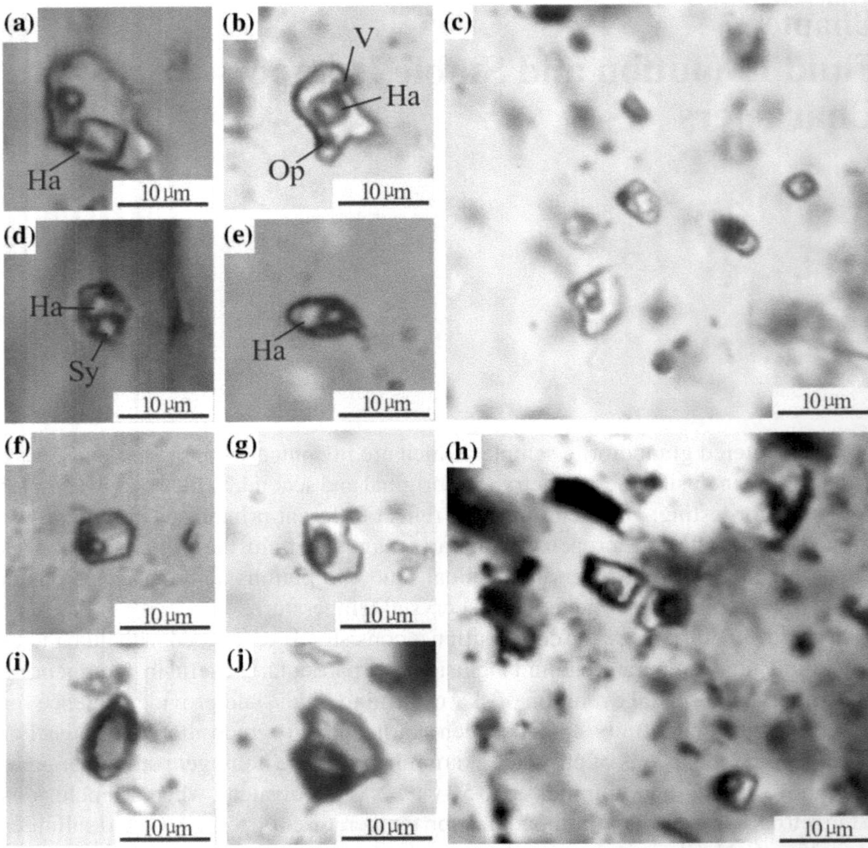

Fig. 7.1 Fluid inclusion (FI) photomicrographs of the Galinge granodiorite at room temperature under transmitted light. **a** 2404-34-22 LHa-type FIs containing ~20 vol% spherical bubble, the cubic transparent daughter mineral is halite. **b** 2404-30-2 LHa-type FIs containing ~15 vol% ellipsoidal bubble, a cubic transparent daughter mineral (halite) and a triangular opaque daughter mineral (possibly chalcopyrite). **c** 2404-30-1V-type, L-type and LHa-type FIs coexisting in the same plane/cluster. **d** 2404-34-17 LHaSy-type FIs containing ~10 vol% bubble and two cubic transparent daughter minerals (halite and sylvite). **e** 2404-34-24 LHa type FIs containing ~50 vol% spherical bubble and cubic transparent halite. **f** 26006-20-5 negative crystal shape inclusion containing ~20 vol% spherical bubble. **g** 2404-30-18 L-type FIs containing ~50 vol% ellipsoidal bubble. **h** 26006-20-1 irregular inclusions in the altered granodiorite. **i** 2404-30-14V-type vapor-rich FIs. **j** 2404-30-24 L-type FIs containing ~50 vol% ellipsoidal bubble

magma at near solidus conditions. Inclusions in anhedral quartz from the Galinge granodiorite intrusions are classified into four types (Fig. 7.1), i.e., V (vapour phase), L (liquid phase), LHa (vapour phase + liquid phase + halite) and LHaSy (vapour phase + liquid phase + halite + sylvite) type (14, 16, 17).

7.2 Fluid Inclusion Microthermometry

The Galinge granodiorite fluid inclusion microthermometry data are listed in Table 7.1 and Fig. 7.2. The salinity, density and minimum trapping pressure were calculated according to NaCl-H_2O system [3, 11]. Homogeneous temperature (Th) of inclusions in 2404-34 sample was divided into three intervals (Fig. 7.2a): 240–360 °C (peak value of 280 °C); 380–460 °C (peak value of 400 °C); 500–560 °C (peak value of 540 °C), salinity was divided into three intervals (Fig. 7.2b): 6–24% NaCl equiv. (peak value of 6% NaCl equiv.); 36–48% NaCl equiv. (peak value of 36% NaCl equiv.); 60–72% NaCl equiv. (peak value of 66% NaCl equiv.). Homogeneous temperature of inclusions in 2404-30 sample ranges from 240 to 340 °C (peak value of 300 °C) (Fig. 7.2a), salinity was divided into two intervals (Fig. 7.2b): 2–12% NaCl equiv. and 30–42% NaCl equiv. It is similar with 2404-30 that inclusions in 2404-36 sample had a significantly lower Th peak value of 280 °C (ranging from 280 to 340 °C) (Fig. 7.2a), a slightly lower salinity of 12–36% NaCl equiv. (Fig. 7.2b). L inclusions in 26006-20 sample had a Th ranging from 280 to 340 °C (peak value of 280 °C), and salinity of 2–6% NaCl equiv.

7.3 Processes Contributing to the Skarn Mineralization

On the basis of microthermometric analysis, we propose a P-T-X path for ascending magmatic-hydrothermal fluids which led to the formation of hydrothermal metamorphic skarn deposit (Fig. 7.3). In the skarn stage, ascending SiO_2-bearing NaCl saturated magmatic fluids separated to a hypersaline liquid and a vapor phase at 500–550 °C, 300–400 bar and 1–1.5 km paleodepth, at higher lithostatic pressure compared to hydrostatic pressure. Brine and vapor continued to ascend, and replaced the wall rock to form unmineralised skarns. As they cooled to ~400 °C in the retrograde stage, due to great consumption of SiO_2, the fluid predominantly consisted of NaCl saturated liquid. The residual fluid reacted with the earlier skarn to form garnet, pyroxene etc. with minor ore minerals. This was followed by a decline in temperature to between 250 and 380 °C at a hydrostatic pressure of 150–50 bar, corresponding a paleodepth of 0.5–1.5 km. During this stage, the mixing of magmatic water with lower-salinity meteoric waters retrograded the skarn resulting in enriched mineralization. The overall hydrothermal cooling event took place during the geological background reversing, like mentioned above, the post-orogenic stage consists of a transition of compressional environment to extensional environment, while the magmatic fluids spread upward and outward on a larger scale.

Table 7.1 Summary of microthermometryic data of fluid inclusions from the Galinge granodiorites

Sample	Mineral	Type	n	Tm_{halite} (°C)	ρ (g/cm^3)	p_t (Mpa)	wt% NaCl	Tm_{ice} (°C)	$Th_{L\text{-}V}$ (°C)
2404-30	qz	L	16		0.724–0.877	3.52–9.09	2.24–9.21	−6.0 to −1.3	246–323
		LHa	3	245–317	1.048–1.100	2.95–9.25	34.37–37.19		245–329
		V	2		0.8	8.68	9.73	−6.4	285–318
2404-34	qz	L	14		0.77–0.92	6.04–10.95	9.08–20.07	−16.8 to −5.9	242–344
		LHa	18	272–558	1.075–1.220	4.08–39.8	36.12–67.9		272–558
2404-36	qz	L	10		0.863–0.958	4.27–4.63	12.05–21.89	−19.3 to −8.3	244–352
		LHa	2	158–164	1.066–1.068	2.98–3.0	29.98–30.22		240–244
26006-20	qz	L	12		0.685–0.77	6.74–13.03	4.03–4.65	−2.8 to −1.0	281–335

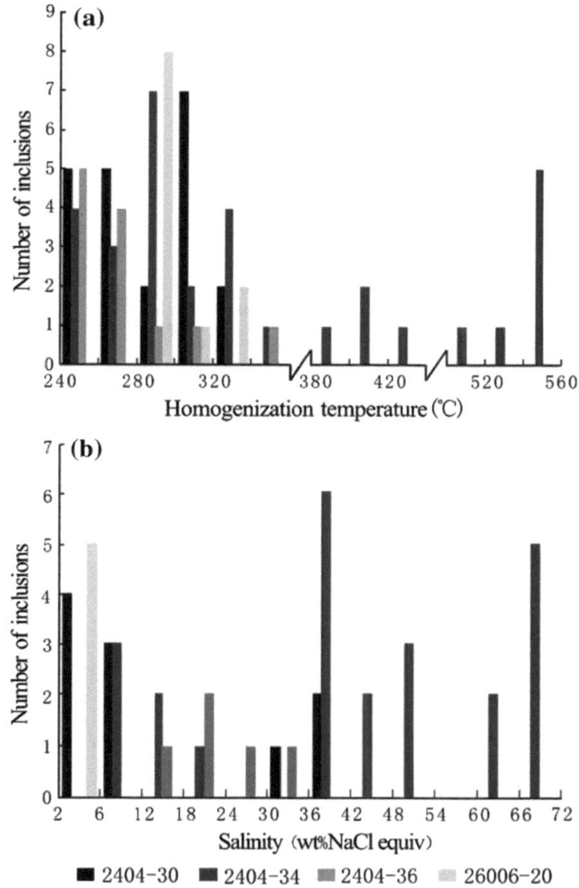

Fig. 7.2 Histogram showing homogenization temperatures and salinities of fluid-inclusion of all samples from Galinge granodiorite. **a** Homogenization temperatures. **b** Total Salinities (expressed as wt% NaCl equiv.)

7.4 Sulfur Isotope Studies

Sulfur isotope compositions were obtained on magnetite, pyrrhotite, pyrite, arsenopyrite, chalcopyrite, galena and sphalerite. The $\delta^{34}S$ values average 1.47‰ in a narrow range between −4.7 and 12.23‰ (Table 7.2). This range suggests that sulfur is dominantly from deep magmatic source or was leached from igneous rocks. Magnetite show a comparatively large range between −2.35 and 12.23‰ [6], which was probably caused by the contamination of crustal sources. The results of pyrite ranges between 1.00 and 4.73‰ [15]. Pyrrhotite and magnetite associated with hedenbergite and hastingsite were hosted in extensively altered mafic volcanic rocks, in

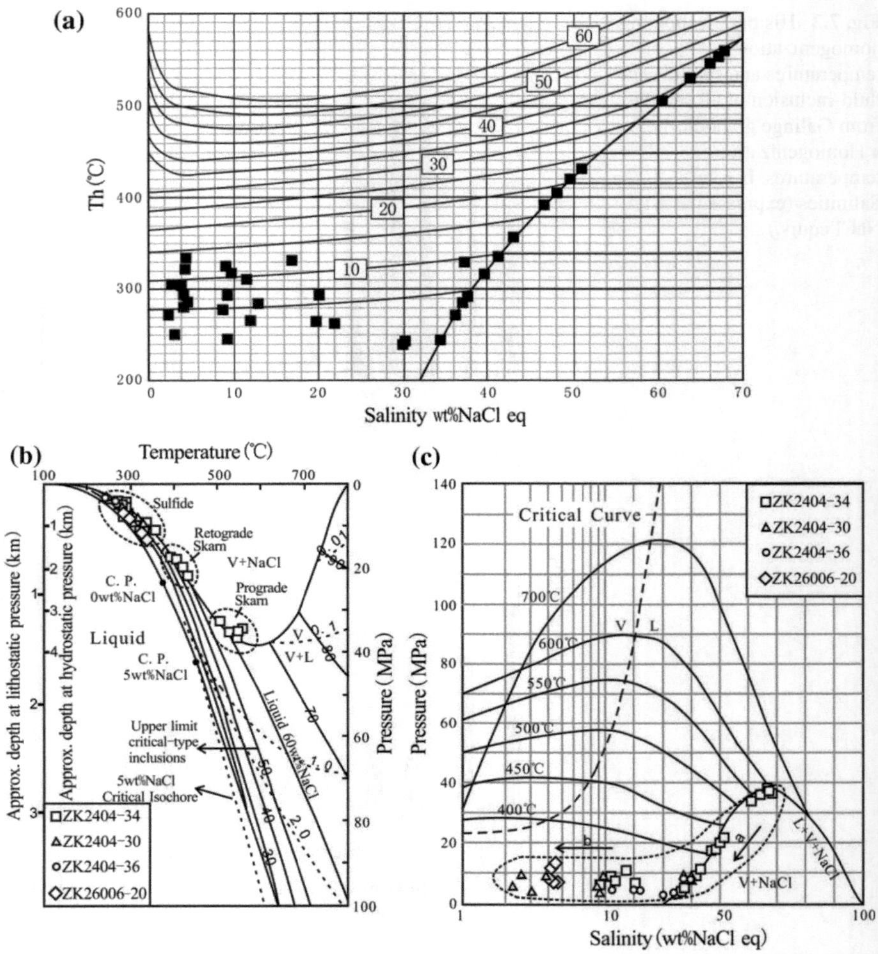

Fig. 7.3 p-t-x phase diagram of the fluid inclusions with halite daughter minerals in NaCl–H_2O system.**a** Summary of microthermometric data, showing salinity in wt% NaCl equiv. versus temperature of final homogenization; salinities determined from halite dissolution and freezing points [3]; fine background lines are isobars (Mpa) of the limiting two-phase and the halite saturation limit of the NaCl–H_2O system. **b** Estimated pressure -temperature diagram and the arrow showing the evolution path of different fluid inclusion types [1, 4, 5]; **c** Estimated pressure-salinity diagram [2, 12]

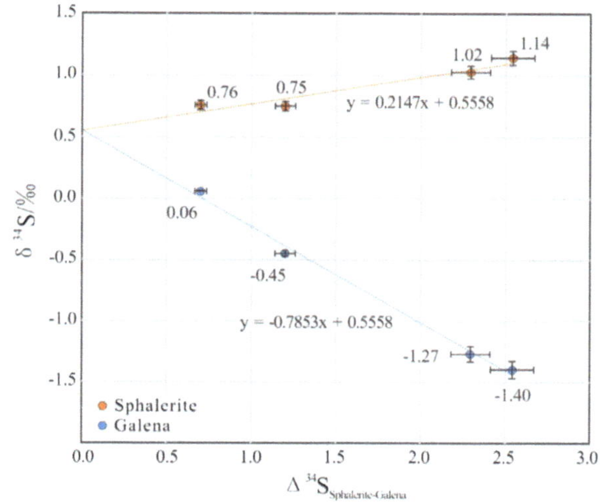

Fig. 7.4 $\delta^{34}S$ fractionation trend between the sphalerite and galena, showing the initial $\delta^{34}S$ values is 0.5558‰. $\Delta^{34}S_{sphalerite\text{-}galena} = \delta^{34}S‰_{sphalerite} - \delta^{34}S‰_{galena}$

which pyrrhotite yields a negative range between −3.8 and −4.7‰. However, both pyrrhotite and arsenopyrite associated with hastingsite in extensively altered mafic volcanic rocks respectively fall in 1.89–1.96‰ $\delta^{34}S$ and 3.52–3.59‰ $\delta^{34}S$. The $\delta^{34}S$ values obtained for chalcopyrite from the drillcores of holes ZK2901 and ZK10604 range from 2.51 to 2.65‰.

Sphalerite and galena from garnet-hedenbergite skarn yield $\delta^{34}S$ range of 0.76–1.8‰ and −1.4 to 1.41‰, respectively. Furthermore, some sphalerite–galena pairs shows consistent fractionation trend (Fig. 7.4), and the initial $\delta^{34}S$ value is 0.5558‰. The fractionation equation of Li and Liu [7] for galena-sphalerite pairs is used to calculate the fractionation temperature, and the temperature obtained from the sphalerite–galena pair ranges from 263 to 748 °C (Table 7.2).

7.5 Lead Isotope Studies

The Pb isotope values of magnetite, galena and pyrrhotite from Galinge skarn deposit are given in Table 7.2. The $^{208}Pb/^{204}Pb$ values average 38.422 for magnetite (range 38.235–38.653), 38.494 for pyrrhotite (range 38.376–38.642) and 38.693 for galena; The $^{207}Pb/^{204}Pb$ values average 15.62 for magnetite (range 15.555–15.659), 15.651 for pyrrhotite (range 15.635–15.69) and 15.688 for galena; The $^{206}Pb/^{204}Pb$ values average 18.582 for magnetite (range 18.445–19.171), 18.568 for pyrrhotite (range 18.552–18.589) and 18.519 for galena. In the lead isotope evolution diagrams (Fig. 7.5), showing designated model reservoirs together with the fields containing their problem average modern composition, all samples plot in the orogen field, which implies that the metallogenic materials are associated with magmatic origination. If the minerals precipitated in equilibrium with pure magmatic water, the

Table 7.2 Summary of Pb and S isotope data for ore minerals from Galinge skarn deposit

Sample no.	Mineral	Mineral Assemblage	$^{208}Pb/^{204}Pb$	$^{207}Pb/^{204}Pb$	$^{206}Pb/^{204}Pb$	t(Ma)	μ	ω	δTh/δU	Δα	Δβ	Δγ	$\delta^{34}S$‰	T(°C)	
2257	Pyrite												4.73		[15]
2258	Pyrite												3		
405	Pyrite												1.5		
408	Pyrite												1		
G-6	Galena		38.693	15.688	18.519	197	9.63	37.79	3.8	82.16	23.92	41.2	−3.08		[6]
G-7	Magnetite		38.372	15.622	18.495	133	9.5	36	3.67	75.76	19.33	29.77	11.62		
G-8	Magnetite		38.44	15.626	18.446	173	9.51	36.57	3.72	76.02	19.76	33.34	−2.35		
G-9	Magnetite		38.253	15.555	18.46	73	9.37	35.1	3.63	69.15	14.71	24	2.8		
G-13	Magnetite		38.635	15.659	18.475	193	9.57	37.52	3.79	79.28	22.01	39.46	0.47		
G-16	Magnetite		38.359	15.645	19.171	−341	9.49	32.85	3.35	117.3	20.95	30.64	3.01		
G-22	Magnetite		38.474	15.615	18.445	160	9.49	36.62	3.73	74.95	18.99	33.69	12.23		
27601-4	Pyrrhotite	Pyrrhotite	38.469	15.635	18.589	81	9.51	36.01	3.66	88.8	20.6	36.58	0.9		This study
27601-6	Pyrrhotite	Pyrrhotite+Magnetite	38.489	15.645	18.553	119	9.54	36.38	3.69	86.69	21.26	37.12	0.1		
4204-9	Pyrrhotite	Pyrrhotite+Magnetite+Pyrite+Hedenbergite	38.376	15.635	18.566	97	9.52	35.76	3.64	87.45	20.6	34.07	- 4.7		
	Pyrite												1.6		
4204-13	Pyrrhotite	Pyrrhotite+Magnetite+Hedenbergite+Hastingsite	38.642	15.69	18.564	168	9.62	37.35	3.76	87.33	24.19	41.24	−3.8		
2901-4	Pyrrhotite	Pyrrohtite+Arsenopyrite+Hastingsite											1.89		
	Arsenopyrite												3.52		
2901-5	Pyrrhotite	Pyrrohtite+Arsenopyrite+Hastingsite											1.96		
	Arsenopyrite												3.59		
2901-14	Chalcopyrite	Pyrrohtite+Chalcopyrite+Garnet+Hastingsite											2.65		

(continued)

Table 7.2 (continued)

Sample no.	Mineral	Mineral Assemblage	$^{208}Pb/^{204}Pb$	$^{207}Pb/^{204}Pb$	$^{206}Pb/^{204}Pb$	t(Ma)	μ	ω	δTh/δU	Δα	Δβ	Δγ	$\delta^{34}S‰$	T(°C)
26006-14	Sphalerite	Sphalerite+Galena+Garnet											1.8	
26006-18	Sphalerite	Sphalerite+Galena											1.2	
26006-22	Sphalerite	Sphalerite+Galena+Hedenbergite											1	
10604-9	Sphalerite	Garnet+Hedenbergite+Sphalerite+Galena+Calcite											0.76	748
	Galena												0.06	
10604-11	Sphalerite	Garnet+Hedenbergite+Sphalerite+Galena											1.14	263
	Galena												−1.4	
10604-12	Sphalerite	Garnet+Hedenbergite+Sphalerite+Galena											0.75	507
	Galena												−0.45	
10604-13	Sphalerite	Sphalerite+Galena+Calcite											1.41	484
	Galena												0.14	
10604-14	Sphalerite	Chalcopyrite+Sphalerite+Galena											1.02	291
	Galena												−1.27	
	Chalcopyrite												2.51	

Note: $\mu = \delta^{238}U/\delta^{204}Pb$; $\omega = \delta^{232}Th/\delta^{204}Pb$; $\Delta\alpha = [\alpha/\alpha M(t)-1]\times 1000$; $\Delta\beta = [\beta/\beta M(t)-1]\times 1000$; $\Delta\gamma = [\gamma/\gamma M(t)-1]\times 1000$; $1000\ln\alpha_{sphalerite\text{-}galena} = \delta^{34}S‰_{sphalerite}-\delta^{34}S‰_{galena} = A(10^6/T^2)$, A = 0.7 (10, 13)

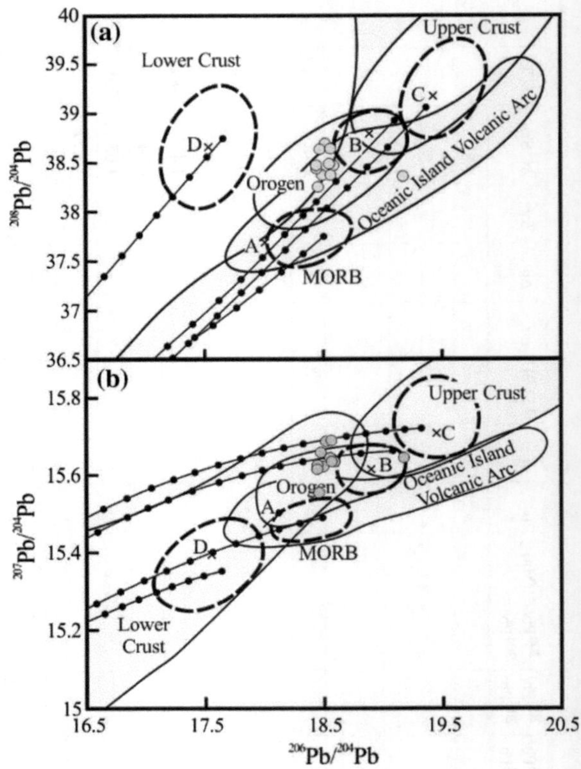

Fig. 7.5 **a**, **b** Diagrams of the Pb-isotopic compositions of magnetite, galena and pyrrhotite from Galinge. Solid lines denote the mantle, upper crust, lower crust, and orogen Pb-isotopic composition field, and dashed lines denote average values of them (A = mantle; B = orogen C = upper crust; and D = lower crust) [18, 19]

mineral isotopes will be higher than if there is a component of either seawater or meteoric water, even if the isotopic composition of magmatic waters can also vary somewhat.

References

1. Baker T, Lang JR (2001) Fluid inclusion characteristics of intrusion-related gold mineralization, Tombstone-Tungsten magmatic belt, Yukon Territory, Canada. Miner Depos 36(6):563–582
2. Bodnar R, Burnham C, Sterner S (1985) Synthetic fluid inclusions in natural quartz. III. Determination of phase equilibrium properties in the system H_2O-NaCl to 1000 °C and 1500 bars. Geochim Cosmochim Acta 49(9):1861–1873
3. Bodnar RJ, Vityk MO (1994) Interpretation of microthermometric data for H_2O-NaCl fluid inclusions. In: Fluid inclusions in minerals: methods and applications, pp 117–130
4. Fournier RO (1999) Hydrothermal processes related to movement of fluid from plastic into brittle rock in the magmatic-epithermal environment. Econ Geol Bull Soc Econ Geol 94(8):1193–1211
5. Hedenquist JW, Arribas A, Reynolds TJ (1998) Evolution of an intrusion-centered hydrothermal system; Far Southeast-Lepanto porphyry and epithermal Cu-Au deposits, Philippines. Econ Geol 93(4):373–404

6. Kong DF, Lai JQ, Liu HC et al (2013) The material sources of Galing iron polymetallic deposit from the evidence of S and Pb isotope. Mordern Mining 535:42–50
7. Li YB, Liu JM (2006) Calculation of sulfur isotope fractionation in sulfides. Geochim Cosmochim Ac 70(7):1789–1795
8. Meinert LD (1992) Skarns and skarn deposits. Geosci Can 19(4):145–162
9. Meinert LD (1993) Igneous petrogenesis and skarn deposits. Miner Depos Model 40:569–583
10. Ohmoto H (1972) Systematics of sulfur and carbon isotopes in hydrothermal ore deposits. Econ Geol 67(5):551–578
11. Roedder E, Bodnar RJ (1980) Geologic pressure determinations from fluid inclusion studies. Annu Rev Earth Pl Sc 8:263
12. Rusk BG, Reed MH, Dilles JH (2008) Fluid inclusion evidence for magmatic-hydrothermal fluid evolution in the porphyry copper-molybdenum deposit at Butte, Montana. Econ Geol 103(2):307–334
13. Rye RO, Ohmoto H (1974) Sulfur and carbon isotopes and ore genesis: a review. Econ Geol 69(6):826–842
14. Wilkinson JJ (2001) Fluid inclusions in hydrothermal ore deposits. Lithos 55(1):229–272
15. Wu TX, Li HL (2009) Geological and geochemical characteristics of the iron polymetallic deposit in the Galinge area, Qinghai province. Bull Mineral Petrol Geochem 28(2):157–161
16. Yu M, Feng CY, Xiao Y, Liu JN, Li DX, Li GC, Ma SC (2013) Features and evolution of metallogenic fluid in Jiadanggen porphyry copper deposit of Gonghe Country. Qinghai Province. Miner Depos 32(1):133–147
17. Yu M, Feng CY, Zhao YM, Li DX, Xiao Y, Liu JN, Li ZF (2014) Fluid inclusion geochemistry in the Kaerqueka copper polymetallic deposit, Qinghai Province and its genetic significances. Acta Geol Sin 88(5):903–917
18. Zartman RE, Doe BR (1981) Plumbotectonics—the model. Tectonophysics 75(1):135–162
19. Zartman RE, Haines SM (1988) The plumbotectonic model for Pb isotopic systematics among major terrestrial reservoirs—a case for bi-directional transport. Geochim Cosmochim Acta 52(6):1327–1339

Chapter 8
Geochronological and Geochemical Constraints on the Galinge Skarn Deposit

8.1 Sampling and Analytical Methods

8.1.1 *Sampling*

Seventeen fresh samples of the Galinge intrusions were collected from the drill cores for petrographic analyses, zircon U–Pb dating and in situ Hf isotopic analysis (Table 8.1). One phlogopite sample coexisting with disseminated/massive magnetite ores (from drill hole ZK0307 at 635.61 m deep) was collected for $^{40}Ar/^{39}Ar$ dating.

8.1.2 *Major and Trace Elements*

Whole rock geochemical analyses were carried out at the Analysis & Testing Center of Beijing Research Institute of Uranium Geology (CNNC). Major elements concentrations were analyzed by X-ray Fluorescence (XRF) spectrometry with Philips PW2404 X-ray fluorescence spectrometer based on GB/T14506.28-93. Trace elements were analyzed by Inductively Coupled Plasma Mass Spectrometry (ICP-MS) based on general rules of DZ/T0223-2001.

8.1.3 *LA-ICP-MS Zircon U–Pb Dating*

Zircons for U–Pb geochronology were separated and handpicked from eight samples including granodiorite, diorite and diorite porphyry dyke. Zircon grains were embedded in epoxy resin and polished with diamond paste. Prior to age determination all

M. Yu, *Metallogenic Mechanism of the Galinge Polymetallic Iron Skarn Deposit, Qiman Tagh Mountains, Qinghai Province*, Springer Theses,
https://doi.org/10.1007/978-981-10-7907-8_8

Table 8.1 The summary of sampling depth, researching methods, and mineral assemblages of Galinge intrusion samples

No.	Drill samples	Depth	Mineral assemblages	Methods
0307-76	Diorite	735 m	Plagioclase (70%) + K-feldspar (10%) + hornblende (5%) + clino-pyroxen (5%) + biotite (10%)	Whole-rock/LA-ICP-MS
0307-79	Diorite	743.73 m		Whole-rock
0307-80	Diorite	748.53		LA-ICP-MS/Hf isotope
0307-81	Diorite	755.46 m		Whole-rock
4003-38	Diorite	859.6		LA-ICP-MS
4003-46	Diorite	919.69 m		Whole-rock
2404-32	Granodiorite	1063.89 m	Plagioclase (40%) + quartz 25%) + K-feldspar (15%) + hornblende (5%) + biotite (10%) + apatite, sphene and zircon (5%)	Whole-rock
2404-34	Granodiorite	1087.59 m		Whole-rock
2404-35	Granodiorite	1092.59 m		Whole-rock
2404-22	Granodiorite	511.6		LA-ICP-MS/Hf isotope
16303-14	Granodiorite	664.38 m		Whole-rock/LA-ICP-MS/Hf isotope
4003-58	Granodiorite	1124 m		Whole-rock
4003-59	Granodiorite	1128 m		LA-ICP-MS
4003-54	Diorite porphyrite dike	1006.19	Plagioclase (80%) + pyroxene (10%) + biotite (5%) + hornblende (5%)	LA-ICP-MS
21605-12	Diorite porphyrite dike	460.5 m		Whole-rock
21605-30	Diorite porphyrite dike	558.97 m		Whole-rock/LA-ICP-MS/Hf isotope
21605-60	Diorite porphyrite dike	690.6 m		Whole-rock

zircon grains were studied by cathode luminescence (CL) imaging to distinguish internal zoning and possible inherited cores. Zircon U–Pb dating was carried out at the Institute of Mineral Resources, Chinese Academy of Geological Sciences (CAGS, Beijing), with a Neptune MC-ICP-MS with self-contained NewWave UP 213 laser ablation system. The spot beam diameter was set to 25 μm, frequency to 10 Hz and energy density to 2.5 J/cm^2. Uranium, Th and Pb concentrations were calibrated by using ^{29}Si as the internal calibrant and M127 as the reference material. The ^{207}Pb/^{206}Pb and ^{206}Pb/^{238}U ratios were corrected by using zircon GJ-1 for the external calibration. Plesovice zircon standards were tested every ten analyses to ensure accuracy [29]. Data were processed with ICPMSDataCal4.3 program [39]. After the data reduction, calculation of weighted mean age and concordia analysis were done with Isoplot 3.0 [42]. Uncertainties of $^{207}Pb/^{206}Pb$, $^{206}Pb/^{238}U$, and $^{207}Pb/^{235}U$ are set at 1σ levels.

8.1.4 Zircon Lu–Hf Isotopes

Zircon Lu–Hf isotope analysis was performed in situ with LA-MC-ICP-MS after age determination. Helium (He) was used as a carrier gas to transport the ablated sample from the laser-ablation cell to the ICP-MS torch via a mixing chamber mixed with Argon (Ar), an eroding beam diameter was set to 55 μm, and zircon GJ1 was used as reference materials with a weighted mean $^{176}Hf/^{177}Hf$ ratio of 0.282004 + 0.000019 (2σ, n = 3) in experiments [29]. $^{176}Lu/^{175}Lu = 0.02658$ and $^{176}Yb/^{173}Yb = 0.796218$ ratios were fixed to adjust the isobaric interferences of ^{176}Lu and ^{176}Yb on ^{176}Hf [10]. For instrumental mass bias correction, Yb isotope ratios were normalized to $^{172}Yb/^{173}Yb = 1.35274$ and Hf isotope ratios to $^{179}Hf/^{177}Hf = 0.7325$ under an exponential law [10]. The mass bias behavior of Lu was supposed to follow that of Yb isotope. Related instrument operation and conditions have been outlined by Hou et al. [30] and Elhlou et al. [16].

8.1.5 Phlogopite $^{40}Ar/^{39}Ar$ Dating

The samples were crushed and purified with a magnetic separator and then cleaned by ultrasonic treatment under ethanol. Purity of the phlogopite grains is above 99%. Sample was wrapped in aluminum foil and loaded into a tube of Al foil, together with 2 or 3 monitor samples. The monitor used in this work is an internal standard: Fangshan biotite (ZBH-25) with reference age = 132.7 ± 1.2 Ma and potassium content = 7.6% [65]. The bottle was irradiated for total 1444 min in the B4 position of a nuclear reactor (The Swimming Pool Reactor, Chinese Institute of Atomic Energy, Beijing). The reactor delivers a neutron flux of 2.6×10^{13} n cm^{-2} s^{-1}. The integrated neutron flux is 2.25×10^{18} n/cm^{-2}.

After the irradiation, the sample and monitors were removed from the quartz bottle and then loaded into the vacuum extraction system. Samples analyzed by the furnace step-heating method utilized a double vacuum resistance furnace. They were baked out for 48 h at 120–150 °C. The released gases were admitted to a purification system. Heating-extraction for each temperature increment and purification lasted 30 min each.

Finally the gases were purified by two Sorb-AC pumps at room temperature. Purified Ar was trapped in an activated charcoal finger at liquid-nitrogen temperature, and then released into the Helix MC Mass Spectrometer to analyze Ar isotopes. Each peak values were selected for 20 data. Measured isotopic ratios were corrected for mass discrimination, atmospheric Ar component, blanks and irradiation induced mass interference. The correction factors of interfering isotopes produced during irradiation were determined by analysis of irradiated K_2SO_4 and CaF_4 pure salts and their values are: $(^{36}Ar/^{37}Ar)_{Ca} = 0.0002389$, $(^{40}Ar/^{39}Ar)_K = 0.004782$, $(^{39}Ar/^{37}Ar)_{Ca} = 0.000806$. All ^{37}Ar was corrected for radiogenic decay. The decay constant λ of ^{40}K is 5.543×10^{-10} a^{-1} [58]. The $^{40}Ar/^{36}Ar$ versus $^{39}Ar/^{36}Ar$ isochron diagram was constructed with ISOPLOT version 2.49 [41].

8.2 Results

8.2.1 Major and Trace Element Geochemistry

Geochemical data of representative samples from the Galinge, Yemaquan and Kaerqueka are listed in Table 8.2. They generally contain similar geochemical features and emplacement age. In the total alkali ($Na_2O + K_2O$) versus silica (TAS) plot [31] (Fig. 8.1a), the compositions of ($Na_2O + K_2O$) show a slight linear increasing trends with increasing SiO_2. Additionally, all the rock compositions fall within the calc-alkaline field. Almost all Galinge rock compositions are metaluminous (A/CNK < 1.0 and A/NK > 0.5) based on the A/CNK versus A/NK diagram (Fig. 8.1b). Moreover, the A/CNK values are decreased from granodiorites to diorite porphyry dykes, but A/NK values shows contrast increasing trend. Additionally, all the samples plot in the calc-alkaline domain in the Modified Alkali Lime Index (MALI: Na_2O + K_2O–CaO) versus SiO_2 discrimination diagram [24] (Fig. 8.1c). According to the geochemical classification on the SiO_2 versus $FeO_t/(FeO_t + MgO)$ discrimination diagram [25] (Fig. 8.1d), most of data plot in the magnesian domain, which is typical for I-type granitic rocks.

Rare earth element (REE) patterns of the granodiorite from Yemaquan, Kaerqueka and Galinge are LREE enriched, flat in the HREE and without significant Eu anomalies ($Eu/Eu^* = 0.54$–0.94) (Fig. 8.2a). Moreover, the $(La/Yb)_N$ ratios of granodiorites are variable between 4.35 and 25.11 (Table 8.2). The REE patterns of the diorites and diorite porphyry dikes show negative Eu anomalies ranging 0.31–0.82 (Fig. 8.2b), and lower $(La/Yb)_N$ ratios ranging 4.43–13.44 (Table 8.2).

Trace elements of the analyzed samples were normalized with primitive mantle values [45] and are plotted on spider diagrams (Fig. 8.2c, d). The Galinge granodiorite, diorite and diorite porphyry dyke have similar trace element patterns, which show that Rb, Th and U are enriched, whereas other elements including plagioclase-compatible Ba and Sr are depleted. Granodiorites from Yemaquan and Kaerqueka have the same features in trace element with those from Galinge. In addition, the garnet-compatible Y shows moderate positive anomaly in the diorite and diorite porphyry dike samples, which reflected that probably little or no garnet was melted.

Table 8.2 Major and trace element data including rare earth elements (RRE) and characteristic parameters of the intrusions from Galinge, Yemaquan, and Kaerqueka

Locations	Galinge												Galinge [2]		
Plutons	Diorite				Granodiorite					Diorite porphyry dyke			Pyroxene diorite		
sample	4003-46	0307-76	0307-79	0307-81	16303-14	2404-34	2404-32	2404-35	4003-58	21605-12	21605-30	21605-60	GLG-6	GLG-7	GLG-8
SiO_2	62.97	58.53	61.54	61.9	60.62	62.48	63.15	61.04	65.58	53.84	52.24	53.7	52.83	52.67	53.54
TiO_2	0.61	0.83	0.68	0.53	0.64	0.6	0.54	0.57	0.56	0.78	0.76	0.87	1.27	1.04	1.09
Al_2O_3	15.63	14.8	15.75	16.3	15.06	14.87	14.55	15.35	15.18	15.58	16.17	16.55	18.26	18.39	18.7
Fe_2O_3	0.67	1.02	0.73	0.76	4.52	4.29	4.49	5.24	1	1.31	1.91	1.8			
FeO	2.92	6.26	4.69	4.66	3.59	3.64	3.93	4.37	3.88	5.1	5.37	4.09	5.99	4	4
MnO	0.114	0.191	0.152	0.203	0.13	0.12	0.13	0.12	0.124	0.139	0.265	0.184	0.17	0.13	0.13
MgO	2.37	3.45	2.77	2.42	2.74	2.26	2.14	2.28	2.06	4.77	3.47	4.53	3.78	3.03	3.2
CaO	4.83	6.69	6.2	4.59	4.86	5	4.19	4.55	4.44	8.45	12.31	10.62	11.05	10.9	11.76
Na_2O	2.5	2.82	2.91	2.14	2.66	2.61	2.31	2.76	2.51	2.73	3.63	3.02	3.14	4.28	4.14
K_2O	4.57	3.71	3.3	3.01	3.68	3.79	4.29	3.54	3.89	3.23	1.04	2.69	2.04	1.13	0.825
P_2O_5	0.116	0.161	0.124	0.1	0.14	0.13	0.11	0.13	0.108	0.152	0.156	0.186	0.43	0.29	0.34
LOI	2.11	1.08	0.66	3.84	2.05	0.86	1.19	1.47	0.54	3.36	2.06	1.4	0.85	3.91	2.08
Total	99.41	99.54	99.49	99.46	100.68	100.7	101	101.4	99.86	99.43	99.37	99.64	99.81	99.76	99.81
Mg#	59.37	49.8	51.53	48.31	57.87	52.78	49.5	48.43	48.87	62.74	53.77	66.6	56	60	61
Cr	39.5	43.4	43.4	29.2	29.2	25.4	33.3	31.4	47.2	277	37.7	119	13.7	14.5	8.9
Ni	8.1	10.8	9.84	2.17	10.8	8.08	10.9	13.6	12.6	28.6	10.1	17.7	11.9	12.9	8.55
Ga	16.4	15.7	16	13.6	18.1	17.9	18.1	18.5	16.8	14.6	16.6	17.8	20.4	24.8	20.8
Rb	174	163	180	115	128	130	175	146	233	118	44.6	132	44.3	34.6	23.7
Sr	219	179	176	118	281	244	259	301	163.6	181	180	178	529	367	516
Y	32.5	32.1	27.2	30.5	32.8	34.8	31.2	32.7	31.2	27.1	27.1	28.8	32.9	38.5	33.1
Nb	9.44	10.7	9.08	8.48	11	11.9	11	10.6	9.65	9.89	8.74	10.3	12.6	16.9	12.6
Ba	1125	471	445	343	752	687	810	835	521	834	267	581	626	291	280
La	47.2	34	29	37.9	21.4	63.9	22.5	37.4	34.8	17	38.6	28.7	28.1	43.9	28

(continued)

Table 8.2 (continued)

Locations	Galinge												Galinge [2]		
Ce	98.8	71.5	62.7	81.6	38.7	124	42.1	70.5	75	40.5	80.4	69	61.8	108	64.3
Pr	9.09	7.05	6.18	8.03	4.93	13.2	5.11	7.77	7.2	5.09	7.44	8.07	8.37	14.1	9
Nd	32	26.2	22.8	28.2	19.9	46.2	19.5	29.5	25.8	19.9	25.3	33.9	37.3	57.1	40.4
Sm	6.18	5.36	4.61	5.31	4.46	7.4	4.65	5.31	5.09	4.11	4.7	8.97	8.72	11.5	9.14
Eu	1.26	1.15	1.03	1.09	1.25	1.25	1.23	1.26	1.01	1.11	1.23	0.94	2.21	2.58	2.41
Gd	5.78	5.57	4.53	5.05	4.76	6.8	4.56	5.49	4.95	4.14	4.73	9.77	7.71	9.71	7.85
Tb	0.85	0.82	0.71	0.79	0.981	1.18	0.922	0.972	0.75	0.64	0.72	1.6	1.27	1.66	1.37
Dy	4.8	4.64	3.92	4.48	5.7	5.92	5.19	5.47	4.33	3.64	4.04	9.4	6.92	8.74	7.17
Ho	0.93	0.93	0.79	0.9	1.18	1.19	1.05	1.16	0.87	0.78	0.82	0.88	1.32	1.67	1.38
Er	2.81	2.79	2.38	2.65	3.66	3.45	3.55	3.37	2.64	2.43	2.38	5.52	3.73	4.68	3.9
Tm	0.48	0.48	0.42	0.44	0.484	0.491	0.507	0.505	0.46	0.47	0.41	0.82	0.55	0.67	0.58
Yb	2.63	2.57	2.23	2.3	3.53	4.03	3.23	3.44	2.45	2.63	2.06	4.65	3.47	4.36	3.47
Lu	0.41	0.38	0.34	0.34	0.547	0.561	0.5	0.478	0.37	0.41	0.3	0.67	0.5	0.57	0.5
Ta	0.89	0.91	0.85	0.87	1.14	1.17	1.13	1.03	0.94	0.78	0.48	0.83	0.68	0.89	0.68
Th	19.9	17.6	22.8	22.9	23.4	27.2	26.9	26.5	26.3	13.2	6.85	9.09	4.73	8.02	6.72
U	7.01	5.49	5.96	6.49	7.37	8.23	8.42	7.95	7.93	4.58	3.43	3.46	1.28	2.86	1.74
Zr	212	203	163	183	189	201	189	158	217	121	162	162			
Hf	5.2	4.87	4.3	4.91	6.09	5.88	5.64	4.88	5.45	3.27	3.95	3.94			
Nb/Ta	10.61	11.76	10.68	9.75	9.65	10.17	9.73	10.29	10.27	12.68	18.21	12.41	18.53	18.99	18.53
Eu/Eu*	0.64	0.64	0.69	0.64	0.83	0.54	0.82	0.71	0.62	0.82	0.80	0.31	0.95	0.92	0.96
$(La/Yb)_N$	12.87	9.49	9.33	11.82	4.35	11.37	5.00	7.80	10.19	4.64	13.44	4.43	5.81	7.22	5.79

(continued)

Table 8.2 (continued)

Locations	Kaerqueka [22]								Yemaquan [40]			
Plutons	Granodiorite								Granodiorite			
sample	KD002	KD003	KD004	KD005	KD029	KD030	KD039	CZK1401-5	YMQ-7	YMQ-8	YMQ-9	YMQ-10
SiO_2	63.47	63.54	63.08	63.8	61.04	61.06	66.08	66.5	66.58	66.69	68.17	66.44
TiO_2	0.67	0.67	0.69	0.66	0.71	0.61	0.5	0.51	0.63	0.31	0.34	0.6
Al_2O_3	16.05	16.52	16.5	16.54	17.94	19.04	15.62	15.6	15.18	15.26	15.45	15.85
Fe_2O_3	1.28	1.67	1.47	1.41	1.77	1.71	1.37	0.1	0.94	1.36	1.6	0.85
FeO	3.52	2.87	3.2	2.96	3.23	2.61	2.62	3.37	3.25	2.67	2.28	3.03
MnO	0.1	0.09	0.09	0.1	0.1	0.08	0.13	0.1	0.07	0.1	0.08	0.07
MgO	2.48	2.24	2.31	2.15	1.74	1.3	1.65	1.57	1.22	0.41	0.5	0.95
CaO	4.37	4.46	4.75	4.25	4.71	4.46	3.59	3.36	2.97	3.58	2.63	3.78
Na_2O	3.43	3.7	3.67	3.47	4.27	4.63	3.2	3.02	3.45	4.35	4.29	3.75
K_2O	2.89	2.83	2.62	2.96	2.66	2.24	3.05	3.03	4.35	4.11	4.58	4.17
P_2O_5	0.16	0.15	0.16	0.15	0.18	0.15	0.12	0.11	0.11	0.1	0.09	0.16
LOI	0.65	0.57	0.55	0.63	0.55	1.23	1.18	1.73				
Total	99.07	99.31	99.09	99.07	99.04	99.12	99.11	99	98.64	98.84	99. 92	99. 49
Mg#	55.91	58.42	56.51	56.66	49.23	47.27	53.13	45.61	40.32	21.65	28.3	36.08
Cr	22.29	20.27	24.28	17.35	12.51	10.8	14.24	11.85		57.62	6.4	
Ni	13.63	13.09	12.23	10.89	5.86	6.1	10.37	7.51		4.29	3.78	
Ga	19.53	20.07	19.53	19.54	23.29	22.5	17.64	18.04	17.87		17.3	
Rb	141.8	126	114.4	80.51	120.1	89.56	114.9	104.1	146	90.82	159	
Sr	338.1	369	359.8	311.8	342	398	228.3	203.1	271	191	265	
Y	16.13	15.75	15.5	14.35	33.67	13.82	18.31	14.76	14.23	11.49	13.5	16.3
Nb	8.87	8.73	8.46	8.31	8.76	9.79	8.76	9.04	8.9	10.45	11.2	
Ba	478.2	505	446.3	482	494.3	694	449.7	341	1671	788.4	682	
La	22.76	19.8	24.78	17.75	25.91	37.11	20.67	18.59	51.8	27.15	48.4	48.8
Ce	46.2	42.4	50.29	39.61	56.24	61.55	39.27	36.2	92.12	50.84	81.7	76.3

(continued)

Table 8.2 (continued)

Locations	Kaerqueka [22]								Yemaquan [40]			
Pr	5.43	4.99	5.67	4.58	7.04	7.06	4.56	4.22	8.94	5.37	8.17	8.24
Nd	20.23	19.97	20.47	17.18	29.26	22.73	17.44	15.13	26.36	20.61	24.5	27.3
Sm	4.11	4.09	4.03	3.62	6.71	3.99	3.63	3.09	4.1	5.01	3.29	4.11
Eu	1.01	1.06	1.03	1.02	1.35	1.6	0.88	0.81	1.15	1.09	0.84	0.76
Gd	3.5	3.47	3.47	3.06	6.66	3.32	3.54	2.79	3.25	2.63	2.68	3.13
Tb	0.53	0.51	0.52	0.46	1.1	0.46	0.56	0.44	0.51	0.38	0.39	0.48
Dy	2.98	2.89	2.92	2.7	6.66	2.59	3.3	2.58	2.64	1.94	2.24	2.38
Ho	0.59	0.57	0.57	0.53	1.34	0.53	0.69	0.53	0.57	0.39	0.47	0.51
Er	1.64	1.58	1.55	1.46	3.59	1.5	1.96	1.47	1.58	1.1	1.38	1.49
Tm	0.25	0.24	0.24	0.22	0.52	0.24	0.31	0.22	0.25	0.16	0.21	0.22
Yb	1.64	1.57	1.55	1.46	3.24	1.65	2	1.48	1.48	1.06	1.6	1.5
Lu	0.25	0.24	0.24	0.22	0.48	0.27	0.3	0.23	0.22	0.18	0.23	0.24
Ta									0.9	0.77	0.99	
Th									16.7	0.21	18.7	
U												
Zr	162.5	153.2	130.8	193.5	108.8	129.8	182.6	125	215	171.3	235	
Hf									5.7	1.42	5.16	
Nb/Ta									9.89	13.57	11.31	
Eu/Eu*	0.81	0.86	0.84	0.94	0.62	1.34	0.75	0.84	0.96	0.92	0.86	0.65
$(La/Yb)_N$	9.95	9.05	11.47	8.72	5.74	16.13	7.41	9.01	25.11	18.37	21.70	23.34

Note: Major elements as oxides in wt.%, trace elements in ppm; Mg# = MgO/(MgO + FeO) (molecular ratio), Eu/Eu* = Eucn/(Smcn • Gdcn)$^{0.5}$; Galinge pyroxene diorite are from Bai et al. 2; Kaerquka data are from Feng et al. [22]; Yemaquan data are from Liu et al. [40]

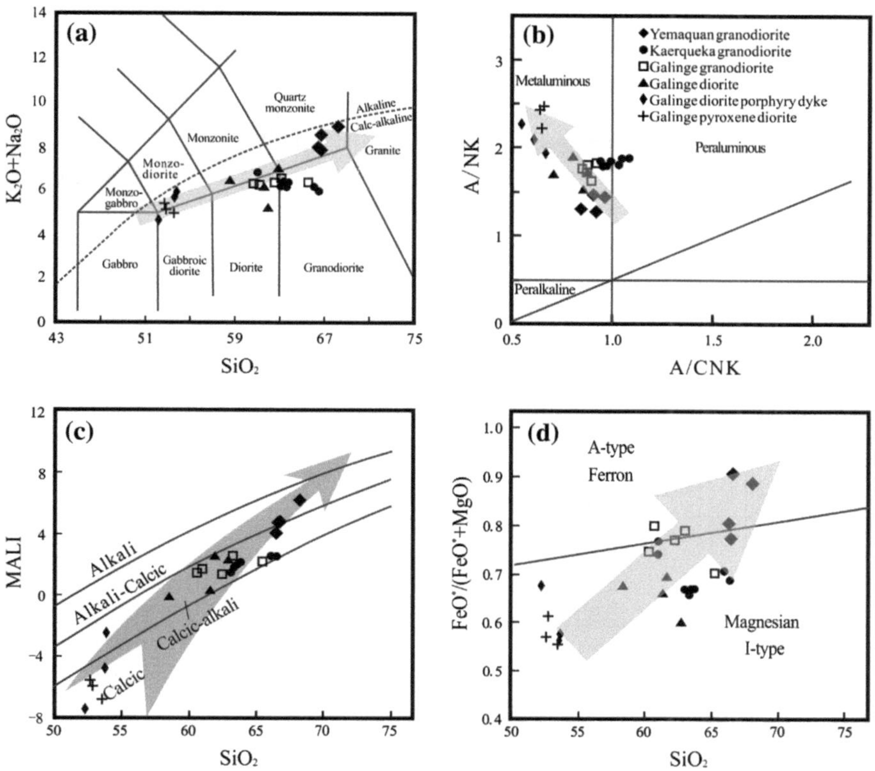

Fig. 8.1 Diagrams of **a** SiO_2–Na_2O + K_2O [31]. **b** A/CNK versus A/NK (A/CNK = molar Al_2O_3/(CaO + Na_2O + K_2O) and A/NK = molar Al_2O_3/(Na_2O + K_2O)) [7, 44]. **c** Modified alkaline index MALI (Na_2O + K_2O–CaO) versus SiO_2 discrimination diagram of Frost et al. [25]. **d** SiO_2 versus FeOt/(FeOt + MgO) discrimination diagram [25]

8.2.2 *Zircon U–Pb Geochronology*

Our results show that the zircons Th/U values range 0.55–1.13 (Mean = 0.73), similar to those of typical igneous zircons (Th/U > 0.1) [70], which implies magmatic origin [11].

All the data points fall along the Concordia (Fig. 8.3). The weighted mean $^{206}Pb/^{238}U$ ages of granodiorite, diorite and diorite porphyry dyke range ca. 229–226 Ma, ca. 223–218 Ma and ca. 226–217 Ma, respectively.

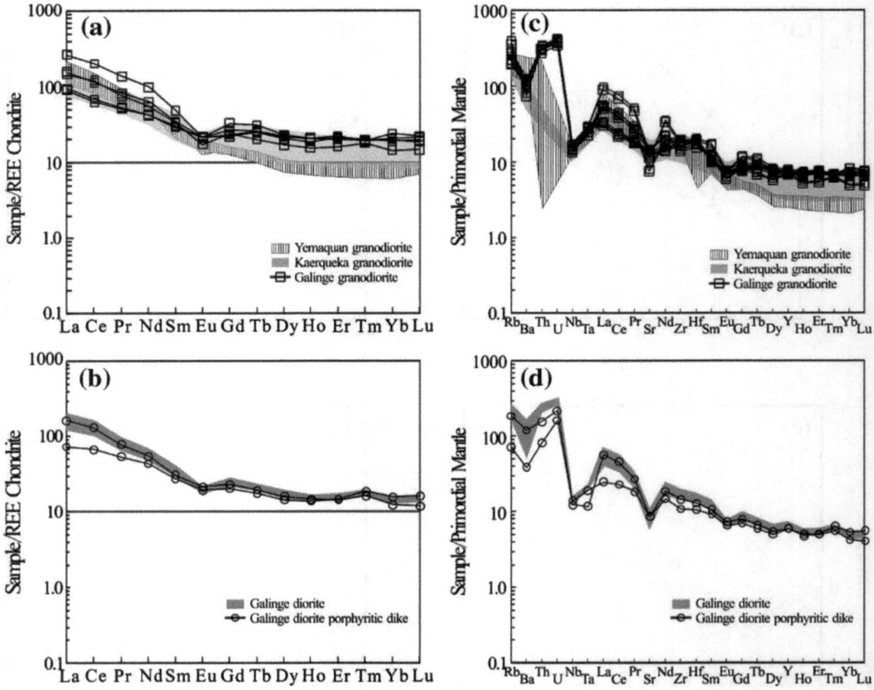

Fig. 8.2 **a, b** Primitive mantle-normalized trace element patterns (normalization values from [45]). **c, d** Chondrite-normalized REE patterns (normalization values from [59]) for the Galinge intrusions and Yemaquan and Kaerqueka granodiorite

8.2.3 Zircon Hf Isotopic Analysis

Initial $^{176}Hf/^{177}Hf$ ratios of all samples range 0.282496–0.282639. A total of 55 spots were analyzed. 53 spots shows negative εHf(t) values ranging from −9.5 to −1.1, whereas the remaining two spots shows a positive εHf(t) values ranging from 0.21 to 0.20 (Fig. 8.4). T_{DM2} model ages vary from 1.248 to 1.571 Ga. The εHf(t) values of Galinge granodiorites range from 0.2 to −4.9, the εHf(t) values of diorites range from −6.0 to −9.5, and the εHf(t) values of diorite porphyry dikes range from −6.0 to −8.6.

8.2.3.1 $^{40}Ar/^{39}Ar$ Dating of Phlogopite

In this study, the ore-forming age is performed on the phlogopite using $^{40}Ar/^{39}Ar$ dating method, which yielded an isochron age of 234.1 ± 3.7 Ma, and a plateau age of 235.8 ± 1.7 Ma. The age spectrum for the phlogopite sample is nearly concordant,

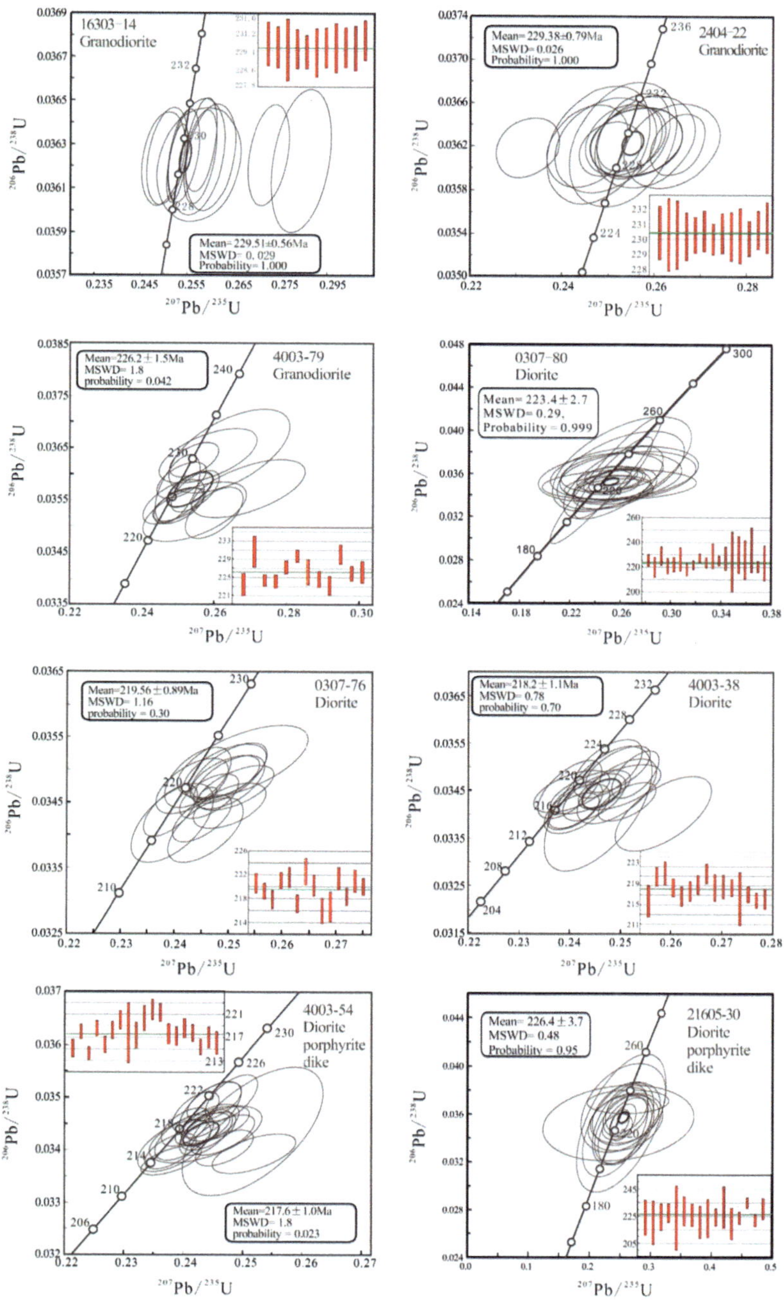

Fig. 8.3 Zircon U–Pb concordia diagram of the Galinge intrusions

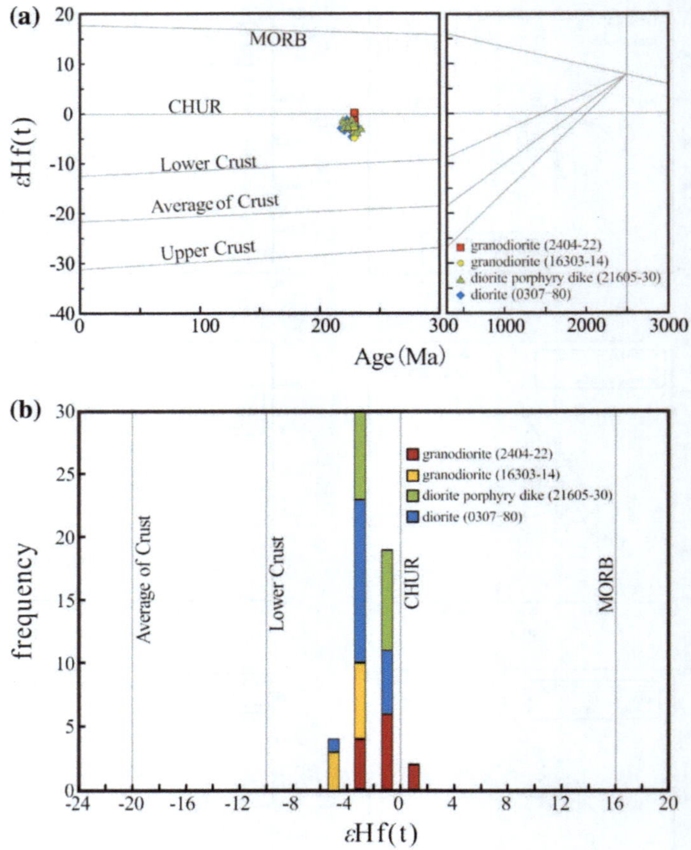

Fig. 8.4 Zircon Hf isotopic compositions of the Galinge intrusions. **a** In situ U–Pb age versus εHf(t). In situ zircon $^{206}Pb/^{238}U$ ages were applied to revise Hf isotopic ratios. εHf(t) were converted by using the corrected Hf isotopic ratios. The MORB array was calculated by Patchett and Tatsumoto [47], and the crustal array was deduced according to Scherer et al. [56]. **b** Histogram of Zircon εHf(t) with the age of 229 Ma for the host granodiorites

with dramatically low initial ages as ~5% gas cumulated, followed by ages which average ~236 Ma for the remainder of the gas cumulated (Table 8.3). Overall, there is a very slight U-shape to the age spectrum, suggesting that excess argon should be considered. The total age is 236.1 Ma. Steps 5–11 (94.46% of the total ^{39}Ar cumulated) and yield a plateau age of 235.8 ± 1.7 Ma (Fig. 8.5a). Steps 5–11 (94.46% of the total ^{39}Ar cumulated) define a very well constrained isochron which yielded an age of 234.1 ± 3.7 Ma (Fig. 8.5b). This isochron indicates that the sample contains a small amount of excess argon (initial $^{40}Ar/^{36}Ar = 302.8 \pm 9.9$). Although all three ages are statistically indistinguishable at the 2σ level, the isochron age is the most accurate, as it accounts for the small amount of excess argon that the sample contains.

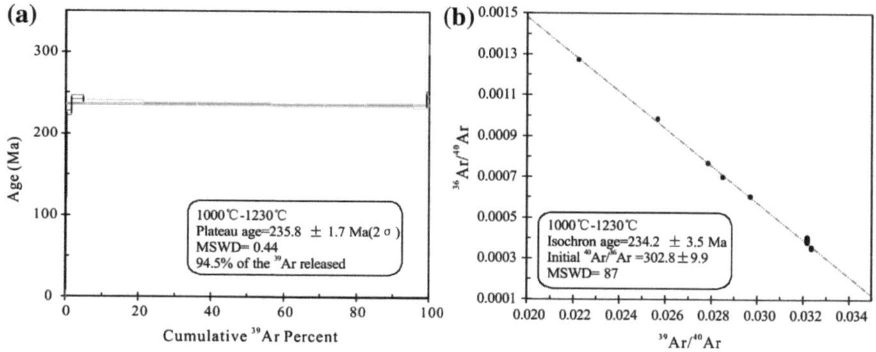

Fig. 8.5 $^{40}Ar/^{39}Ar$ step heating age spectra and $^{36}Ar/^{40}Ar$–$^{39}Ar/^{40}Ar$ isotope correlation diagrams for phlogopite collected from different facies of the contact aurole. In the plateau age diagram (**a**), boxes represent the relative fraction of ^{39}Ar released in heating steps and extend vertically to $\pm 1\sigma$. In the isotope correlation diagrams (**b**), circles indicate the gas compositions for plateau steps, which were incorporated into the data regression. Error ellipses represent $\pm 1\sigma$

Ca/K ratios are, with the exception of the first four steps, uniformly low, which may be influenced by degassing of a minor secondary mineral phase, or minor inclusions of a higher Ca phase.

8.3 Discussion

8.3.1 Late Triassic Magmatism and Mineralization

In this study, we have obtained zircon U–Pb ages of 229–226 Ma for granodiorite, 223–218 Ma for diorite, and 226–217 Ma for diorite porphyry dyke. Thus, the Galinge diorite and diorite porphyry dike postdate the granodiorite. Previous studies show that Late Triassic (237–204 Ma) was an important tectonic active period in which the Qiman Tagh was under a post-collisional compressive to extensional transition tectonics as evidenced by the extensive mantle-derived magmatism [22, 43, 66]. Granodiorite, monzogranite, and syenogranite in the Qiman Tagh orogen are identified as calc-alkaline and alkaline granitoid plutons [20], which is a common and distinctive feature for post-collisional magma evolution [3, 72]. The histogram of syenogranite, monzogranite and granodiorite ages shows that the intrusion of granodiorites mainly occurred in ca. 236–220 Ma, the monzogranite in ca. 230–215 Ma, and the syenogranite in ca. 204–224 Ma (Fig. 8.6). This age distribution pattern coincides well with typical post-collisional granitic magmatism in a tensional regime.

More precise metallogenic ages were reported from the Qiman Tagh belt recently (Table 8.4). It is an explosive period for Fe–Cu–Mo–Pb–Zn–Au mineralization in the

Table 8.3 40Ar/39Ar stepwise heating analytical data for phlogopite from the Galinge skarn deposit

T(°C)	$(^{40}Ar/^{39}Ar)_m$	$(^{36}Ar/^{39}Ar)_m$	$(^{37}Ar/^{39}Ar)_m$	$(^{38}Ar/^{39}Ar)_m$	^{40}Ar(%)	F	^{39}Ar ($\times 10^{-14}$ mol)	^{39}Ar(Cum.) (%)	Ca/K	Age(Ma)	±1s(Ma)
700	710.3401	2.4017	4.3525	0.4918	0.13	0.9428	0.02	0.05	7.504	9	102
800	95.8637	0.2778	0.4511	0.0669	14.41	13.8151	0.08	0.26	0.778	121.3	5.8
900	74.9916	0.1646	0.101	0.0443	35.16	26.3711	0.42	1.46	0.174	224.9	2.5
960	59.8343	0.1051	0.0441	0.0327	48.09	28.7727	1.17	4.79	0.076	244.1	2.3
1000	45.0163	0.0573	0.0213	0.0237	62.37	28.0755	5.72	21.12	0.037	238.5	2.2
1040	35.083	0.0246	0.0343	0.0178	79.27	27.8124	5.3	36.24	0.059	236.4	2.2
1080	35.9326	0.0276	0.0186	0.0182	77.3	27.7758	7.15	56.64	0.032	236.1	2.2
1110	38.9716	0.0383	0.0074	0.0204	70.94	27.6454	5.09	71.16	0.013	235.1	2.2
1150	33.672	0.0203	0.0116	0.0169	82.13	27.6555	4.03	82.65	0.02	235.2	2.2
1190	30.919	0.0109	0.0128	0.015	89.56	27.6924	4.51	95.52	0.022	235.5	2.2
1230	31.092	0.0122	0.0902	0.0151	88.36	27.4763	1.31	99.25	0.155	233.8	2.2
1300	34.4899	0.0172	0	0.0144	85.27	29.4088	0.23	99.92	0	249.1	3.3
1400	67.633	0.1085	0	0.0288	52.6	35.5776	0.03	100	0	297	14

Note: W = 27.48mg, J = 0.005035; m represents the ratios of measured isotope, total age = 236. 1 Ma, F = $^{40}Ar^{*}/^{39}Ar$, cum. = cumulate

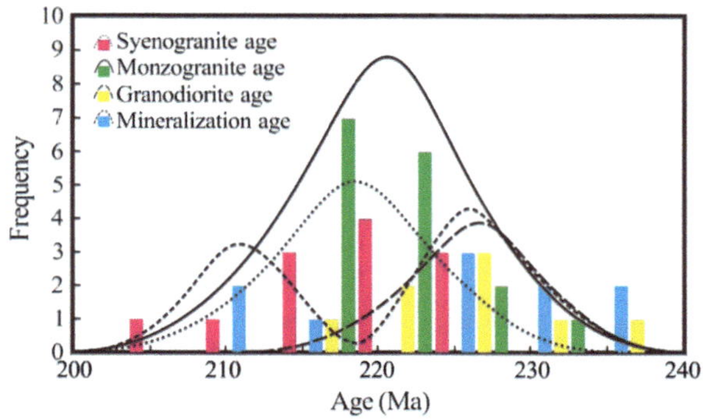

Fig. 8.6 Probability density plot for the Middle to Late Triassic intrusions in Qiman Tagh

Qiman Tagh metallogenetic belt from 240 to 210 Ma based on isotope dating. Besides, many important skarn deposits, e.g., Galinge, Yemaquan, Hutouya, Tawenchahan, Kaerqueka, Weibao and Kendekeke, some porphyry and epithermal deposits were developed, e.g., Ulan Uzhur, Yazigou, Changshan and Lalingzaohuo. All of these porphyry-related skarns and epithermal deposits are spatially and temporally associated with the Triassic granodiorite, monzogranite and syenogranite rocks (Table 8.4). The Galinge skarn Fe mineralization age (phlogopite $^{40}Ar/^{39}Ar = 234 \pm 3.7$ Ma) is largely coeval with the Galinge granodiorite age, and suggests possible genetic link between them. Tawenchahan skarn deposit, spatially closed to the Galinge deposit, gets a mineralization age of 229.9 ± 3.5 Ma through $^{40}Ar/^{39}Ar$ method [62]. In addition, the Suolajier and Hutouya skarn deposit get a mineralization age range of ca. 225–239 Ma (Table 8.4) [19, 21]. These skarn deposits are intensively developed in a range of ca. 225–240 Ma, and strongly associated with monzogranite and granodiorite according to temporal relations (Fig. 8.6). In addition, two important deposits associated with the alkaline syenogranite are the Yugouzi Fe–Cu (Mo) and Yazigou Zn–Pb–Cu porphyry-skarn deposits (Table 8.4). Zhou et al. [75] confirmed that the metallogenic age of Yugouzi skarn deposit is 210.1 ± 4.8 Ma through Re–Os dating of molybdenite, and the associated Yugouzi syenogranite age is 210 ± 0.6 Ma [4]. Age of the Yazigou porphyry mineralization is 224.7 ± 3.4 Ma [27], which reflects that the syenogranite porphyry (224.0 ± 4.6 Ma) was the metallogenic source [37]. These porphyry or epithermal deposits associated with syenogranite mainly range in ca. 210–224 Ma. Therefore, we conclude here the Qiman Tagh mineralization belt can be divided in two mineralization stage: 225–240 Ma for skarn mineralization and 210–224 Ma for porphyry or epithermal mineralization stage. Furthermore, the Galinge deposit is a typical skarn deposit associated with the granodiorite-derived hydrothermal fluids formed in a post-collisional environment, rather than SEDEX-type as previously suggested [9, 69, 71].

Table 8.4 The metallogenic age of porphyry-skarn deposit from Qiman Tagh

Deposit	Type	Associate igneous rocks	Minerals	Method	Age (Ma)	Reference
Galinge	Skarn	Granodiorite	Phlogopite	$^{40}Ar–^{39}Ar$	235.8 ± 1.7	This study
Tawenchahan	Skarn	Granodiorite	Muscovite	$^{40}Ar–^{39}Ar$	229.9 ± 3.5	[62]
Suolajier	Skarn	Granodiorite	Molybdenite	$^{187}Re–^{187}Os$	239.0 ± 11.0	[19]
Hutouya	Skarn	Granodiorite and monzogranite	Molybdenite	$^{187}Re–^{187}Os$	225.0 ± 4.0	[21]
			Molybdenite	$^{187}Re–^{187}Os$	230.1 ± 4.7	
Yazigou	Porphyry	Porphyritic syenogranite and monzogranite	Molybdenite	$^{187}Re–^{187}Os$	224.7 ± 3.4	[27]
Lalingzaohuo	Porphyry-skarn	Granodiorite	Molybdenite	$^{187}Re–^{187}Os$	240.8 ± 4.0	[67]
		Syenogranite	Molybdenite	$^{187}Re–^{187}Os$	214.5 ± 4.9	
Changshan	Porphyry-skarn	Porphyraceous monzogranite	Molybdenite	$^{187}Re–^{187}Os$	228	[20, 74]
		Syenogranite	Molybdenite	$^{187}Re–^{187}Os$	218	
Yugouzi	Porphyry-skarn	Syenogranite	Molybdenite	$^{187}Re–^{187}Os$	210.1 ± 4.8	[75]

8.3.2 Petrogenesis of the Galinge Granitoids

The Galinge intrusions belong to the Late Triassic Qiman Tagh post-collisional intrusions. The granodiorite from Galinge, Yemaquan and Kaerqueka are metaluminous and high-K calc-alkaline. High-K calc-alkaline granitoids are widely present in various geodynamic environments and can be derived from diverse magma sources [3, 5, 35]. Many researchers support that depressions melting of the thickened continental crust is the main sources for post-collisional magmas [28, 36, 51, 57, 61]. But some scholars considered that partial melting of mantle component is a plausible source [8, 13, 14].

Generally, plagioclase fractionation could lead to a strongly negative Eu anomaly [53] (Fig. 8.7). However, the Eu/Eu^* values decrease with increasing partial melting of amphibole [46] (Fig. 8.7). The moderate Eu anomalies ($Eu/Eu^* = 0.54–0.94$) of all samples are were likely to be mainly caused by the partial melting of plagioclase and amphibole together as shown in the Fig. 8.7. The decreased Ba/Yb ratios from the granodiorite to the diorite porphyry dikes may have been influenced by the partial melting degree of amphibole, and the partial melting degree of amphibole for the granodioritic magma is lower than the degree for the diorite magma, and the degree for the pyroxene diorite is the highest (Fig. 8.7). Therefore, we assumed here the acidic-mafic signature of magma may be controlled by the partial melting degree of amphiboles in lower crust. Experimental studies focusing on the melting behavior of natural amphibolite facies rocks have confirmed that granitic liquids can be produced by melting reactions that begin with the dehydration of amphibole [52, 55, 68].

All the Galinge intrusions show a metaluminous signature with a relatively low A/CNK range (A/CNK < 1.0). Partial melting experiments indicate that low alumina melts ($Al_2O_3 < 15\%$) are partial melting products of lower crustal rocks [50].

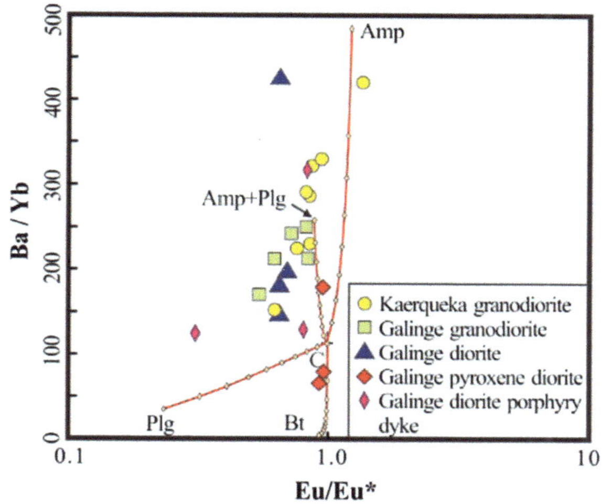

Fig. 8.7 Eu/Eu* versus Ba/Yb diagram for the various Galinge and Kaerqueka granitoids compared to the biotite (Bt), plagioclase (Plg) and amphibole (Amp) partial melting trends which were calculated using the partition coefficients of Keskin [34]. Tick marks represent the fraction of partial melting versus various minerals, which was calculated using the petromodeler software [18]. The starting points stands for the composition of bulk continental crust [60]. The linear trends for the Galinge intrusions indicate that the evolution of magma is dominantly controlled by the plagioclase and amphibole

In contrast, partial melting of MORBs under subduction pressure and temperature conditions (20 kbar and 1100 °C) produced high-alumina ($A1_2O_3 = 17.2\%$) dacitic magma with higher degrees of partial melting (10–20%) [51]. In addition, magmas generated by the partial melting of mafic amphibolite facies rocks in lower crust are strongly to moderately peraluminous (A/CNK > 1) up to amphibole-out boundary, neutral at the boundary (A/CNK = 1.0), and increasingly metaluminous (A/CNK < 1.0) beyond it [17, 49]. Therefore, the A/CNK values gradually decreasing from granodiorites to diorite porphyry dikes indicates that amphibolite dehydration melting degrees from intermediate-acidic granodiorite to mafic diorite porphyry dykes become higher and higher.

In the Nb/Ta versus Zr/Sm diagram (Fig. 8.8), the compositions of the Galinge intrusions show more similar characters with Archean continental crust that contains a relatively low Nb/Ta and high Zr/Sm ratio ranges. Furthermore, the Nb/Ta ratios of the Galinge intrusion increase progressively from the granodiorite to the diorite porphyry dykes, and this trend is also compatible with increasing partial melting degree of amphibolite facies rocks. Generally, low Nb/Ta and high Zr/Sm ratios are caused by the low degree melting of rutile-free amphibolite [23]. In contrast, the melt yielded by eclogite or amphibole eclogite generally contains high Nb/Ta ratios (Fig. 8.8). Therefore, partial melting of amphibolite facies rocks in lower crust is more convincing for the formation of the Galinge intrusions. In addition, at most

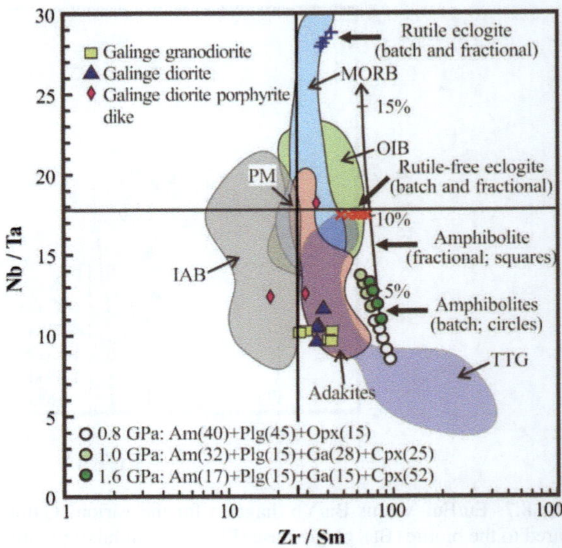

Fig. 8.8 Nb/Ta versus Zr/Sm diagram for the Galinge intrusions. The average Nb/Ta values from various tectonic environment are: 15.7 for adakite from Drummond et al. [15]; 16.7 for MORB from Kamber et al. [33]; 15.9 ± 0.6 for OIB source region from Pfänder et al. [48]; 11 for Archean continental crust from Taylor and Mclennan [60]. In addition, rutile-bearing eclogite produced by slab-melting is an additional reservoir with high Nb/Ta ratios [26, 54]

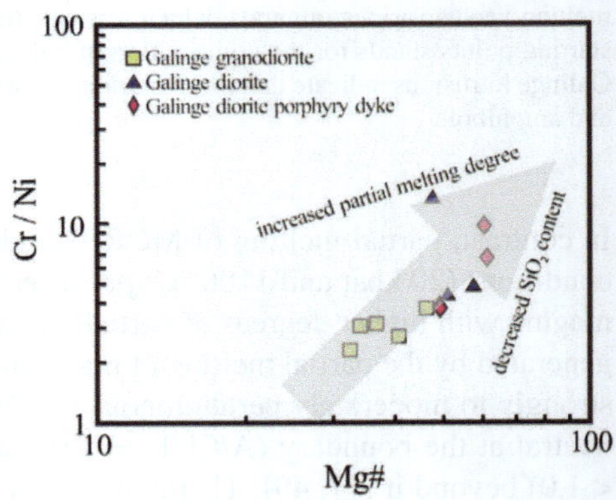

Fig. 8.9 Cr/Ni versus Mg# diagram of the Galinge granodiorite, diorite and diorite porphyry dikes

convergent margins thickened crust cannot reach the metamorphic conditions of eclogite [12], and the metamorphic basement of the Galinge region does not contain eclogite [38, 73].

The higher Mg# and Cr/Ni content in the diorite porphyry dikes and diorites (Fig. 8.9) than those of the granodiorite implies that they were the partial melting degree may be related with the SiO_2 content in melts. Experiments by Brown and Fyfe [6] suggest that SiO_2 content of melts derived from partial melting of basalts decreased with the increasing pressure, which indicate that the liquids would evolve be more dioritic at higher pressures. Therefore, accompanying the increasing pressure and partial melting degree, the melts generated by the amphibole dehydration melting may be tend to be more dioritic component.

Zircon Hf isotopic analyses yielded a wide range of Hf isotopic composition. Most zircons have negative εHf(t) values between −4.87 and −1.05. Figure 8.4 shows that almost all data points plot in the region between the lower crustal line and the chondritic reference line (the range of εHf(t) from +1 to −1). Although granitoids with εHf(t) $\approx$ 0 might be produced by mixing between older depleted mantle and enriched crust derivatives, such mixing would probably result in a broad range of initial isotope ratios from highly positive to strongly negative [1]. Therefore, the narrow εHf(t) range of the Galinge intrusions indicates that their parental magmas may have been derived predominately from the lower crust (Fig. 8.4a, b). In addition, a few of samples have weakly positive εHf(t) values of the Galinge granodiorites. As reviewed by Johnson et al. [32], the presence of garnet as a residual phase during partial melting influences strongly εHf(t) values. Dehydration melting of amphibolite facies in the lower crust produces significant amount of garnet in the residues, and therefore the resultant melt evolves to anomalously low εHf values over time. The relatively low $^{176}Lu/^{177}Hf$ and $^{176}Hf/^{177}Hf$ may have been related to the residual garnet, which has the greatest potential for decoupling Hf and Nb isotopic systems [63, 64].

8.4 Conclusion

(1) In this study, we obtained zircon U–Pb ages of 229–226 Ma for granodiorite, 223–218 Ma for diorite, and 226–217 Ma for diorite porphyry dyke.
(2) The Galinge skarn Fe mineralization age (phlogopite $^{40}Ar/^{39}Ar = 234 \pm 3.7$ Ma) is largely coeval with the Galinge granodiorite age, which is consistent with the regional granitoid and porphyry-related metallogenic ages and suggest possible genetic link between them.
(3) The Qiman Tagh mineralization belt can be divided into two mineralization stages: 225–240 Ma for skarn mineralization and 210–224 Ma for porphyry or epithermal mineralization stage.
(4) The different geochemical characters indicate that the fractionation between the melt and residuals were attributed to the melting of amphibole and plagioclase and the presence of garnet in the residues. Partial melting of amphibolite-facies rocks may have generated the Galinge granitoids. The partial melting degree is associated with the SiO_2 content of generated melts.

References

1. Amelin Y, Lee D, Halliday AN et al (1999) Nature of the Earth's earliest crust from hafnium isotopes in single detrital zircons. Nature 399(6733):252–255
2. Bai YN, Sun FY, Jian Y, Liu HC, Zhang DM (2016) Zircon U-Pb geochronology and geochemistry of pyroxene diorite in Galinge iron-polymetallic deposit. East Kunlun. Global Geol 35(1):17–27

3. Barbarin B (1999) A review of the relationships between granitoid types, their origins and their geodynamic environments. Lithos 46(3):605–626
4. Bing Q, Gao YB, Li K et al (2015) Zircon U-Pb-Hf isotopes and whole rock geochemistry constraints on the petrogenesis of iron-rare metal mineralization related alkaline granitic intrusive rock in Yugouzi area, eastern Kunlun, Xinjiang. Acta Petrol Sin 31(9):2508–2520
5. Bonin B (1990) From orogenic to anorogenic settings: evolution of granitoid suites after a major orogenesis. Geol J 25(3–4):261–270
6. Brown GC, Fyfe WS (1970) The production of granitic melts during ultrametamorphism. Contrib Mineral Petrol 28(4):310–318
7. Chappell BW (1999) Aluminium saturation in I- and S-type granites and the characterization of fractionated haplogranites. Lithos 46(3):535–551
8. Chappell BW, White A (1992) I- and S-type granites in the Lachlan Fold Belt. Geol Soc Am Spec Pap 272:1–26
9. Chen SS, Fu YX, Bao GY et al (2009) Characteristics and genesis of the Galinge polymetallic Fe deposit in west segment of East Kunlun Moutain, Qinghai province. Miner Resour Geol 23(6):542–546
10. Chu NC, Taylor RN, Chavagnac V et al (2002) Hf isotope ratio analysis using multi-collector inductively coupled plasma mass spectrometry: an evaluation of isobaric interference corrections. J Anal Atom Spectrom 17(12):1567–1574
11. Connelly JN (2001) Degree of preservation of igneous zonation in zircon as a signpost for concordancy in U/Pb geochronology. Chem Geol 172(1):25–39
12. Defant MJ, Drummond MS (1990) Derivation of some modern arc magmas by melting of young subducted lithosphere. Nature 347(6294):662–665
13. Didier J, Barbarin B (1991) The different types of enclaves in granites-nomenclature. Enclaves Granite Petrol 13:19–24
14. Didier J, Renouf JT (1973) Granites and their enclaves: the bearing of enclaves on the origin of granites. Elsevier, Amsterdam
15. Drummond MS, Defant MJ, Kepezhinskas PK (1996) Petrogenesis of slab-derived trondhjemite–tonalite–dacite/adakite magmas. Trans R Soc Edinb: Earth Sci 87(1–2):205–215
16. Elhlou S, Belousova E, Griffin WL et al (2006) Trace element and isotopic composition of GJ-red zircon standard by laser ablation. Geochim Cosmochim Acta 70(18):A158
17. Ellis DJ, Thompson AB (1986) Subsolidus and partial melting reactions in the quartz-excess $CaO + MgO + Al_2O_3 + SiO_2 + H_2O$ system under water-excess and water-deficient conditions to 10 kb: some implications for the origin of peraluminous melts from mafic rocks. J Petrol 27(1):91–121
18. Ersoy EY (2013) PETROMODELER (Petrological Modeler): a Microsoft® Excel© spreadsheet program for modelling melting, mixing, crystallization and assimilation processes in magmatic systems. Turk J Earth Sci 22(1)
19. Feng CY, Li DS, Qu WD et al (2009) Re-Os Iisotopic dating of molybdenite from the Suolajier skarn-type copper-molybdenum deposit of Qimantage Mountain in Qinghai Province and its geological significance. Rock Mineral Anal 28(3):223–227
20. Feng CY, Li DS,Wu ZS, Li JH, Zhang ZY, Zhang AK, Shu XF, Su SS (2010) Major types, time-space distribution and metallogenesis of polymetallic deposits in the Qimantage metallogenic belt, eastern Kunlun area. Northwestern Geol 43(4):10–17
21. Feng CY, Wang XP, Shu XF et al (2011) Isotopic chronology of the Hutouya skarn lead-zinc polymetallic ore district in Qimatage area of Qinghai Province and its geological significance. J Jilin Univ (Earth Sci Ed) 41(6):1806–1818
22. Feng CY, Wang S, Li GC et al (2012) Middle to Late Triassic granitoids in the Qimantage area, Qinghai Province, China: chronology, geochemistry and metallogenic significances. Acta Petrol Sin 28(2):665–678
23. Foley SF, Tiepolo M, Vannucci R (2002) Growth of early continental crust controlled by melting of amphibolite in subduction zones. Nature 417(6891):837–840
24. Frost CD, Frost BR (2011) On ferroan (A-type) granitoids: their compositional variability and modes of origin. J Petrol 39–53

25. Frost BR, Barnes CG, Collins WJ et al (2001) A geochemical classification for granitic rocks. J Petrol 42(11):2033–2048
26. Green TH (1995) Significance of Nb/Ta as an indicator of geochemical processes in the crust-mantle system. Chem Geol 120(3):347–359
27. He SY, Li DS, Li LL et al (2009) Re-Os age of molybdenite from the Yazigou copper (molybdenum) mineralized area in eastern Kunlun of Qinghai Province, and its geological significance. Geotecton Metallog 33(2):236–242
28. Holloway JR, Ford CE (1975) Fluid-absent melting of the fluoro-hydroxy amphibole pargasite to 35 kilobars. Earth Planet Sci Lett 25(1):44–48
29. Hou K, Li Y, Tian YY (2009) In situ U-Pb zircon dating using laser ablation-multi ion counting-ICP-MS. Mineral Depos 28(4):481–492
30. Hou KJ, Li YH, Zou TR et al (2007) Laser ablation-MC-ICP-MS technique for Hf isotope microanalysis of zircon and its geological applications. Acta Petrol Sin 23(10):2595–2604
31. Irvine T, Baragar W (1971) A guide to the chemical classification of the common volcanic rocks. Can J Earth Sci 8(5):523–548
32. Johnson CM, Shirey SB, Barovich KM (1996) New approaches to crustal evolution studies and the origin of granitic rocks: what can the Lu–Hf and Re–Os isotope systems tell us? Geol Soc Am Spec Pap 315:339–352
33. Kamber BS, Ewart A, Collerson KD, Bruce MC, McDonald GD (2002) Fluid-mobile trace element constraints on the role of slab melting and implications for Archaean crustal growth models. Contrib Mineral Petr 144(1):38–56
34. Keskin M (1994) Genesis of collision-related volcanism on the Erzurum-kars plateau, North eastern Turkey. Durham University, pp 1–363
35. Kouchi A, Sunagawa I (1983) Mixing basaltic and dacitic magmas by forced convection. Nature 304(5926):527–528
36. Le Breton N, Thompson AB (1988) Fluid-absent (dehydration) melting of biotite in metapelites in the early stages of crustal anatexis. Contrib Mineral Petrol 99(2):226–237
37. Li SJ, Sun FY, Feng CY et al (2008) Geochronological study on Yazigou polymetallic deposit in Eastern Kunlun, Qinghai Province. Acta Geol Sin 20(7):949–955
38. Liu YC, Ye ZF (1998) A new cognition on high grade-metamorphic rocks in Jinshuikou Area, Eastern Kunlun. Qinghai Geol 1:18–26
39. Liu YS, Hu ZC, Gao S et al (2008) In situ analysis of major and trace elements of anhydrous minerals by LA-ICP-MS without applying an internal standard. Chem Geol 257(1):34–43
40. Liu JN, Feng CY, Zhao YM, Li DX, Xiao Y, Zhou JH, Ma YS (2013) Characteristics of intrusive rock, metasomatites, mineralization and alteration in Yemaquan skarn Fe-Zn polymetallic deposit, Qinghai Province. Miner Depos 1:008
41. Ludwig KR (2001) User's manual for Isoplot/Ex v. 2.47. A geochronological toolkit for Microsoft Excel. BGC Special Publication 1a, Berkeley, 55 p
42. Ludwig KR (2003) User's manual for Isoplot 3.00: a geochronological toolkit for Microsoft Excel. Berkeley Geochronological Center, Special Publication
43. Luo ZH, Ke S, Cao YQ et al (2002) Late Indosinian mantle-derived magmatism in the East Kunlun. Geol Bull China 21(6):292–297
44. Maniar PD, Piccoli PM (1989) Tectonic discrimination of granitoids. Geol Soc Am Bull 101(5):635–643
45. McDonough WF, Sun S, Ringwood AE et al (1992) Potassium, rubidium, and cesium in the Earth and Moon and the evolution of the mantle of the Earth. Geochim Cosmochim Acta 56(3):1001–1012
46. Nicholls IA, Harris KL (1980) Experimental rare earth element partition coefficients for garnet, clinopyroxene and amphibole coexisting with andesitic and basaltic liquids. Geochim Cosmochim Acta 44(2):287–308
47. Patchett PJ, Tatsumoto M (1981) A routine high-precision method for Lu-Hf isotope geochemistry and chronology. Contributions to Mineralogy and Petrology 75(3):263–267
48. Pfänder JA, Münker C, Stracke A, Mezger K (2007) Nb/Ta and Zr/Hf in ocean island basalts—implications for crust–mantle differentiation and the fate of Niobium. Earth Planet Sc Lett 254(1):158–172

49. Rapp RP (1995) Amphibole-out phase boundary in partially melted metabasalt, its control over liquid fraction and composition, and source permeability. J Geophys Res: Solid Earth (1978–2012) 100(B8):15601–15610
50. Rapp RP, Watson WB (1991) Partial melting of amphibolite/eclogite and the origin of tonalite-trondhjemitic magmas. Precambrian Res 51(1):1–25
51. Rapp RP, Watson EB (1995) Dehydration melting of metabasalt at 8–32 kbar: implications for continental growth and crust-mantle recycling. J Petrol 36(4):891–931
52. Rapp RP, Watson EB, Miller CF (1991) Partial melting of amphibolite/eclogite and the origin of Archean trondhjemites and tonalites. Precambrian Res 51(1):1–25
53. Rudnick RL, Fountain DM (1995) Nature and composition of the continental crust: a lower crustal perspective. Rev Geophys 33(3):267–309
54. Rudnick RL, Barth M, Horn I, McDonough WF (2000) Rutile-bearing refractory eclogites: missing link between continents and depleted mantle. Science 287(5451):278–281
55. Rushmer T (1991) Partial melting of two amphibolites: contrasting experimental results under fluid-absent conditions. Contrib Mineral Petrol 107(1):41–59
56. Scherer EE, Whitehouse MJ, Münker C (2007) Zircon as a monitor of crustal growth. Elements 3(1):19–24
57. Skjerlie KP, Douce AEP, Johnston AD (1993) Fluid absent melting of a layered crustal protolith: implications for the generation of anatectic granites. Contrib Mineral Petrol 114(3):365–378
58. Steiger RH, Jäger E (1977) Subcommission on geochronology: convention on the use of decay constants in geo-and cosmochronology. Earth Planet Sci Lett 36(3):359–362
59. Sun SS, McDonough WF (1989) Chemical and isotopic systematics of oceanic basalts: implications for mantle composition and processes. Geol Soc Lond Spec Publ 42(1):313–345
60. Taylor SR, Mclennan SM (1995) The geochemical evolution of the continental crust. Rev Geophysics 33(2):241–265
61. Thompson AB (1982) Dehydration melting of pelitic rocks and the generation of H_2O-undersaturated granitic liquids. Am J Sci 282(10):1567–1595
62. Tian CS, Feng CY, Li JH et al (2013) 40Ar-39Ar geochronology of Tawenchahan Fe-polymetallic deposit in Qimantag Mountain of Qinghai Province and its geological implications. Miner Depos 32(1):169–176
63. Vervoort JD, Patchett JP (1996) Behavior of hafnium and neodymium isotopes in the crust: Constraints from Precambrian crustally derived granites. Geochim Cosmochim Acta 60(19):3717–3733
64. Vervoort JD, Patchett PJ, Gehrels GE et al (1996) Constraints on early Earth differentiation from hafnium and neodymium isotopes. Nature 379:624–627
65. Wang F, Zhu R, Yang L et al (2008) $^{40}Ar/^{39}Ar$ analyses on Quaternary K-Ar standard BB-24: evaluations. Int J Mass Spectrom 270(1):16–22
66. Wang BZ, Luo ZH, Li HY et al (2009) Petrotectonic assemblages and temporal-spatial framework of the Late Paleozoic-Early Mesozoic intrusions in the Qimantage Corridor of the East Kunlun belt. Geol China 36(4):769–782
67. Wang FC, Chen J, Xie ZY et al (2013) Geological features and Re-Os isotopic dating of the Lalingzaohuomolybdenum polymetallic deposit in East Kunlun. Geol China 40(4):1209–1217
68. Wolf MB, Wyllie PJ (1994) Dehydration-melting of amphibolite at 10 kbar: the effects of temperature and time. Contrib Mineral Petrol 115(4):369–383
69. Wu TX, Li HL (2009) Geological and Geochemical Characteristics of the Ironpolymetallic deposit in the Galinge area, Qinghai province. Bull Mineral Petrol Geochem 28(2):157–161
70. Wu F, Yang Y, Xie L et al (2006) Hf isotopic compositions of the standard zircons and baddeleyites used in U-Pb geochronology. Chem Geol 234(1):105–126
71. Wu XX, Bao GY, Yi YC et al (2007) The study on the genesis and geological characteristics of Galinge high grade iron deposit of Qinghai province. Gold Sci Technol 15(4):36–40
72. Yu M, Feng CY, Zhao YM et al (2015) Genesis of post-collisional calc-alkaline and alkaline granitoids in Qiman Tagh, East Kunlun, China. Lithos 239:45–59
73. Zhang JX, Meng FC, Wan YS et al (2003) Early Paleozoic tectono-thermal event of the Jinshuikou Group on the southern margin of Qaidam: Zircon U-Pb SHRIMP age evidence. Geol Bull China 22(6):397–404

74. Zhao SF, WU ZS, Zhang AK, Liu GL (2014) Geological Features, Deposit Genesis and Prospecting Potential of Changshan Molybdenum Deposit in Qimantage, Qinghai Province. Northwestern Geol 47(1):179–187
75. Zhou JH, Feng CY, Wang H et al (2014) Re-Os dating of molybdenite from the Yugouzi Fe-Cu (Mo) deposit in Qimantage, eastern Kunlun and its geological implications. Geol Explor 50(1):0001–0007

Chapter 9
Genesis of Post-collisional Calc-Alkaline and Alkaline Granitoids in Qiman Tagh

Qiman Tagh fold belt (or the North Kunlun belt) has been subjected to intense deformation, metamorphism and is characterized by voluminous magmatism throughout geological time. WNW-NW-trending compressive or transpressive faults are the main structures in the Qiman Tagh (Fig. 9.1). The Middle to Late Triassic was the most important period in the geological evolution of the Qiman Tagh orogen (Fig. 9.1) [30, 77], during which the Qiman Tagh orogen was in a post-collisional setting, witnessing a transition from a compressional to an extensional environment. The period is characterized by the occurrence of both mantle-derived rocks and voluminous intermediate to felsic intrusive rocks, including syenogranite, monzogranite, granodiorite, tonalite and diorite [30, 48]. Feng et al. [30] showed that the emplacement of post-collisional calc-alkaline granitoids was between 237 and 204 Ma. All granitoid types contain various quantities of mafic microgranular enclaves (MME) [30], which is interpreted to be a result of magmas mixing of different sources [3, 60].

Post-collisional magmatism is characterized by voluminous granitoid intrusions [6]. Post-collisional tectonics are complex and involve crustal-scale shears, extensional or transtensional faults. Therefore, whether magmas are formed by partial melting of the upper mantle or thickened lower crust is hard to distinguish [35]. Nevertheless, the two most abundant types of post-collisional granitoids are high-K calc-alkaline granitoids and silica-oversaturated alkaline to peralkaline granitoids [26, 47]. Alkaline granitoids may occur as large plutons with sizes comparable with those of calc-alkaline granitoids, and the two are often spatially and temporally closely associated. Calc-alkaline granitoids have often a mixed crustal and mantle source, but alkaline granitoids are poorly understood in terms of their origin [2]. There are three models accounting for the formation of alkaline magmas: (1) Via fractional crystallization of mantle-derived mafic to intermediate magmas [5, 7]; (2) Via partial melting of the residual source that has already undergone a granitic magma extraction in the deeper crust [19, 79]; (3) Metasomatism by magmatic-derived fluids enriched in carbon compounds and halogens is proposed [1, 38].

M. Yu, *Metallogenic Mechanism of the Galinge Polymetallic Iron Skarn Deposit, Qiman Tagh Mountains, Qinghai Province*, Springer Theses,
https://doi.org/10.1007/978-981-10-7907-8_9

Various research has been carried out to understand magmatism in the East Kunlun. However, evolution history of the Middle to Late Triassic Qiman Tagh magmatism remains unresolved. This paper aims to: (1) Systematically obtain field, petrogeochemical and geochronological data of the calc-alkaline and alkaline granitoids (Galinge, Yemaquan, Kaerqueka, Xiaohongshan, Jingren, Hutouya, Kaerqueka, and Ulan Uzhuer) that are spatially and temporally linked; (2) Discuss the genesis and evolution of the two distinctive calc-alkaline (CA) and alkaline (Alk) rock suites; (3) Explain the genetic processes that form the various post-collisional granitoid types.

9.1 Late Triassic Post-collisional Granites

Qiman Tagh is composed of two major magmatic belts: One subparallel to the SW-trending Altun fault and is dominated by Ordovician and Silurian igneous rocks. The other is NW-trending and is mainly composed of Hercynian and Indosinian igneous rocks (Fig. 9.1). The Indosinian igneous rocks are the major subject of this study, and consist mainly of voluminous intermediate to felsic granitoids including syenogranite, monzogranite, granodiorite, tonalite and diorite. Detailed petrographic descriptions are listed in Table 9.1. The syenogranites contain higher contents of K-feldspar than monzogranites and granodiorites. Table 9.2 summarizes the published high-precision U–Pb zircon ages of the Late Triassic (237–204 Ma) Qiman Tagh granitoids and diorites, as well as some whole rocks Rb–Sr ages.

9.1.1 Major and Trace Elements

In this paper, a total of 52 samples were collected for analyses, which include some representative granitic rocks from different locations in Qiman Tagh [17, 30, 45, 86,

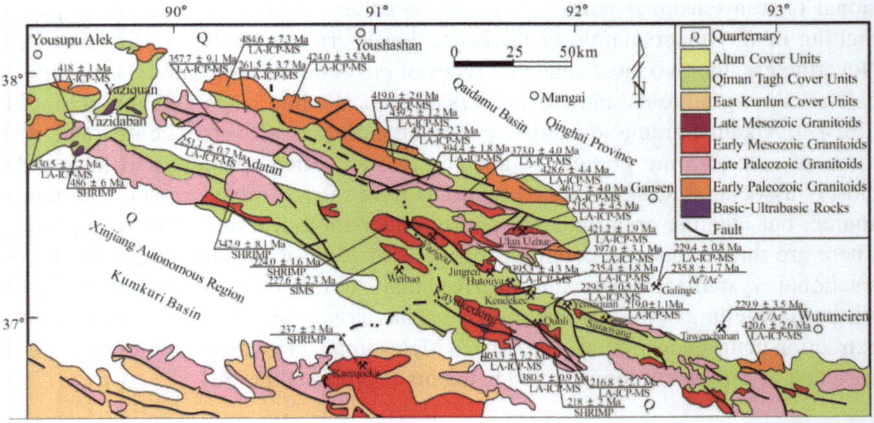

Fig. 9.1 Geologic map showing the structural geology for Indosinian convergent zone along the southern margin of the East Kunlun Mountain [27, 43, 84]

Table 9.1 The location and petrographical contents of Qiman Tagh middle to late Triassic granitoids

Location	Lithology	Position	Occurrence	Description	Major minerals	Literature
Jingren	syenogranite		Stock	Flesh pink color, medium-fine grained structure	Kfs (50%) + Plg (15%) + Qtz (30%) + Bio (5%)	[44]
Ulan Uzhur	syenogranite		Dike	Flesh pink color, porphyritic structure	Kfs (60%) + Plg (15%) + Qtz (25%)	[68]
Xiaohongshan	syenogranite	Lng: 91° 52′; Lat: 37° 30′	Batholith	Flesh pink color, medium-fine grained granitic structure	Kfs (55%) + Plg (13%) + Qtz (30%) + Bio (1%)	[17]
Kaerqueka	monzogranite	Lng: 91° 01.1′; Lat: 36° 45. 8′	Batholith	Light flesh pink color, medium-coarse grained porphyaceous structure	Kfs (35%) + Plg (30%) + Qtz (20%) + Bio (15%)	[30]
Hutouya	monzogranite	Lng: 91° 37′23″; Lat: 37° 04′ 09″	Stock	Light gray color, medium-fine grained granitic structure	Kfs (35%) + Plg (25%) + Qtz (30%) + Bio (10%)	[29]
Yemaquan	monzogranite	M1 and M6	Stock	Light gray color, medium-fine grained granitic structure	Kfs (40%) + Plg (35%) + Qtz (25%) + Amp (5%) + Bio (5%)	[45]
Yemaquan	granodiorite	M9, M10 and M13	Stock, apophyse	Light gray color, medium-fine grained granitic structure	Plg (40%) + Kfs (20%) + Qtz (15%) + Amp (15%) + Bio (5%)	[45]
Kaerqueka	granodiorite	Lng: 91° 02.13′; Lat: 36° 43.6′	Stock	Gray color, medium-fine grained granitic structure	Plg (50%) + Kfs (20%) + Qtz (20%) + Amp (5%) + Bio (5%)	[78]
Galinge	granodiorite	Lng: 92° 21′ 09″; Lat: 37° 08′ 02″	Batholith	Gray color, medium-fine grained granitic structure	Plg (45%) + Kfs (25%) + Qtz (20%) + Amp (5%) + Bio (5%)	[86]

M1 represents the number of magmatic anomaly area in deposit

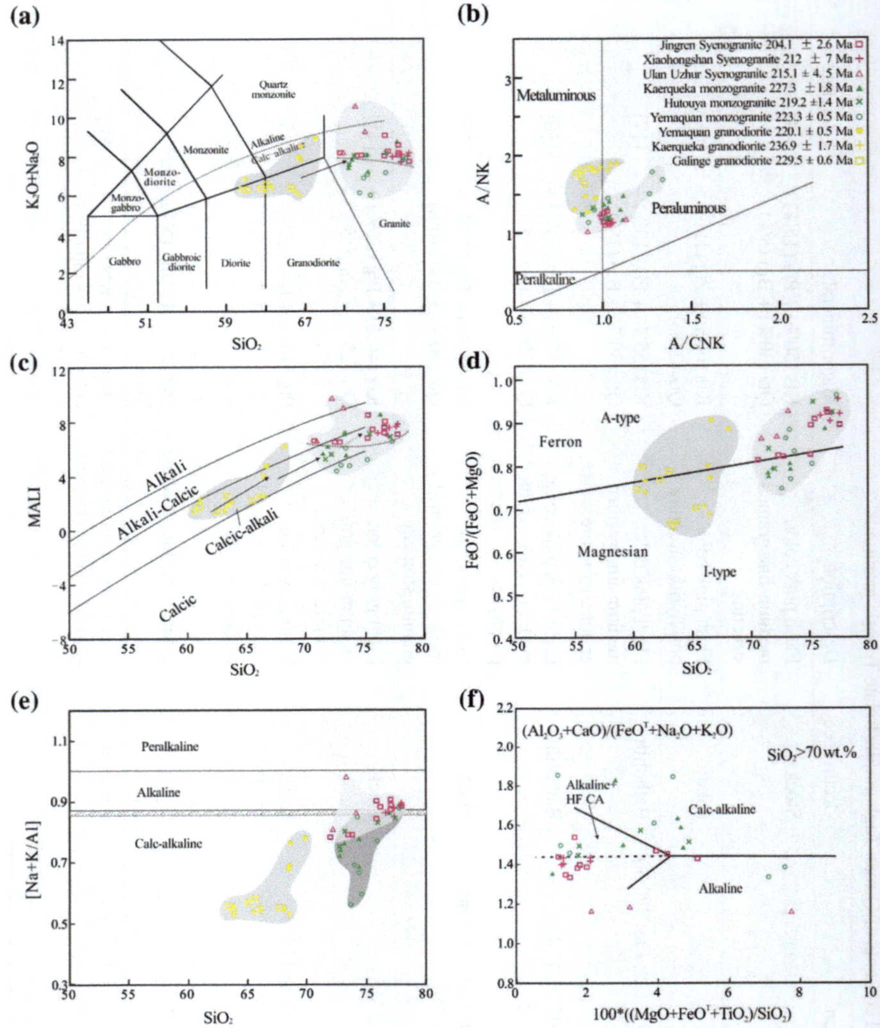

Fig. 9.2 **a** Total alkali silica diagram (TAS) representing the boundary ($\sigma = 4$) between alkaline and calc-alkaline [37]. **b** A/CNK versus A/NK diagram: A/CNK = $Al_2O_3/(K_2O + Na_2O + CaO)$ (molar ratios); A/NK = $Al_2O_3/(K_2O + Na_2O)$ (molar ratios) [15, 49, 72]. **c** Plots of the granitoids in the modified alkali lime index MALI ($Na_2O + K_2O–CaO$) versus SiO_2 discrimination diagram of Frost and Frost [31] showing an alkali-calcic nature for rocks. **d** Plots of the Qiman Tagh granitoids and the enclaves in the SiO_2 versus FeOt/(FeOt + MgO) discrimination diagram [32], showing a ferroan suite for the rocks. **e** SiO_2 versus NK/A (agpaitic index, mol%), peralkaline rocks have an agpaitic index that exceed 1, while other alkaline rocks are generally above 0.87 [46]. **f** 100 * $((MgO + FeO_t + TiO_2)/SiO_2)$ versus $(Al_2O_3 + CaO)/(FeO_t + Na_2O + K_2O)$, after Sylvester [73]. HF CA = high fractionated calc-alkaline granite

Table 9.2 Summery of age datas of granitoids and gabbros from the East Kunlun orogen

Pluton	Lithology	Method	Ages (Ma)	References
Kaerqueka	Monzogranite	SHRIMP U–Pb	227.3 ± 1.8	[30]
	Granodiorite	SHRIMP U–Pb	236.9 ± 1.7	[78]
Yazigou	Porphyritic Moyite	SHRIMP U–Pb	224.0 ± 1.6	[43]
Yemaquan	Granodiorite	LA-ICP-MS	220.1 ± 0.5	[88]
	Monzogranite	LA-ICP-MS	223.3 ± 0.5	
Kendekeke	Granodiorite	LA-ICP-MS	229.5 ± 0.5	[83]
	Monzogranite	LA-ICP-MS	230.5 ± 4.2	[82]
Changshan	Granite	SHRIMP U-Pb	220 ± 1	[30]
	Syenogranite	SHRIMP U-Pb	219.9 ± 1.3	
Hutouya	Granodiorite	LA-ICP-MS	235.4 ± 1.8	[29]
	Monzogranite	SHRIMP U-Pb	219.2 ± 1.4	
Jingren	Syenogranite	SHRIMP U-Pb	204.1 ± 2.6	[44]
Galinge	Granodorite	LA-ICP-MS	229.4 ± 0.8	[86]
	Diorite	LA-ICP-MS	217.6 ± 1.0	[86]
	Granite	SIMS	226. 4 ± 1. 7	[30]
Maxingdaban	Monzogranite	LA-ICP-MS	218 ± 2	[80]
Chulutaohai	Granodiorite	SIMS	226. 4 ± 1. 7	[30]
Xiaohongshan	Syenogranite	Rb-Sr	212 ± 7	[17]
Ulan Uzhur	Monzogranite Porphyry	SHRIMP U-Pb	215.1 ± 4. 5	[68]

88]. In the total alkali ($Na_2O + K_2O$) versus silica (TAS) plot (Fig. 9.2a), the distinction between Alk and CA granitoids is not only clear at SiO_2 content, but also at alkali concentrations. Alkali concentrations for the syenogranites are slightly higher than the monzogranites. In addition, all the rock compositions fall within the calc-alkaline field, and they have a positive correlation with SiO_2. Most syenogranite compositions are peraluminous based on the A/CNK versus A/NK diagram (Fig. 9.2b), the monzogranites do not have strictly distinguished boundary between metaluminous and peraluminous fields, however, almost all granodiorite samples are situated in the metaluminous field. In the modified alkali lime index (MALI: $Na_2O + K_2O–CaO$) versus SiO_2 discrimination diagram [31] (Fig. 9.2c), all the samples plot in the calc-alkaline domain and increase against the content of SiO_2. In the SiO_2 versus $FeO_t/(FeO_t + MgO)$ diagram [32] (Fig. 9.2d), all syenogranites plot in the ferroan domain, which was also recognized as the A-type, whilst most monzogranites plot in the same domain as the syenogranites. Nevertheless, the granodiorites plot predominantly in the magnesian field, typical for I-type granitoids. Most Alk syenogranites are characterized by agpaitic index ($(Na_2O + K_2O)/Al_2O_3$), mol% ratio) ranging from 0.8 to

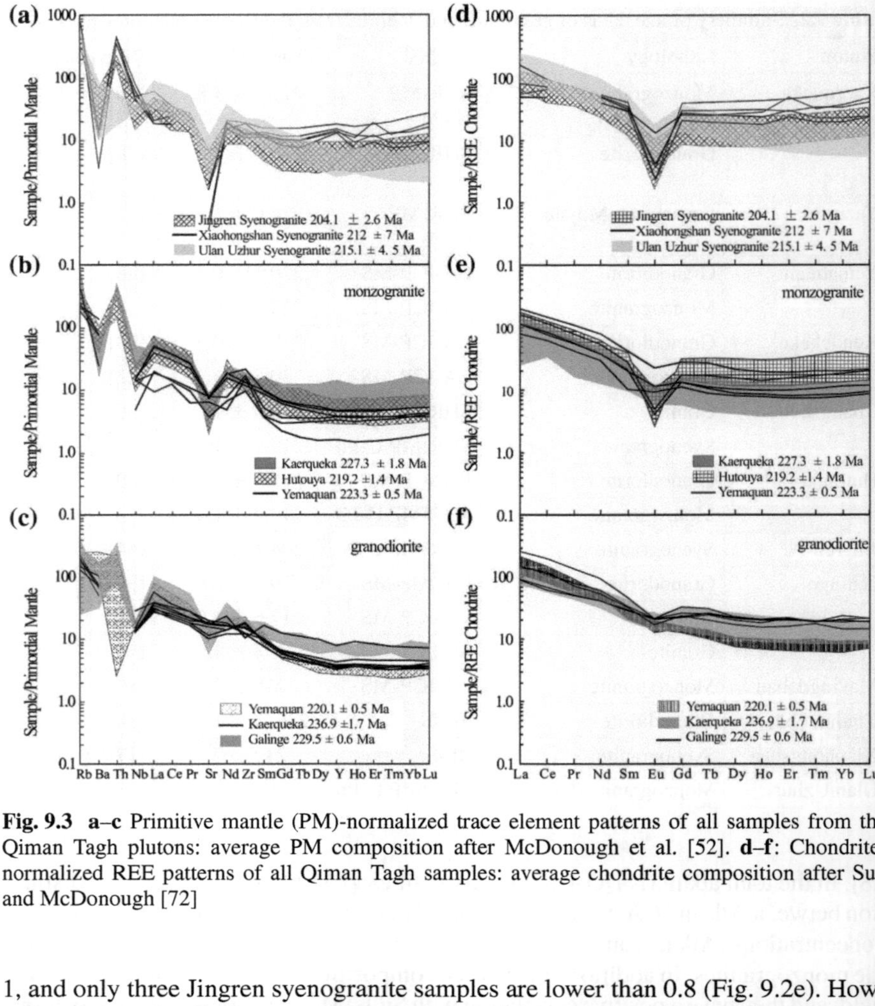

Fig. 9.3 a–c Primitive mantle (PM)-normalized trace element patterns of all samples from the Qiman Tagh plutons: average PM composition after McDonough et al. [52]. **d–f**: Chondrite-normalized REE patterns of all Qiman Tagh samples: average chondrite composition after Sun and McDonough [72]

1, and only three Jingren syenogranite samples are lower than 0.8 (Fig. 9.2e). However, all the calc-alkaline granitoids are below 0.87, which is the boundary between alkaline and calc-alkaline. The $(Al_2O_3 + CaO)/(FeO_t + Na_2O + K_2O)$ versus $100 * (MgO + FeO_t + TiO_2)/SiO_2$ diagram also distinguishes the alkaline syenogranites from the calc-alkaline monzogranites (Fig. 9.2f).

In the primitive mantle normalized multi-element diagram (Fig. 9.3a–c) [52], both of the Qiman Tagh CA and Alk granitoids are enriched in large-ion lithophile elements (LILEs; e.g., Rb and Th) and depleted in high field-strength elements (HFSEs; e.g., Pb, Ti, Y, Hf, Zr, Nb and Ta) (Fig. 9.3a).

Despite the fact that alkaline syenogranites have similar chondrite-normalized REE patterns as calc-alkaline granitoids, which are characterized by the enrichment of HREEs (Fig. 9.3d–f), the negative Eu anomalies of the alkaline syenogranites are more pronounced than that of the calc-alkaline granitoids. The most ΣREE of

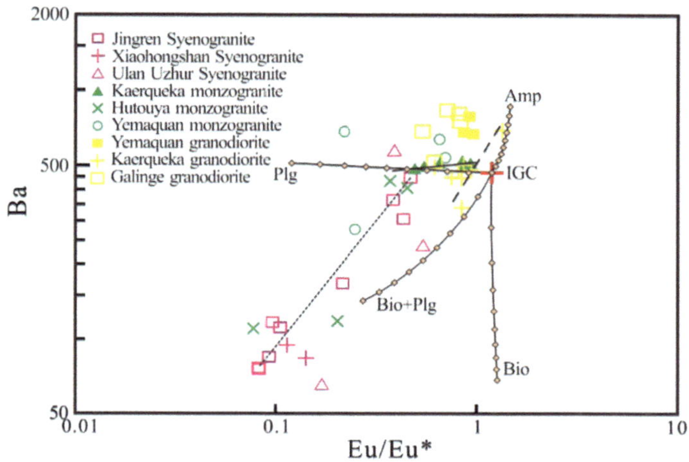

Fig. 9.4 Eu/Eu* versus Ba (ppm). Biotite (Bio), plagioclase (Plg), amphibole (Amp) and biotite + plagioclase (Bio + Plg) trends calculated using the partition coefficients of (1994). Tick marks represent the fraction of partial melting versus various minerals calculated using the petromodeler software [24]. The starting points stand for the composition of intermediate granulite [65]. The solid line and dashed line shows the fractionation trends for monzogranitic magma and syenogranitic- and granodioritic magma, respectively

alkaline syenogranites is above 400 ppm, of which most ΣLREE/ΣHREE ratios are variable from 1.06 to 4.23. Moreover, the $(La/Yb)_N$ ratios are ranging from 1.6 to 11.24, which means that the sources of these rocks were probably fractionated with LREE enrichment. The CA monzogranites in REE distribution patterns show moderate Eu negative anomalies (0.08–0.86), the ΣREE generally exceed 400 ppm similar to the Alk rocks, and the ΣLREE/ΣHREE ratios range from 1.06 to 4.2. The CA granodiorites rarely have significant Eu negative anomalies.

The significance of the plagioclase, K-feldspar, biotite and amphibole fractionation is illustrated in the Eu/Eu* versus Ba diagram (Fig. 9.4). The Eu/Eu* values of Qiman Tagh syenogranites range from 0.06 to 0.81, the monzogranites range from 0.08 to 0.86, and the granodiorites range from 0.54 to 1.34 (Table 9.3). The Ba content decreases sharply from the I- to A-type granitoids. Different magma types have different fractionation trends: The evolution of syenogranitic magma is influenced mainly by the fractionation of biotite, K-feldspar and plagioclase, whilst that of monzogranite is mainly controlled by plagioclase fractionation. In contrast, granodioritic magma evolution shows a different trend from that of syenogranite, and is probably controlled by amphibole fractionation.

The syenogranites from Xiaohongshan, Jingren and Ulan Uzhur intrusions have lower $Mg^{\#}$ values that range from 12.37 to 43.5 (mean = 29.32), whereas the monzogranites from Hutouya, Kaerqueka and Yemaquan have moderate $Mg^{\#}$ values ranging from 13.94 to 47.37 (mean = 34.3), and the granodiorites from Kaerqueka, Yemaquan and Galinge have higher $Mg^{\#}$ values ranging from 21.65 to 58.42 (mean = 47.44) (Table 9.3). A prominent feature of mantle assimilation and fractional crystallization

Table 9.3 The Sr–Nd isotopic data of syenogranite, monzogranite and granodiorite from Qiman Tagh

Location	Jingren				Kaerqueka			Kaerqueka					Yemaquan	
Pluton	Syenogranite				Monzogranite			Granodiorite					Granodiorite	
Sample	JY08	JY31	JY32	JY35	KD007	KD008	KD009	KD002	KD003	KD029	KD037	CZK1401-5	YMQ-TW1	YMQ-TW3
87Sr/86Sr (i)	0.71389	0.71062	0.71036	0.71058	0.71192	0.71162	0.71266	0.71066	0.71068	0.71147	0.7112	0.71103	0.711592	0.712451
143Nd/144Nd (i)	0.512143	0.512156	0.512153	0.512154	0.512086	0.512101	0.512062	0.512043	0.512058	0.512038	0.51202	0.512047	0.51232	0.512206
εSr (i)	−463.88	−100.3	−143.9	−1043.49	−18.39	−29.56	−79.37	25.59	36.64	46.51	21.65	17.96	19.78	39.82
εNd (i)	−8.04	−6.96	−6.9	−8.22	−8.17	−7.82	−9.97	−9.38	−9.11	−9.95	−9.92	−9.32	−2.96	−5.55
Age (Ma)	226	226	226	226	227	227	227	237	237	237	237	237	220	220

The Yemaquan Sr–Nd isotopic data are from Zhang [87], and the Kaerqueka and Jingren Sr–Nd isotopic data are from Feng et al. [30]

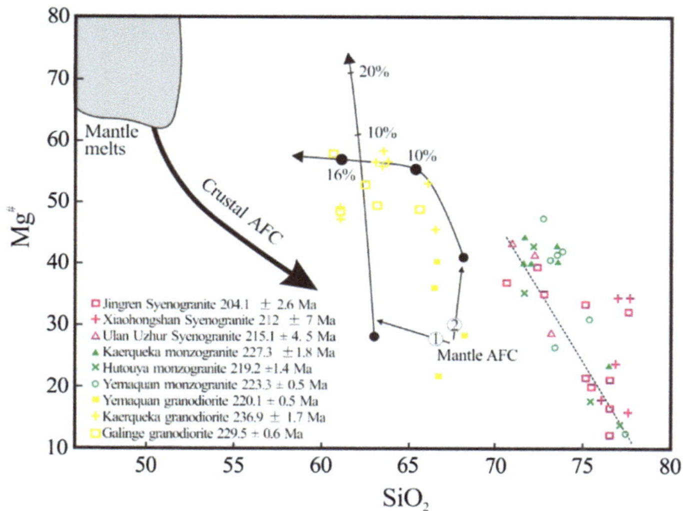

Fig. 9.5 Mg# versus SiO_2 diagram for the various Qiman Tagh granitoids compared to experimentally forming high pressure partial melts of garnet-amphibolite and eclogite, mantle melts. Mantle AFC curves are from Smithies [69] and show the proportion of the assimilated peridotite. The crustal AFC curves are from Stern and Kilian [70]

(AFC) pattern is depicted in the Mg# versus SiO_2 diagram (Fig. 9.5). The granodiorites from Kaerqueka and Galinge with $SiO_2 < 70\%$ have Mg# > ~45 (up to 58.42). Such data plot approximately along the mantle AFC curves, which point towards a mantle contamination [40, 70]. However, the granodiorites from Yemaquan have relatively lower Mg# values (Mg# = 21.65–40.32), which may have been the consequence of putative partial melting without the contamination of mantle melt [61]. The Mg# values of syenogranite show a linear decreasing trend with increasing SiO_2. Likewise, most of monzogranites have medium Mg# values along weak linear trends.

9.1.2 *Chemical Discrimination*

Diagrams of Ga/A1 versus certain major and trace elements (Fig. 9.6) can distinguish the CA and Alk granitic suites. A-type granites are distinguished from I-type granites by their higher Nb, Ga, Y and Rb, and lower Al, Mg, Ca, Sr and Ba [19, 81]. Although subalkaline compositions in some suites tend to overlap the compositions of other granite types, the more alkaline compositions are identified as A-type [79]. In Fig. 9.6, the CA granodiorites are located in the I- and S-type fields, and most of the Alk syenogranite compositions are identified as A-type. Whereas some CA monzogranite compositions tend to exceed the rectangular boundary and appear to be mixed with the Alk syenogranites, the Rb/Sr versus Y/Sr plot (Fig. 9.7) can minimize

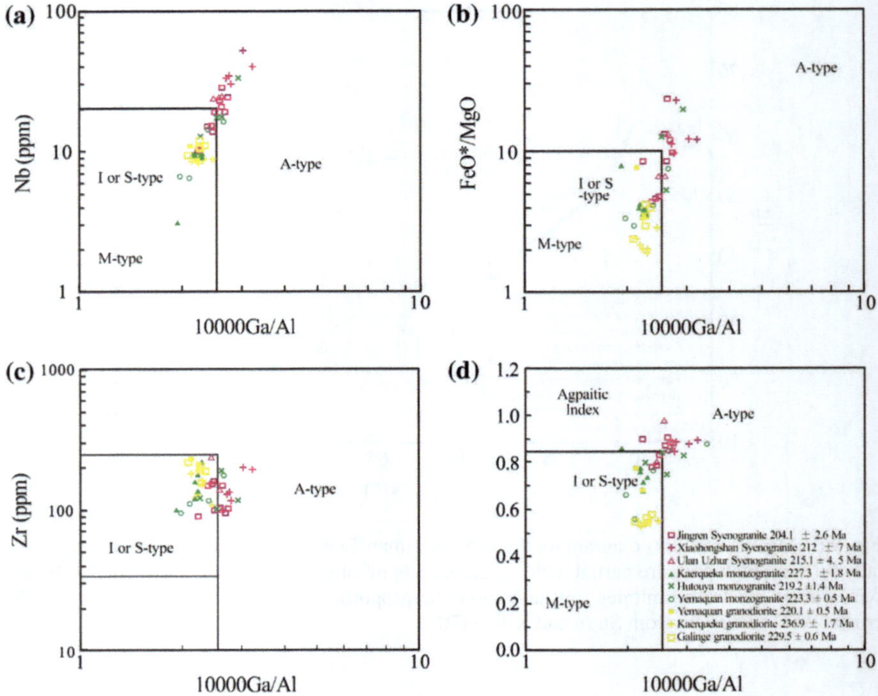

Fig. 9.6 a–d 10,000 Ga/A1 versus Nb, FeO^t/MgO, Zr and agpaitic index plots of CA and Alk granitic rocks for comparison purposes, I, S and M-type granites (in rectangular boxes) [79]

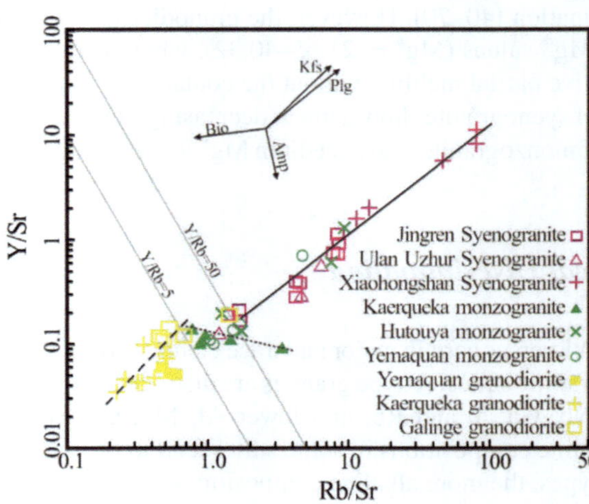

Fig. 9.7 Rb/Sr versus Y/Sr diagram. The solid line, thin dashed line and broad dashed line show the evolution trend of Alk syenogranites, monzogranites and granodiorites, respectively. Bio = Biotite, Amp = Amphibole, Plg = Plagioclase, Kfs = K-feldspar

the overlapped range between the monzogranite and syenogranite composition. CA series are almost distinguishable from the Alk series by Y/Rb ratios (granodiorite: Y/Rb < 5, monzogranite: 5 < Y/Rb < 50, syenogranite: 50 < Y/Rb).

9.2 Genesis of the Qiman Tagh Calc-Alkaline and Alkaline Magmas

Major elements geochemistry indicates that the Qiman Tagh granodiorites and monzogranites belong to the metaluminous, high-K calc-alkaline suite. High-K calc-alkaline granitoids (KCG) are present in various geodynamic environments [2]. They occur either during periods of stasis in a collision event, or during post-collisional compressive-extensional transitions [42]. Therefore, KCGs are abundant in orogenic belts and usually coexist with peraluminous alkaline granitoids. The syenogranites from Xiaohongshan, Jingren and Ulan Uzhur geochemically resemble peraluminous alkaline granites.

There is no consensus on the origin of A-type granitoids, but lower crustal partial melting is one of the most widely accepted models [19]. Alternatively, Creaser et al. [20] suggested that partial melting of crustal igneous sources (prior to which there was no melt depletion) followed by granulite-facies metamorphism may generate A-type granitoids. The A-type magma genesis has also been attributed to partial or complete melting of an alkali-metasomatized crust [51]. Furthermore, an oceanic island basalt and continental crust mixed source is also proposed by Eby [21, 22]. Crystallization of mantle-derived, parental magmas [75] and intense fractionation of I-type magmas [79] are some of the other propositions. In spite of many different genetic models for the A-type granitoids, they share similar chemical features and are derived from similar crustal melts.

Our study suggests that the partial melting of the residual material in the lower crust may explain the generation of A-type granitoids in Qiman Tagh. Firstly, the A-type melts are dry or almost anhydrous, which means fractionation from I-type melts is unlikely to be a cause because it would generate hydrous components [19]. In addition, experimental work [18] favors the formation of A-type granites by partial melting of an essentially anhydrous source and at elevated temperatures (>830 °C), which is higher than their calc-alkaline counterparts. Hence, the Qiman Tagh Alk granitoids were unlikely to be the differentiation products of CA granites [19]. In contrast to the alkaline granites, the calc-alkaline granites can be derived from anatexis of the lower crust by vapor-absent dehydration reactions of amphibole at temperatures of 750–850 °C [47].

A-type granites result mainly from partial melting of F and/or Cl enriched and dry granulitic residue in the lower crust after extraction of hydrous granite [16, 79]. Experimental studies [62, 66] indicate that dehydration melting of amphibolites may produce melts of felsic to intermediate calc-alkaline compositions (forming high-Al trondhjemite-tonalite, granodiorite, quartz diorite and diorite), leaving behind a granulite (plagioclase + clinopyroxene + othopyroxene + ilmenite + garnet) residue (at 8–12 kb) and garnet granulite to eclogite residues (at 12–32 kb). While amphibole and biotite remain as stable residual phases with F-rich minerals in the restite [36], less Cl would be retained in biotite and amphibole [53]. Therefore, the breakdown of the residual biotite or amphibole from the last melting event could release high amounts of volatile, and the following melting events must occur at higher temperature than the first melting event.

Granulite-facies metamorphism occurs in the lower crust where temperatures generally exceed 750 °C (and sometimes >1100 °C) [10]. Lower crustal thickening during collisions contributes to the formation of eclogite-facies rocks, when the continental crust is subjected to temperatures >900 °C at depths >40 km [34]. The middle crust is predominantly composed of amphibolite-facies rocks, containing significant K, Th and U [65]. As mentioned above, different mineral phase equilibria at confining physico-chemical conditions of metamorphism control the different geochemical parameters of melts, and thus resulting in different mineral phases.

The modal batch melting model is applied in this partial melting process study, which assumes that the melt remains in chemical equilibrium with restite during melting, and the proportion of minerals that undergo melting is the same as that in the source [89]. Note that crystal-liquid separation is generally imperfect in nature. Nonetheless, batch melting model is particularly useful in explaining the fractionation of trace elements between source rock and their restite compositions. The trace element partitioning coefficients for different rock-forming minerals from Keskin [41] are used in the melting model. The starting point of partial melting stands for the composition of intermediate granulite as per Rudnick and Fountain [65] or average crustal composition as per Taylor and McLennan [74].

In Fig. 9.4, the fractionation between plagioclase and melt will lead to a strong Eu anomaly [65]. However, biotite scarcely make any influence on the Eu/Eu^* values due to the almost equal partitioning coefficients of Sm, Eu and Gd. Furthermore, the Eu/Eu^* values will slightly decrease with the increasing fractionation of amphibole during partial melting. Based on the different fractionated trends of plagioclase, biotite and amphibole, the syenogranite evolution is controlled by the melting of plagioclase and biotite, the monzogranite is controlled by the melting of plagioclase, and the granodiorite is controlled by the melting of amphibole.

Brown and Fyfe [11] experimentally simulated a gradual melting progression, and stated that the melt compositions become enriched in the albite component as pressure increases, and indicated that the orthoclase contents of the melts increase with increasing phengite content associated with fusion. In addition, low Mg content is associated with muscovite melting. The solidus reaction at 825 ± 25 °C is that of phengite. Phengite is likely to be stable in the relatively MgO-poor and CaO- and Al_2O_3-rich basaltic system, whereas biotite and orthopyroxene are largely stable in picritic basalt (MgO-rich and CaO- and Al_2O_3-poor), and biotite, instead of phengite, may be stabilized even in rocks of basaltic compositions [54]. The melting of phengite or biotite can dramatically increase the A/CNK and K_2O contents in the bulk composition of the melt, and hence syenogranites have higher A/CNK values and K_2O contents than granodiorites and monzogranites.

Above the amphibole out boundary, clinopyroxene is the sole ferromagnesian mineral [67]. In silicate melts of granitic and basaltic compositions, tetrahedral and octahedral sites predominate and transition metal ions can enter into both types of sites [14]. The strong preference of Ni^{2+}, Co^{2+}, Cr^{3+} and V^{3+} for octahedral sites in minerals leads to their enrichment in certain pyroxenes, which are refractory residues during partial melting, because they have high crystal field stabilization energies [12]. In addition, Mg^{2+} has a compatible ionic radius for the M1 site of pyroxene which closely approximates octahedral symmetry [13]. Consequently, the melting of amphibole, leaving clinopyroxene in the residual melt, results in further depletion in $Mg^{\#}$, Ni, Co, Cr and V, as found in A-type granites.

Primitive mantle-normalized multi-element diagrams (Fig. 9.3a–c) show apparent negative anomalies of Ba, Nb, Pb, Ti, as well as LILE enrichments and HFSE depletions. LILE enrichments and HFSE depletions are commonly interpreted to be characteristics of island arc magmatism [56], which suggests that the CA and Alk silicic magma may have formed by partial melting of the juvenile crust formed in an island arc setting [26]. The almost flat REE distribution patterns with weak negative Eu anomalies (Fig. 9.3e–f) may be attributed to amphibole fractionation in partial melting [5]. The elevated LREE/HREE may be attributed to the presence of accessory minerals such as titanite, zircon and apatite [50, 76]. Esna-Ashari et al. [25] showed that there is a correlation between the plagioclase/amphibole ratio and the shape of REE patterns. Trace and rare earth elements (REEs) properties of Qiman Tagh granidoids resemble those of typical granites formed in a volcanic arc and within-plate settings [58].

A notable feature of the Kaerqueka and Galinge granodiorites is that the $Mg^{\#}$ (>45) are considerably higher than that of the Qiman Tagh monzogranites and syenogranites ($Mg^{\#} < 45$) (Fig. 9.5). It is noteworthy that the experimentally yielded partial melts of basalt have a maximum $Mg^{\#}$ value of ~45 [61]. Therefore, melts with $Mg^{\#} >$ ~50 is likely to have been contaminated by a more mafic material than basalt [69]. Experimental studies demonstrated that the addition of ~10% peridotite can increase the $Mg^{\#}$ of adakites from ~44 to ~55 (Fig. 9.5) [64], which can be achieved in nature via mantle assimilation and fractional crystallization (AFC) [85]. Consequently, the compositional range for the Triassic Qiman Tagh granodiorites has suggested both crustal partial melting and the addition of a small amount of peridotite.

The Qiman Tagh I-type and A-type granites exhibit moderately negative $\varepsilon_{Nd}(T)$ values (−9.97 to −2.96) and comparatively moderate initial $^{87}Sr/^{86}Sr$ values (0.71036–0.71389) (Table 9.3), which further indicate that they were mainly derived from ancient crustal sources and probably mixed with mantle material [30, 87]. A simple modeling for source contamination (Fig. 9.8) indicates that mixing of the upper mantle peridotite (along with about 5% of subducted continental mass) with rocks of middle to lower crustal characteristics, would provide the Sr–Nd isotopic compositions of the Qiman Tagh granites [39].

In the tectonic discrimination diagrams for granitoids [59], almost all studied granitic rocks plot within the field of post-collisional granites (Fig. 9.9a). Post-collisional Alk syenogranites are compositionally similar to within-plate granitoids (WPG), whereas the post-collisional CA granites show affinities towards the volcanic

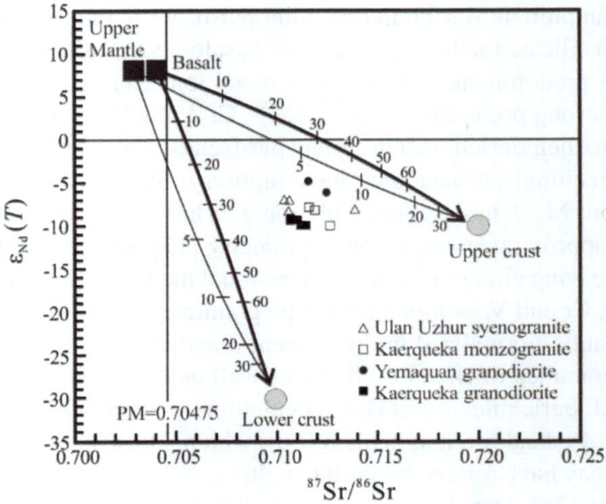

Fig. 9.8 $\varepsilon_{Nd}(T)$ versus initial $^{87}Sr/^{86}Sr$ plot for a variety of CA and Alk granitic rocks from the Qiman Tagh orogen. Mixing calculations for source and melt contaminations are based on Jahn et al. [39]. Upper mantle and upper continental crust element data are from Taylor and McLennan [74], lower crust data is from Rudnick and Fountain [65]

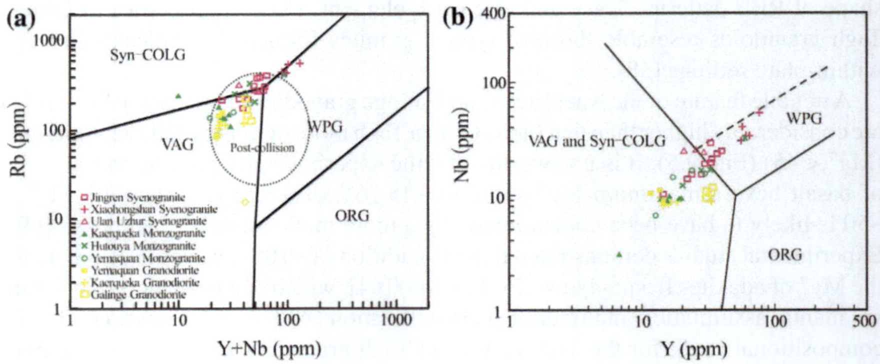

Fig. 9.9 Geotectonic discrimination diagram of all the samples. **a** Rb–(Yb + Ta) discrimination diagram of all samples [59]. **b** Ta–Yb discrimination diagram of all samples [59] Ocean Ridges Granites (ORG), Within-Plate Granites (WPG), Volcanic Arc Granites (VAG), Syn-collisional granites (Syn-COLG), Oceanic Arcs (OA), Continental Arcs (CA), Active Continental Margins (ACM), Within-Plate Volcanic Zones (WPVZ), WPB Within-Plate Basalts, Mid-Oceanic Ridge Basalts (MORB)

arc granitoids (VAG) and syn-collisional granites, possibly suggesting contamination by previous arc magmatic rocks (Fig. 9.9a, b).

Typical Alk granites can be subdivided into two groups based on Y/Nb ratio (Fig. 9.10) [21]: A_1 granites (Y/Nb < 1.2) are formed in anorogenic settings, e.g., in continental rifts; whereas A_2 granites (Y/Nb > 1.2) are formed in post-collisional

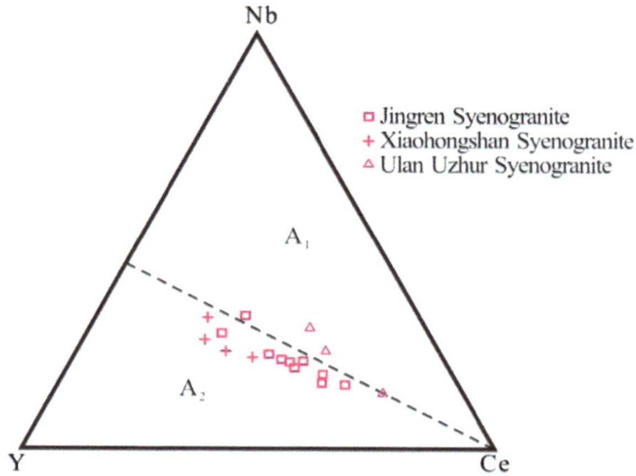

Fig. 9.10 A_1 and A_2 granitoids discrimination diagram for the Alk granites in Qiman Tagh [22]. Dashed line corresponds to Y/Nb ratio of 1/2. A_1 (anorogenic) granitoids are from rift, plume and hotspots; A_2 (post-collisional) granitoids are from post-orogenic and anorogenic environments

setting [22]. In the A_1 and A_2 granitoids discrimination diagram [22], most of the Qiman Tagh Alk granites fall into the A_2 (post-collisional granite) field.

Almost all of the Qiman Tagh granitoid suites plot in the relatively high Y/Nb ratio spectrum approaching the ratios of average lower crust (Y/Nb = 2.407, Ce/Nb = 4.699, Ba/La = 21.194) [33] (Fig. 9.11), which is associated with the generation of the A_2 group granitoids from previously melted crust [19]. It is noteworthy that although there is a substantial variation in Ba/La and Ce/Nb ratios throughout the evolution of the Qiman Tagh granitoid suites, the Y/Nb ratios remain remarkably constant, which implies all of the magmas may have the same type of crustal source [21]. In addition, Ce/Y ratios gradually decrease from granodiorites to the syenogranites (Fig. 9.11b), which is produced by the melting of the amphibole with contrastingly high content of Y and HREE. The granodiorites have remarkably higher Ba/La ratios, which is probably associated with the partial melting of the biotite since it contains relatively high Ba.

The continental La/Nb ratio can be used to constrain the proportion of crust formed in intraplate versus convergent margin settings [65]. The average La/Nb ratio in the lower crust is 2.6 ± 0.8 [4] and the values in the upper crust is 1.2 [74] (Fig. 9.12). It is generally depleted in Nb within the continental crust relative to highly incompatible La and therefore, intraplate magmas have La/Nb generally <1.0, whereas convergent margin magmas typically have high La/Nb ratios [72]. Since Nb is slightly more incompatible than La during partial melting, the La/Nb ratios decrease throughout the evolution of the Qiman Tagh granitoid suites from CA granitoids to Alk granitoids.

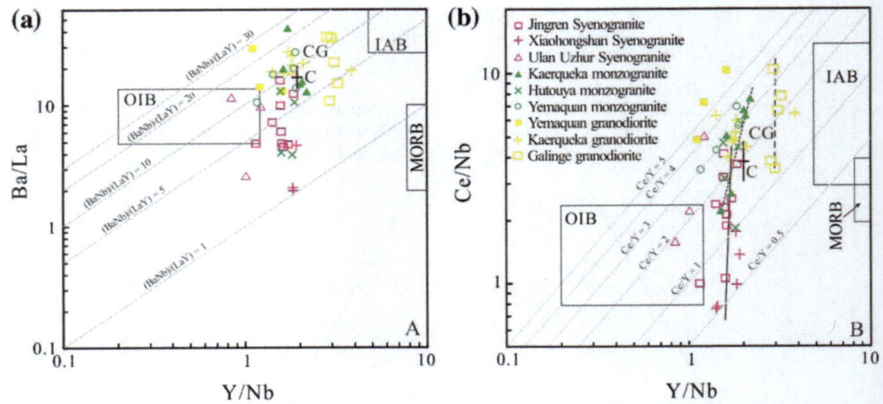

Fig. 9.11 **a** Y/Nb versus Ba/La diagram. **b** Y/Nb versus Ce/Nb [21]. OIB = field for Oceanic Island Basalts, IAB = field for Island Arc Basalts, CG = syn-Collisional Granite fields (from the representative analyses in Pearce et al. [59]. C = average crustal ratios from Taylor and McLennan [74]. The solid line in diagram (**b**) shows the evolution trend of Alk syenogranites, the thin dashed line shows the evolution trend of monzogranites, and the broad dashed line shows the evolution trend of granodiorites

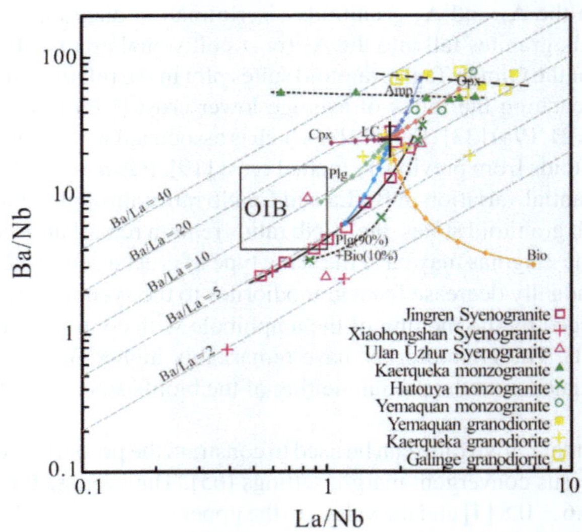

Fig. 9.12 La/Nb versus Ba/Nb diagram. OIB = field for Oceanic Island Basalts from Pearce et al. [59]. LC = average lower crustal ratios from Taylor and McLennan [74]. The solid line represents almost all of the Qiman Tagh granitoid suites fall in the region of CG and average crust between the OIB field and IAB field. The solid line, thin dashed line and broad dashed line represent the evolution trend of Alk syenogranites, monzogranites and granodiorites, respectively

9.3 Middle- to Late Triassic Post-collisional Magmatism in Qiman Tagh

In Qiman Tagh, intrusion of granodiorites, monzogranites and syenogranites occurred mainly in ca. 236.9–220.11 Ma, ca. 230.5–215.1 Ma and ca. 224–204.1 Ma, respectively (Fig. 9.13).

The presence of peraluminous and alkaline granitoids in Qiman Tagh signifies the end of the orogenic stage and the start of the anorogenic intraplate stage [8, 23, 57], which coincides well with typical post-collisional magmatic trend from calc-alkaline I-type granitoids to alkaline A-type granitoids. In Qiman Tagh, the I-type granodiorite and monzogranite melts were first produced by the melting of amphibolite-facies rocks. With the progressive temperature rise in the crust, the Alk granitic melts were formed by the partial melting of the residual garnet-granulite-facies rocks in the lower crust (Fig. 9.14). Our petrogenetic model provides an alternative explanation for the close space-time association of the CA and Alk granitoids in Qiman Tagh, and the model may also be applicable to many other CA and Alk granitoid occurrences along other orogenic belts around the world [11].

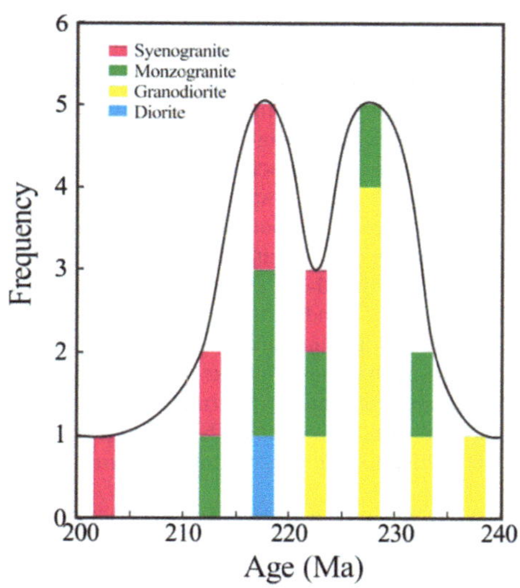

Fig. 9.13 Age compilation histogram for the Qiman Tagh granitoids and diorite [28, 80]

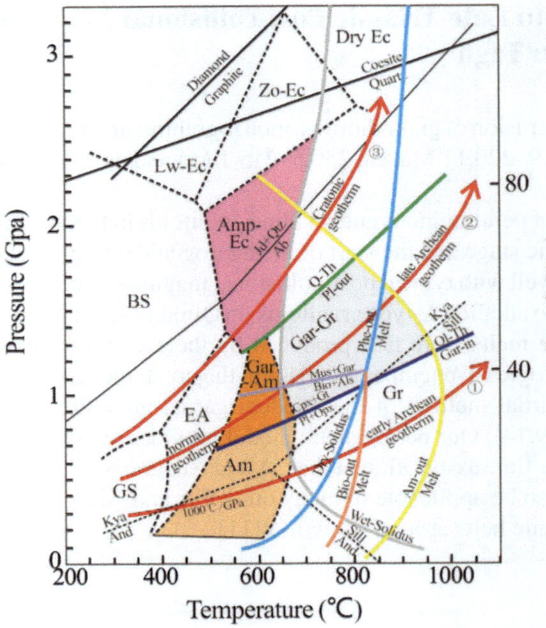

Fig. 9.14 Granulite-facies was subdivided into high-pressure granulite (garnet-bearing granulite) and low pressure granulite (two pyroxene-granulite) on the basis of the schematic reaction Opx + Ca-rich + Pl = Gt + Cpx + Q + Na-rich. The high-pressure amphibolites was separated with the low-pressure amphibolite on the basis of the reaction Hb + Pl = Di + Gt. The sillimanite-kyalite transition marks the boundary between low-P granulites (Gr) and high-P granulites (Gar-Gr). The 'garnet-in' in olivine tholeiite and 'plagioclase-out' in quartz tholeiite is from O'Brien and Rotzler [55]. Garnet is stable above the 'Gt-in' curves, and the plagioclase disappears above the 'Pl-out' curves. The geotherms, red lines, showing the P-T conditions for early Archean geotherm and late Archean geotherm from Rapp et al. [63], and cratonic geotherms from Boyd and Gurney [9], pass out the intervals between the wet-solidus lines and amphibole-out curves. The P-T field between the normal geotherm and cratonic geotherm is a "cooler thermal regime", and the field between the normal geotherm and early archean geotherm is a "warmer thermal regime" [71]. Gt-Am = Garnet Amphibolite; Am = Amphibolite; Amp-Ec = Amphibolite eclogite; EA = Epidote Amphibolite; GS = Greenschist; BS = Blueschist; Lw-Ec = Lawsonite Eclogite; Zo-Ec = Zoisite Eclogite

9.4 Conclusion

(1) Post-collisional magmatism in the Qiman Tagh region is characterized by the emplacement of calc-alkaline (CA) I-type and alkaline (Alk) A-type granitoids. The two granitoid types are spatially- and temporally closely related.

(2) In spite of the close spatial-temporal link, the Qiman Tagh CA granitoids show completely different major and trace elements geochemistry from the Qiman Tagh Alk plutons. Our new geochemical data demonstrate that the source of these granitoids were derived from the crust rather than from the mantle.

(3) The Qiman Tagh CA (I-type) granodiorite and monzogranite were probably generated by dehydration melting of amphibolite-facies rocks in the lower crust, and amphibole fractionation had shaped their geochemical characters.
(4) The Qiman Tagh (A-type) syenogranite may have been produced by the partial melting of high-pressure granulite-facies rocks, which had previously undergone partial melting as amphibolite-facies rocks. The fusion of phengite and plagioclase may have controlled their geochemical characters.

References

1. Bailey DK (1980) Volcanism, earth degassing and replenished lithosphere mantle. Phil Trans R Soc Lond A, Math Phys Sci 297(1431):309–322
2. Barbarin B (1999) A review of the relationships between granitoid types, their origins and their geodynamic environments. Lithos 46(3):605–626
3. Barbarin B, Didier J (1992) Genesis and evolution of mafic microgranular enclaves through various types of interaction between coexisting felsic and mafic magmas. Trans R Soc Edinb: Earth Sci 83(1–2):145–153
4. Barth MG, McDonough WF, Rudnick RL (2000) Tracking the budget of Nb and Ta in the continental crust. Chem Geol 165(3):197–213
5. Blundy J, Wood B (2003) Mineral-melt partitioning of uranium, thorium and their daughters. Rev Mineral Geochem 52(1):59–123
6. Bonin B (2004) Do coeval mafic and felsic magmas in post-collisional to within-plate regimes necessarily imply two contrasting, mantle and crustal, sources? A review. Lithos 78(1):1–24
7. Bonin B, Giret A (1990) Plutonic alkaline series: Daly gap and intermediate compositions for liquids filling up crustal magma chambers. Schweizerische Mineralogische und Petrographische Mitteilungen (Bull Suisse Minéral et Pétrogr) 70(2):175–187
8. Bonin B, Azzouni-Sekkal A, Bussy FO et al (1998) Alkali-calcic and alkaline post-orogenic (PO) granite magmatism: petrologic constraints and geodynamic settings. Lithos 45(1):45–70
9. Boyd FR, Gurney JJ (1986) Diamonds and the African lithosphere. Science 232(4749):472–477
10. Brown M (2008) Characteristic thermal regimes of plate tectonics and their metamorphic imprint throughout Earth history: When did Earth first adopt a plate tectonics mode of behavior? When Did Plate Tectonics Begin on Planet Earth? 440:97
11. Brown GC, Fyfe WS (1970) The production of granitic melts during ultrametamorphism. Contrib Mineral Petrol 28(4):310–318
12. Burns RG (1973) The partitioning of trace transition elements in crystal structures: a provocative review with applications to mantle geochemistry. Geochim Cosmochim Acta 37(11):2395–2403
13. Burns RG (1993) Mineralogical applications of crystal field theory, vol 5. Cambridge University Press
14. Burns RG, Fyfe WS (1967) Crystal-field theory and the geochemistry of transition elements. Res Geochem 2:259–285
15. Chappell BW (1999) Aluminium saturation in I- and S-type granites and the characterization of fractionated haplogranites. Lithos 46(3):535–551
16. Chen B, Jahn B (2004) Genesis of post-collisional granitoids and basement nature of the Junggar Terrane, NW China: Nd–Sr isotope and trace element evidence. J Asian Earth Sci 23(5):691–703
17. Chen DL, Liu L, Che ZC et al (2001) Determination and preliminary study of Indosinian aluminous A-type granites in the Qimantag area, southeastern Xinjiang. Geochimica 30(6):540–546
18. Clemens JD, Holloway JR, White AJR (1986) Origin of an A-type granite: experimental constraints. Am Mineral 71(3–4):317–324

19. Collins WJ, Beams SD, White AJR et al (1982) Nature and origin of A-type granites with particular reference to southeastern Australia. Contrib Mineral Petrol 80(2):189–200
20. Creaser RA, Price RC, Wormald RJ (1991) A-type granites revisited: assessment of a residual-source model. Geology 19(2):163–166
21. Eby GN (1990) The A-type granitoids: a review of their occurrence and chemical characteristics and speculations on their petrogenesis. Lithos 26(1):115–134
22. Eby GN (1992) Chemical subdivision of the A-type granitoids: petrogenetic and tectonic implications. Geology 20(7):641–644
23. Eklund O, Konopelko D, Rutanen H et al (1998) 1.8 Ga Svecofennian post-collisional shoshonitic magmatism in the Fennoscandian shield. Lithos 45(1):87–108
24. Ersoy EY (2013) PETROMODELER (Petrological Modeler): a Microsoft® Excel© spreadsheet program for modelling melting, mixing, crystallization and assimilation processes in magmatic systems. Turk J Earth Sci 22(1):
25. Esna-Ashari A, Tiepolo M, Valizadeh M et al (2012) Geochemistry and zircon U-Pb geochronology of Aligoodarz granitoid complex, Sanandaj-Sirjan Zone, Iran. J Asian Earth Sci 43(1):11–22
26. Eyal M, Litvinovsky B, Jahn BM et al (2010) Origin and evolution of post-collisional magmatism: coeval Neoproterozoic calc-alkaline and alkaline suites of the Sinai Peninsula. Chem Geol 269(3–4):153–179
27. Fan JJ, Li C, Xie CM et al (2015) Petrology and U-Pb zircon geochronology of bimodal volcanic rocks from the Maierze Group, northern Tibet: constraints on the timing of closure of the Banggong-Nujiang Ocean. Lithos 227:148–160
28. Feng CY, Li DS, Wu ZS, Li JH, Zhang ZY, Zhang AK, Shu XF, Su SS (2010) Major types, time-space distribution and metallogenesis of polymetallic deposits in the Qimantage metallogenic belt, eastern Kunlun area. Northwestern Geol 43:10–17
29. Feng CY, Wang XP, Shu XF, Zhang AK, Xiao Y, Liu JN, Ma SC, Li GC, Li DX (2011) Isotopic chronology of the Hutouya skarn lead-zinc polymetallic ore district in Qimatage area of Qinghai Province and its geological significance. J Jilin Univ (Earth Sci Ed) 41(6):1806–1818
30. Feng CY, Wang S, Li GC et al (2012) Middle to Late Triassic granitoids in the Qimantage area, Qinghai Province, China: chronology, geochemistry and metallogenic significances. Acta Petrol Sin 28(2):665–678
31. Frost CD, Frost BR (2011) On ferroan (A-type) granitoids: their compositional variability and modes of origin. J Petrol 39–53
32. Frost BR, Barnes CG, Collins WJ et al (2001) A geochemical classification for granitic rocks. J Petrol 42(11):2033–2048
33. Hans Wedepohl K (1995) The composition of the continental crust. Geochim Cosmochim Acta 59(7):1217–1232
34. Harley SL (2008) Refining the P-T records of UHT crustal metamorphism. J Metamorph Geol 26(2):125–154
35. Harris NBW, Kelley S, Okay AI (1994) Post-collision magmatism and tectonics in northwest Anatolia. Contrib Mineral Petr 117(3):241–252
36. Holloway JR, Ford CE (1975) Fluid-absent melting of the fluoro-hydroxy amphibole pargasite to 35 kilobars. Earth Planet Sci Lett 25(1):44–48
37. Irvine T, Baragar W (1971) A guide to the chemical classification of the common volcanic rocks. Can J Earth Sci 8(5):523–548
38. Jahn BM, Wu F, Capdevila R, Martineau F, Zhao Z, Wang Y (2001) Highly evolved juvenile granites with tetrad REE patterns: the Woduhe and Baerzhe granites from the Great Xing'an Mountains in NE China. Lithos 59(4):171–198
39. Jahn B, Wu F, Lo C et al (1999) Crust-mantle interaction induced by deep subduction of the continental crust: geochemical and Sr-Nd isotopic evidence from post-collisional mafic-ultramafic intrusions of the northern Dabie complex, central China. Chem Geol 157(1):119–146
40. Kay RW (1978) Aleutian magnesian andesites: melts from subducted Pacific Ocean crust. J Volcanol Geotherm Res 4(1):117–132
41. Keskin M (1994) Genesis of collision-related volcanism on the Erzurum-kars plateau, North eastern Turkey. Durham University, pp 1–363

42. Lameyre J (1988) Granite settings and tectonics. Rend Soc It Miner Petrol 43:215–236
43. Li C, Zhai QG, Dong YS et al (2008) Oceanic crust on the northern margin of Gondwana: evidence from Early Paleozoic ophiolite in central Qiangtang, Qinghai-Tibet Plateau. Geol Bull China 27:1605–1612
44. Liu YH, Mo XX, Yu XH, Zhang XT, Xu GW (2006) Zircon SHRIMP U-Pb dating of the Jingren granite, Yemaquan region of the east Kunlun and its geological significance. Acta Petrol Sin 22(10):2457–2463
45. Liu JN, Feng CY, Zhao YM et al (2013) Characteristics of intrusive rock, metasomatites, mineralization and alteration in Yemaquan skarn Fe-Zn polymetallic deposit, Qinghai Province. Miner Depos 1:008
46. Liégeois J, Black R (1987) Alkaline magmatism subsequent to collision in the Pan-African belt of the Adrar des Iforas (Mali). Geol Soc Lond Spec Publ 30(1):381–401
47. Liégeois J, Navez J, Hertogen J et al (1998) Contrasting origin of post-collisional high-K calc-alkaline and shoshonitic versus alkaline and peralkaline granitoids. The use of sliding normalization. Lithos 45(1):1–28
48. Luo ZH, Ke S, Cao YQ, Deng JF, Zhan HW (2002) Late Indosinian mantle-derived magmatism in the East Kunlun. Geol Bull China 21(6):292–297
49. Maniar PD, Piccoli PM (1989) Tectonic discrimination of granitoids. Geol Soc Am Bull 101(5):635–643
50. Mao JW, Li HY, Pei RF (1995) Nd-Sr isotopic and petrogenetic studies of the Qianlishan granite stock, Hunan Province. Miner Depos 14(3):235–242
51. Martin RF (2006) A-type granites of crustal origin ultimately result from open-system fenitization-type reactions in an extensional environment. Lithos 91(1):125–136
52. McDonough WF, Sun S, Ringwood AE et al (1992) Potassium, rubidium, and cesium in the Earth and Moon and the evolution of the mantle of the Earth. Geochim Cosmochim Acta 56(3):1001–1012
53. Munoz JL, Ludington SD (1974) Fluoride-hydroxyl exchange in biotite. Am J Sci 274(4):396–413
54. Nakamura D (2003) Stability of phengite and biotite in eclogites and characteristics of biotite- or orthopyroxene-bearing eclogites. Contrib Mineral Petrol 145(5):550–567
55. O'Brien PJ, Rotzler J (2003) High-pressure granulites: formation, recovery of peak conditions and implications for tectonics. J Metamorph Geol 21(1):3–20
56. Parada MA, Nyström JO, Levi B (1999) Multiple sources for the Coastal Batholith of central Chile (31-34 S): geochemical and Sr-Nd isotopic evidence and tectonic implications. Lithos 46(3):505–521
57. Pearce JA (1996) Sources and settings of granitic rocks. Episodes 19(4):120–125
58. Pearce JA, Parkinson IJ (1993) Trace element models for mantle melting: application to volcanic arc petrogenesis. Geol Soc Lond Spec Publ 76(1):373–403
59. Pearce JA, Harris NB, Tindle AG (1984) Trace element discrimination diagrams for the tectonic interpretation of granitic rocks. J Petrol 25(4):956–983
60. Poli GE, Tommasini S (1991) Model for the origin and significance of microgranular enclaves in calc-alkaline granitoids. J Petrol 32(3):657–666
61. Rapp PR (1997) Heterogeneous source regions for Archean granitoids. In: de Wit MJ, Ashwal LD (eds) Greenstone belts. Oxford University Press, Oxford
62. Rapp RP, Watson EB (1995) Dehydration melting of metabasalt at 8–32 kbar: implications for continental growth and crust-mantle recycling. J Petrol 36(4):891–931
63. Rapp RP, Watson EB, Miller CF (1991) Partial melting of amphibolite/eclogite and the origin of Archean trondhjemites and tonalites. Precambrian Res 51(1):1–25
64. Rapp RP, Shimizu N, Norman MD et al (1999) Reaction between slab-derived melts and peridotite in the mantle wedge: experimental constraints at 3.8 GPa. Chem Geol 160(4):335–356
65. Rudnick RL, Fountain DM (1995) Nature and composition of the continental crust: a lower crustal perspective. Rev Geophys 33(3):267–309
66. Rushmer T (1991) Partial melting of two amphibolites: contrasting experimental results under fluid-absent conditions. Contrib Mineral Petrol 107(1):41–59

67. Rutter MJ, Wyllie PJ (1988) Melting of vapour-absent tonalite at 10 kbar to simulate dehydration-melting in the deep crust. Nature 331(6152):159–160
68. She HQ, Zhang DQ, Jing XY, Guan J, Zhu HP, Feng CY, Li DX (2007) Geological characteristics and genesis of the Ulan Uzhur porphyry copper deposit in Qinghai. Geol China 2:013
69. Smithies RH (2000) The Archaean tonalite-trondhjemite-granodiorite (TTG) series is not an analogue of Cenozoic adakite. Earth Planet Sci Lett 182(1):115–125
70. Stern CR, Kilian R (1996) Role of the subducted slab, mantle wedge and continental crust in the generation of adakites from the Andean Austral volcanic zone. Contrib Mineral Petrol 123(3):263–281
71. Stüwe K (2007) Geodynamics of the lithosphere: an introduction. Springer
72. Sun SS, McDonough WF (1989) Chemical and isotopic systematics of oceanic basalts: implications for mantle composition and processes. Geol Soc Lond Spec Publ 42(1):313–345
73. Sylvester PJ (1989) Post-collisional alkaline granites. J Geol 97(3):261–280
74. Taylor SR, McLennan SM (1985) The continental crust: its composition and evolution. Blackwell Scientific, Oxford, pp 1–312
75. Turner SP, Foden JD, Morrison RS (1992) Derivation of some A-type magmas by fractionation of basaltic magma: an example from the Padthaway Ridge, South Australia. Lithos 28(2):151–179
76. Vervoort JD, Patchett PJ, Gehrels GE et al (1996) Constraints on early Earth differentiation from hafnium and neodymium isotopes. Nature 379:624–627
77. Wang BZ, Luo ZH, Li HY, Shen HW, Hu XL (2009) Petrotectonic assemblages and temporal-spatial framework of the Late Paleozoic-Early Mesozoic intrusions in the Qimantage Corridor of the East Kunlun belt. Geol China 36(4):769–782
78. Wang S, Feng CY, Li SJ, Jiang JH, Li DS, Su SS (2009) Zircon SHRIMP U-Pb dating of granodiorite in the Kaerqueka polymetallic ore deposit, Qinmantage Mountain, Qinghai Province, and its geological implications. Geol China 36(1):74–84
79. Whalen JB, Currie KL, Chappell BW (1987) A-type granites: geochemical characteristics, discrimination and petrogenesis. Contrib Mineral Petrol 95(4):407–419
80. Wu YZ, Qiao GB, Chen DH (2011) A preliminary study on relationship between tectonic magmatism and mineralization in Qimantage area, Eastern Kunlun Mountains. Geotecton Metallog 35:232–241
81. Wu F, Sun D, Li H et al (2002) A-type granites in northeastern China: age and geochemical constraints on their petrogenesis. Chem Geol 187(1):143–173
82. Xi RG, Xiao PX, Wu YZ (2010) The Geological significances, composition and age of the monzonitic granite in Kendekeke iron mine. Northwestern Geol 43(4):195–202
83. Xiao Y, Feng CY, Liu JN, Yu M, Zhou JH, Li DX, Zhao YM (2013) LA-MC-ICP-MS zircon U-Pb dating and sulfur isotope characteristics of Kendekeke Fe-polymetallic deposit. Qinghai Province. Miner Depos 32(1):177–186
84. Xu ZQ, Yang JS, Li HB et al (2006) The Qinghai-Tibet plateau and continental dynamics: a review on terrain tectonics, collisional orogenesis, and processes and mechanisms for the rise of the plateau. Geol China 33(2):221–238
85. Yogodzinski GM, Kay RW, Volynets ON et al (1995) Magnesian andesite in the western Aleutian Komandorsky region: implications for slab melting and processes in the mantle wedge. Geol Soc Am Bull 107(5):505–519
86. Yu M (2013) Geochemistry and zonation of the Galinge iron deposit, Qinghai province. China University of Geosciences (Beijing), Beijing
87. Zhang AK (2012) Studies on late Paleozoic-early Mesozoic magmatism and mineralization in Yemaquan area, Qinghai province. China University of Geosciences, Beijing, pp 1–155
88. Zhang AK, Liu GL, Mo XX et al (2012) Relationship between tectonic settings and metallogenesis of Late Paleozoic-Early Mesozoic intrusive rock in Qimantage, Qinghai Province. Northwestern Geol 1:005
89. Zou H (1998) Trace element fractionation during modal and nonmodal dynamic melting and open-system melting: a mathematical treatment. Geochim Cosmochim Acta 62(11):1937–1945

Chapter 10
Metallogenic Mechanism of the Galinge Polymetallic Iron Skarn Deposit, Qiman Tagh Mountains, Qinghai Province

The Qiman Tagh Orogenic Belt (QTOB), located along the northern part of the Qinghai-Tibet plateau, was constructed through protracted accretion and collision of a collage of terranes during subduction and closure of the Qiman Tagh Ocean, a branch of Paleo-Tethys Ocean from the Neoproterozoic to Early Mesozoic. The orogen is located between the Qaidam Basin and Kumukuri Basin, and cut by the Altun Fault to the west. The early Neoproterozoic (ca. 1000–820 Ma) ages from this orogen suggest a link with the formation of the supercontinent Rodinia.

The QTOB is tectonically divided into the North Qiman Tagh Terrane (NQT) and the South Qiman Tagh Terrane (SQT). The NQT developed as an active continental margin, and preserves abundant Early Paleozoic granitoids, which possibly formed through the melting of old basement, and a series of mafic–ultramafic rocks considered as VA (volcanic arc) type ophiolites. In contrast, the SQT witnessed intra-oceanic subduction, where SSZ (supra-subduction zone) type ophiolites are documented together with island arc tholeiite (IAT) and calc-alkaline lavas, in a primary oceanic island arc environment during the Early Paleozoic. With continued subduction, the young island arc was transformed into a mature island arc with thickened crust. This region preserves typical evidence for sedimentation and volcanism in the initial stages of volcanic arc development. The collision between the SQT and NQT occurred probably in the Late Silurian (ca. 422 Ma) and continued until ca. 398 Ma, as evidenced from the ages of the abundant within-plate granitic magmatism in the NQT that formed after 398 Ma. In the SQT, voluminous oceanic island arc granitoids formed during the Early-Middle Devonian (ca. 418–389 Ma), with contrasting geochemical features as to those in the NQT. The SQT is interpreted as an exotic terrane that has been incorporated into the continental margin and contributed significantly to the continental growth in this orogenic belt. A trench jam might explain the large gap (ca. 357–251 Ma) of granitoid magmatism.

The final closure of the Paleo Tethyan Qiman Tagh Ocean might have occurred in the Late Permian, and resulted in the accretion of the Kumukuri microcontinent; which formed in response to the orocline formation of western Qiman Tagh Orogen

M. Yu, *Metallogenic Mechanism of the Galinge Polymetallic Iron Skarn Deposit, Qiman Tagh Mountains, Qinghai Province*, Springer Theses,
https://doi.org/10.1007/978-981-10-7907-8_10

and the rotation of the western South Qiman Tagh Terranes. A series of Y-depleted granitoids formed during Early-Middle Triassic (before 237 Ma), which might be associated to the partial melting of thickened lower crust induced by the oceanic lithosphere delamination. Subsequently, a series of calc-alkaline and alkaline granitoids generated through melting of older crustal material which were emplaced in the SQT, and their formation is interpreted to be linked with the transition from post-collision to within-plate settings. Our model is not only suitable to trace the tectonic evolution of the Qiman Tagh orogen, but also valid for the plate tectonic setting orogens in the modern earth. The completely different geotectonic setting between the SQT and NQT response to the scientific issue why most of skarn deposits are distributed in the SQT, which are related to the Triassic granitoids only occurred in the SQT.

The Triassic post-collisional magmatism of Qiman Tagh is characterized by the intrusion of voluminous intermediate to felsic granitoids, including syenogranite, monzogranite, granodiorite, tonalite and diorite. The granitoids can be divided into two magmatic suites: Calc-alkaline (CA) and alkaline (Alk), which were emplaced from ~236 to ~204 Ma. The CA suite contains metaluminous granodiorites and monzogranites. Typical Qiman Tagh CA granodiorites show moderately fractionated REE patterns ((La/Yb)N = 4.35–25.11) with significant negative Eu anomalies ($Eu/Eu^* =$ 0.54–1.34), and the primitive mantle-normalized spidergrams show strong depletion of Nb and Sr. The Qiman Tagh CA monzogranites show similar fractionated REE patterns ((La/Yb)N = 2.70–13.5) with less prominent negative Eu anomalies, and the chondrite-normalized spidergrams show strongly depleted Ba, Nb and Sr. The Alk suite, including syenogranite, is highly potassic (K_2O/Na_2O = 1.09–3.56) and peraluminous (A/CNK = 0.91–1.06). Compared to typical Qiman Tagh CA granodiorites, the Qiman Tagh Alk granitoids can be distinguished by their higher Rb, Nb, Ga/Al, FeO^*/MgO, Y/Sr and Rb/Sr, as well as their lower Mg#, MgO, CaO, Al_2O_3, Sr, Co, V, Eu/Eu^*, Ba/Nb, La/Nb, Ba/La and Ce/Nb. The Qiman Tagh CA rocks were most likely to be derived from the partial melting of garnet-amphibolite-facies rocks in the lower crust, leaving behind anhydrous granulite-facies rocks with plagioclase and garnet in the residue. The Alk rocks may have formed by the continued partial melting of granulite-facies rocks at elevated temperatures (>830 °C). The various geochemical features.

The Qiman Tagh orogen, the westward extending parts of the East Kunlun orogenic system, hosts a variety of deposit types including intrusion-related iron-polymetallic skarn deposit, epithermal base and precious metal veins, magmatic Cu–Ni sulfide deposit, banded iron formation (BIF) and orogenic gold deposit. The intrusive rocks related with skarns in the areas are similar in terms of age, alteration, emplacement and morphology. Although the most of magma associations are characteristic of calc-alkaline and metaluminous, however, their chemical properties have subtle differences. The magma associated with magnetite ores are generally oxidized, but associated with sphalerite-galena ores are characteristically reductive (Fig. 10.1). Magnetite ores occurred in skarns formed as stratabound ores and irregular lenses located close to the contact. The H–O isotope compositions of skarn-forming fluids are consistent with the result of fluid inclusion studies, indicating the mixing of

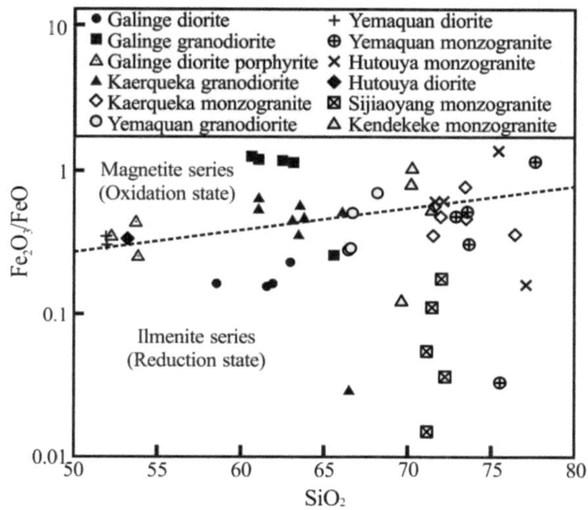

Fig. 10.1 Plot of the granitoids in the SiO_2 versus Fe_2O_3/FeO discrimination diagram [18], showing a magnetite- and ilmenite series for the rocks

magmatic fluids and meteoric water [6, 12, 20, 36]. The S and Pb isotope studies of sulfide reflect a magmatic source [6, 35, 40].

Economically important skarn deposits occur within the Upper Triassic magmatic metallogenetic belt of Qiman Tagh, part of East Kunlun orogenic system. Fe, Pb, Zn and Cu deposits are among the most common mineralization within this belt. The Galinge Fe–Pb–Zn–Cu skarn deposit displays both some general similarities with other skarn deposits within Qiman Tagh belt. Despite this, they also show a large number of differences in terms of some aspects.

In the Qiman Tagh metallogenic belt, Fe, Zn, Pb, Cu and Au skarns and epithermal Cu and Mo deposits are spatially and temporally associated with Triassic granitoid rocks which include granodiorite, monzogranite and syenogranite (Table 10.1). The plutonic associations related to mineralization have various ages between middle-Triassic (237–226 Ma) and late-Triassic (226–204 Ma) (Fig. 10.2) [10, 37].

The two most important deposits associated with the alkaline syenogranite are the Yugouzi Fe–Cu (Mo) and Yazigou Zn–Pb–Cu porphyry-skarn deposit (Table 10.1). Zhou et al. [44] confirmed that the metallogenic age of Yugouzi skarn deposit associated with syenogranites is 210.1 ± 4.8 Ma. The age of porphyry mineralization in Yazigou reflects that the porphyry syenogranite (224.0 ± 4.6 Ma) was the metallogenic mother rock. In the Galinge magnesian zonation, the phlogopite coexisting with magnetite yields 234.2 ± 3.5 Ma isochron age which could be regarded as an age of magnetite formation [37]. The metallogenetic age of Galinge, Tawenchahan, Weibao, Binggou and Suolajier skarn deposit show the granodiorite pluton was principal heat and fluid source. Nevertheless, Wuyihe, Hutouya, Yemaquan, Kayakedeng Tagh and Kendekeke deposit are not only associated with granodiorite, but deeply related with monzogranite (Table 10.1). Nevertheless, the Galinge granodiorite contains relatively higher Fe_2O_3/FeO ratios than the intrusions from other Qiman Tagh

Table 10.1 The metallogenic age of porphyry-skarn deposit from Qiman Tagh

Deposit	Type	Associate igneous rocks	Minerals	Method	Age (Ma)	Reference
Galinge	Skarn	Granodiorite	Phlogopite	$^{40}Ar–^{39}Ar$	235.8 ± 1.7	This study
Tawenchahan	Skarn	Granodiorite	Muscovite	$^{40}Ar–^{39}Ar$	229.9 ± 3.5	[32]
Suolajier	Skarn	Granodiorite	Molybdenite	$^{187}Re–^{187}Os$	239.0 ± 11.0	[7]
Hutouya	Skarn	Granodiorite and monzogranite	Molybdenite	$^{187}Re–^{187}Os$	225.0 ± 4.0	[9]
			Molybdenite	$^{187}Re–^{187}Os$	230.1 ± 4.7	
Yazigou	Porphyry	Porphyritic syenogranite and monzogranite	Molybdenite	$^{187}Re–^{187}Os$	224.7 ± 3.4	[13]
Lalingzaohuo	Porphyry-skarn	Granodiorite	Molybdenite	$^{187}Re–^{187}Os$	240.8 ± 4.0	[34]
		Syenogranite	Molybdenite	$^{187}Re–^{187}Os$	214.5 ± 4.9	
Changshan	Porphyry-skarn	Porphyraceous monzogranite	Molybdenite	$^{187}Re–^{187}Os$	228	[8, 43]
		Syenogranite	Molybdenite	$^{187}Re–^{187}Os$	218	
Yugouzi	Porphyry-skarn	Syenogranite	Molybdenite	$^{187}Re–^{187}Os$	210.1 ± 4.8	[44]

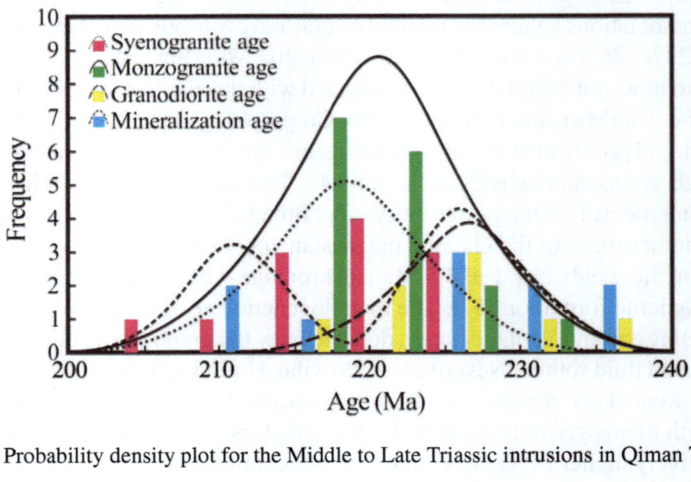

Fig. 10.2 Probability density plot for the Middle to Late Triassic intrusions in Qiman Tagh

Table 10.2 The calculated physical-chemical conditions of Galinge intrusions on the basis of the chemical composition of igneous amphiboles

Location	Galinge							Hutouya		
Sample	Granodiorite		Granodiorite		Diorite			Monzogranite		
	16303-14-2-1	16303-14-3-1	2404-35-2-2	2404-35-2-3	0307-76-1-1	0307-76-1-2	0307-76-1-3	HTY1-6	HTY1-7	HTY1-9
T (°C)	726	718	693	708	656	718	728	838	820	850
P (MPa)	57	50	48	54	26	52	45	156	144	145
OD (km)	2	1.8	1.7	1.9	0.9	1.8	1.6	5.5	5.1	5.1
CD (km)	2.1	1.9	1.8	2	1	2	1.7	5.9	5.4	5.5
ΔNNO	1.9	1.6	1.7	1.6	2.1	2	2.3	0.1	−0.1	0.2
log fO_2	−13.7	−14.1	−14.7	−14.4	−15.4	−13.8	−13.2	−12.9	−13.5	−12.6
H_2O_{melt} (wt%)	4.3	5	4.2	3.7	3.5	5	5.3	3.7	4.2	3.8

Location	Kaerqueka B		Kaerqueka C				
Sample	Garanodiorite		Granodiorite			Xenolith	
	KE01	KE04	KE42	KE48	KE49	KE44	KE47
T (°C)	827	814	782	781	796	792	779
P (MPa)	176	146	127	128	139		
OD (km)	6.2	5.1	4.5	4.5	4.9		
CD (km)	6.7	5.5	4.8	4.8	5.2		
ΔNNO	−0.2	0.1	0.2	0.2	0.0		
log fO_2	−13.4	−13.4	−14.1	−14.1	−13.9		
H_2O_{melt} (wt%)	6.4	5.8	6.4	6.0	6.0		

The calculation of T-P-H_2O_{melt}-fO_2 conditions from the composition of calcic amphiboles of calc-alkaline rocks is on the basis of Ridolfi et al. [25], which operates within Microsoft Excel (Office 2003). It gives reliable results in the ranges of 1064–766 °C, 73–1000 MPa, $-0.3 < \Delta NNO < 2.5$, H_2O_{melt} 3.4–10.6 wt%

deposit (Fig. 10.1). This is why abundant Fe accumulated in the Galinge skarn deposit because the strong oxidation state in the melt. Beside this, the Galinge skarn deposit differs from the other skarns with respect to the much higher Cu, Mo and precious metal content.

The Galinge, Yemaquan and Tawenchahan deposits show a Fe–Pb–Zn–Cu geochemical signature in which magnetite, sphalerite, galena and chalcopyrite are the dominant ore phases. The Kaerqueka, Ulan Uzhur and Yazigou porphyry-skarn deposit are associated mainly with Cu and Mo rather than Fe ore. Chalcopyrite is the dominant Cu-bearing phase and is associated with pyrite and molybdenite in Kaerqueka. Fe–Co–Bi–Cu–Pb–Zn–Au–Ag are the main phases in Kendekeke skarn deposit [35]. The Fe-ore accompanied with Pb–Zn–Ag–Cu–Au is the main phases in the southern ore belt of Kendekeke, and Co–Bi–Au–Ag ore phases, including native bismuth, Bi-tellurides, bismuthinite, linnaeite, erythrine, and arsenopyrite were found in post-skarn sulfide mineralization stage in northern ore belt [35]. Co–Bi–Au–Ag mineralization in Kendekeke are most related with gabbro rather than granodiorite [41]. Weibao is one significant Pb–Zn–Cu rich ore deposit which is associated with banded skarn with high pyroxene/garnet ratios. Furthermore, there is a remarkable mineralized zonation in Weibao, in which the Cu mineralization is covered by the Pb–Zn mineralization. Qunli is a typical oxidized Fe skarn deposit mainly composed of magnetite, hematite, and specularite, it lacks sulfides or precious metals. Sijiaoyang exhibits a well-developed sulfide-rich ore dominated by Zn and Pb, and calc-silicate skarn dominated by garnet and pyroxene. The Galinge deposit is the most similar to Tawenchahan in terms of mineralogy, lithological controls and intrusive rocks (Table 10.1).

The compositional and physical contrast between prograde and retrograde skarn, in addition to fluid property (e.g.: T, P, pH and Salinity) changes, is responsible for the magnetite precipitation, e.g., a temperature drop and volatile consumption in the fluids may have been involved. The Galinge skarn deposit uniquely develops a significant earlier Mg-rich skarn associated with magnetite. The FeO content of the Galinge olivines shows a contra relation with the magnetite mineralization degrees in magnesian skarn systems. In addition, ranges in pyroxene and garnet compositions are much wider in Galinge than in Yemaquan and Kaerqueka due to the enrichment in Mn in pyroxene and Al in garnets, where skarn garnet and pyroxene composition shows a restricted range. The skarn-stage alteration was of considerably high temperature range of 460–560 °C, and of high $\Delta \log fO_2$(HM) range of -12 to 0, which are favorable for iron ions to be partially incorporated into andradite and hedenbergite rather than magnetite in the hydrothermal fluids. We propose that the hydrothermal fluids at Galinge were initially derived from a magmatic source. Beside this, the Galinge deposit develops so many hydrated minerals in the retrograde stage, which are almost yielded by variably superimposing upon early anhydrate minerals at a relatively high fOH^- condition as a consequence of increasing pH value. The formation and disappearance of tremolite, chondrodite, phlogopite, serpentine, epidote and prehnite has tentatively constrained a mineralization temperature range of 360–460 °C and $\Delta \log fO_2$(HM) range of -15 to -3 for the retrograde-stage alteration. The Mg-rich magnetite from the II ore domain is mostly associated with serpentine which is an alteration product of forsterite, tremolite and diopside [11, 42]. Serpentine replacing olivine increases MgO content of the magnetite.

Fluorine-bearing chondrodite, tremolite and phlogopite indicate that contact metamorphism was accompanied by fluorine metasomatism [33]. In addition, the Galinge deposit developed abundant Cl-rich hastingsite and B-rich tourmaline and axinite in the IV and V ore domains [39, 38]. This is interpreted as a volatile- (F, Cl and B) rich core situated in the central, proximal part of the ore field, and to be formed immediately above the fluid plume emerging from the source intrusion [5]. The volatile-bearing hydrothermal fluid carrying F, Cl and B transported abundant metals into the wall rocks over long distances. However, the other skarn deposits from Qiman Tagh didn't exist this remarkable enrichment of volatile phase. These are primary factors of why the Galinge deposit is larger than the others in this belt.

Here we need to discuss where and how Cl_2 are from, if they are captured by the melt formed by the partial melting of the lower crust, or generated by the crystallization of melt, or in any cases both together. The answers to these questions are critical to explain why volatile components are saturated in the fluid. Generally, in the early high temperature and oxygen fugacity and strong acidic melt, chloride are mainly occurred in oxides including ClO_3^-, ClO_2^- and ClO^-. But all of these acid radical ions are not stable in the acid fluid, and reacted with HCl to form Cl_2 via the reactions:

$$
\begin{aligned}
ClO_3^-(\text{melt}) + 5Cl^-(\text{fluid}) + 6H^+ &= 3Cl_2(\text{fluid}) + 3H_2O \\
ClO_2^-(\text{melt}) + 3Cl^-(\text{fluid}) + 4H^+ &= 2Cl_2(\text{fluid}) + 2H_2O \\
ClO^-(\text{melt}) + Cl^-(\text{fluid}) + 2H^+ &= Cl_2(\text{fluid}) + H_2O.
\end{aligned}
$$

As a consequence, the magmatic hydrothermal fluid is volatile saturated. But actually the occurrences of Cl in the siliceous-aluminous magma are complexes combined with Al^{3+}, which will react with SiO_4^{4-} at an increasing oxygen fugacity condition:

$$14SiO_4^{4-} + 4AlCl_3 + 52H^+ + 3O_2 = 4AlSi_3O_8^- + 2SiO_2 + 6Cl_2 + 26H_2O,$$

and accompanying the formation of plagioclase, quartz and water. Therefore, lots of Cl_2 bearing fluid will be produced during the crystallization of magma. Likewise, the crystallization of biotite and amphibole in the magma will also produce lots of Cl_2 in the same way.

Generally, average ligand number of the iron chloride complex in the hydrothermal fluid system is close to 2 at lower temperatures. Dominance of the species $FeCl_2^\circ$ is assumed for the iron transportation. At lower temperatures or higher pressures, increased ionization would occur and the species $FeCl^+$ should become increasingly important. Similarly , at very high chloride concentrations, species such as $FeCl_3^-$

and $FeCl_4^{-2}$ could be expected to increase in importance, and conversely, at lower concentrations, $FeCl^+$ should again be favored over $FeCl_2^{\circ}$. If higher chloride complexes become dominant, a large enhancement of metal solubility results, proportional to the chloride concentration raised to the power of the chloride ligand number. Dominance of the neutral dichloride species is therefore assumed for lead and zinc, as well as for iron, at 500 °C and 1 kbar, whereas for copper the monochloride $CuCl^{\circ}$ is assumed. When the temperature is higher than 300 °C, the main iron chloride complexes are ($Na(FeCl_4)$ or $Na_2(FeCl_4)$) in the fluid, which are equilibrated as: $FeCl_4^- = FeCl_3 + Cl^-$ and $FeCl_4^{2-} = FeCl_2 + 2Cl^-$. As metal concentration elevates and its complexes become a supplier as well as a consumer of chloride, chloride becomes more strongly distributed toward H^+. When acid fluid moves into carbonate rock strata, the chloride complexes will breakdown as a consequence of the pH value increasing:

$$3FeCl_4^{2-} + 6CaCO_3 + 3H_2O = Fe_3O_4 + 6CaCl_2 + 5CO_2 + CO + 6OH^-$$
$$6FeCl_4^- + 12CaCO_3 + 3H_2O = 2Fe_3O_4 + 12CaCl_2 + 12CO_2 + 1/2O_2 + 6OH^-,$$

or as well as the high saline and temperature fluid mix with low saline and temperature sky water:

$$3FeCl_4^{2-} + 4H_2O = Fe_3O_4 + 6HCl + 6Cl^- + H_2 \uparrow$$
$$6FeCl_4^- + 8H_2O + H_2 = 2Fe_3O_4 + 18HCl + 6Cl^-$$

The behavior of siderophile and chalcophile metals during subsequent transportation from the sources depends critically on oxidation state (fO_2) and sulfur fugacity (fS_2), because these parameters control the stability and abundance of sulfide phases in the fluids. At relatively high oxygen fugacity and temperature conditions, SO_2 and sulfate dominates in fluids, and iron is transported as Fe^{3+} in fluid, which corresponds to the SO_4^{2-} and hematite stable area in Fig. 10.3. Hematite is an significantly phase once sulfate reduction lowers the pH sufficiently and the oxidation state is controlled by the magnetite-hematite buffer [24, 30]. Therefore, ferrous iron oxidized to ferric iron is a causal results of sulfate reduction: $2Fe^{2+}Cl_2 + 3SO_4^{2-} + 12H^+ = 2Fe^{3+} + 3SO_2\uparrow + 2Cl_2\uparrow + 6H_2O$. Whereas at a low oxygen fugacity, hydrosulphide complexes mainly existed in fluids [23, 24, 29]. Therefore, it is impossible to form any sulfides at a relatively high oxygen fugacity ($fO_2 >$ c. 10^{-19}) condition. But with decrease of oxygen fugacity ($fO_2 <$ c. 10^{-19}), Fe^{2+} contents become more and more in the fluid, and thus magnetite precipitate as a consequence. When the fugacity of sulfur in the fluid increases ($fS_2 > 10^{-8}$), the SO_2 will be transferred to H_2SO_4 and H_2S as a consequence of hydrolysis reaction: $4SO_2 + 4H_2O = 3H_2SO_4 + H_2S$, and thus lots of magnetites will be replaced by pyrrhotites. When the fugacity of sulfur in the fluid progressively increase ($fS_2 > 10^{-5}$), most of magnetites and pyrrhotites will be replaced by pyrites (Fig. 10.3). In most cases, the concentration of H_2S is independent with the varying pH values of the fluid, but is controlled by the hydrolysis equilibration [15].

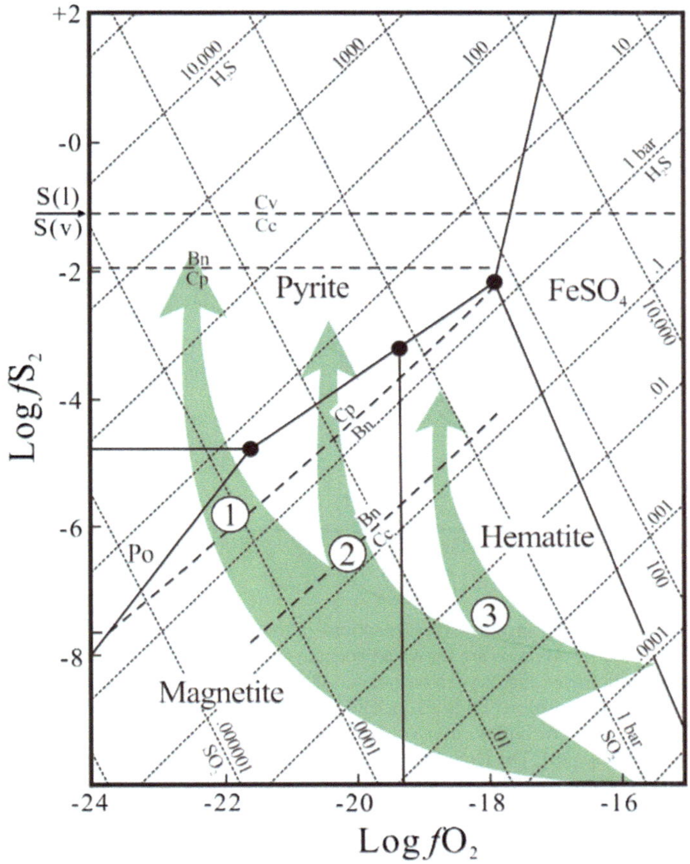

Fig. 10.3 Buffered relations for S_2 and O_2 fugacities in the Fe–S–O (solid lines) and Cu–Fe–S–O (dashed lines) systems at 500 °C and 1kbar total pressure, based on the SUPCRT, a package software for calculating thermodynamic properties. Dotted lines indicate SO_2 and H_2S fugacities and the arrow indicates the sulfur condensation point; the three evolution paths represents Po = Pyrrhotite, Cp = Chalcopyrite, Bn = Bornite, Cc = Chalcocite, and Cv = Covellite. Only a portion of the Bn/Cc boundary is shown

The primary oxygen fugacity condition of magma is very important to effect the magnetite precipitation. Evolved arc magmas and associated fluids are often characterized by variations in oxidation state over typical ranges for arc magmas (ΔFMQ = 0 to +2) [16, 21, 27]. Richards [24] suggested that the redox state of the porphyry deposit systems will be controlled between the pyrite-magnetite/pyrrhotite (PPM) and magnetite-hematite (MH) buffer (ΔFMQ = +3 to +5). Therefore, porphyry deposits are more oxidized than their parental magmas (ΔFMQ = +1 to + 2). Sun et al. [29] proposed that magmatic oxygen fugacity could controls the redox

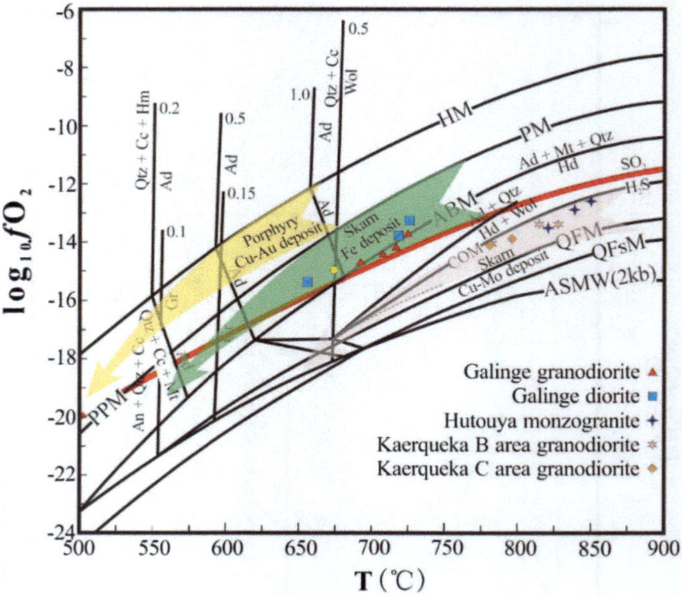

Fig. 10.4 Log fO_2-T diagram for several reactions in the system Ca–Fe–Si–O–H [19, 31]. Mole fractions of CO_2 in C–O–H fluids are shown adjacent to garnet and wollastonite decarbonation reactions. The geothermometer and oxygen barometer of granodiorite and diorite are calculated on the basis of amphibole composition (Table 10.2). The curve labeled HM defines the hematite-magnetite oxygen buffer [22, 28]; The buffer curve of pyrrhotite-pyrite-magnetite (PPM) was defined by the Robie and Hemingway [26]; The H_2S–SO_2 buffer line is controlled by the reaction of $4SO_2 + 4H_2O = 3H_2SO_4 + H_2S$; The curve labeled ABM defines the amphibole-biotite-magnetite oxygen buffer. While those curves marked with QFM and QFsM referred to the quartz-fayalite-magnetite and quartz-ferrosilite-magnetite assemblage oxygen buffers under conditions of one atmosphere pressure [1]. The curve labeled ASMW (2 kb) denotes the annite-sanidine-magnetite-H_2O buffer under conditions of $P_{total} = P_{H_2O} = 2$ kbars [14]. The curves labeled COM represents the assemblage coexisting with augite and othopyroxene [3]

state of sulfur in exsolved fluids and magma. In the Galinge iron deposit, the oxygen fugacity-temperature evolution path of the ore-forming granodiorite magma is almost parallel to the ABM oxygen buffer line and above the H_2S–SO_2 buffer line (Fig. 10.4). When the temperature dropped to 550 °C, the oxygen fugacity decreased to c. 10^{-19} which is in agreement with the oxygen fugacity condition of magnetite precipitation. However, in the Kaerqueka and Hutouya skarn Cu ± Mo deposits, the evolution path is approximately closed to the COM oxygen buffer line (Fig. 10.4). The oxygen fugacity of Kaerqueka and Hutouya magma is apparently lower than that of Galinge deposit when the temperature reached the same level.

Thus, magnetite in skarn deposits factually precipitated as a result of the disproportionation of aqueous SO_2 [24]: 4 $SO_{2(aq)}$ + 4 H_2O = $H_2S_{(aq)}$ + 3 $H_2SO_{4(aq)}$. This reaction is usually effected by the variation in fO_2 of magma and evolved fluid, which was constrained by the sulfur-sulfur oxide buffer, or approximately ΔQFM + 2 (Fig. 10.4) [4]. Above the sulfur-sulfur oxide buffer, hydrothermal magnetite precipitation in skarn deposit can be described in terms of reaction with aqueous SO_2: $9Fe^{2+}Cl_{2(aq)} + SO_{2(aq)} + 10H_2O = 3Fe^{2+}Fe_2{}^{3+}O_4 + H_2S_{(aq)} + 18HCl_{(aq)}$. And it is similar to the common occurrence of magnetite in porphyry copper deposits.

One mechanism to trigger precipitation of sulphides from sulfur-poor brines is to elevate the sulfur activity [2] through the generation of H_2S by disproportionation of SO_2 upon cooling [17]. Therefore, sulfide precipitation will take place provided that the oxygen fugacity below the sulfur-sulfur oxide buffer. In the relatively low oxygen fugacity fluid environment, ore-forming fluid is dominated by S^{2-}, which promotes formation of sulfides in such an environment. As depicted in Fig. 10.4, along the evolution path of skarn Fe deposits, magnetite primarily forms above sulfur-sulfur buffer in the prograde stage. After magnetite precipitation, sulfites forms when oxidation states fall below sulfur-sulfur oxide buffer. The evolution path of skarn Fe deposits is completely different with that of skarn Cu ± Mo deposits in which the redox state is controlled between the sulfur-sulfur oxide and QMF buffer (ΔQFM = + 1 to + 2). Therefore, chalcopyrite precipitation in skarn systems can be described in terms of reaction with newly formed H_2S and H_2SO_4, not SO_2:

$$8CuCl_{(aq)} + 8Fe^{2+}Cl_{2(aq)} + 15H_2S_{(aq)} + H_2SO_{4(aq)} = 8CuFe^{3+}S_2 + 24HCl(aq) + 4H_2O$$

The mafic volcanic rocks provide a hydrothermal fluid-buffered systems for the acquisition, transport, and deposition of metals. The rock buffer reactions is considered as a function of silicate mineral hydrolysis and base metal ions dissolution, which can be combined with the variability of fO_2, fS_2, and total chloride. Solubility is increased by higher chloride and by lower fS_2, with relatively little change with change in fO_2. The reduced sulfur remains the dominate sulfur species. At high chloride concentrations, high dissolved metal to reduced sulfur ratios may soar. Among the reasons for all these changes, a foremost crucial point is that H_2S concentration is independent of pH, and therefore, fH_2S is controlled by aqueous equilibria as similar as fO_2. Increasing acidity by increasing total chloride, therefore, simply increases the dissolution of magnetite from the volcanic rocks, not the concentration of H_2S. The result of the dissolution is a sulfide-deficient solution [15].

Based on the temporal-spatial summary on the tectonic, magma, fluid and mineralization, we discuss the metallogenic mechanism and model of the Galinge skarn deposit as followed (Fig. 10.5).

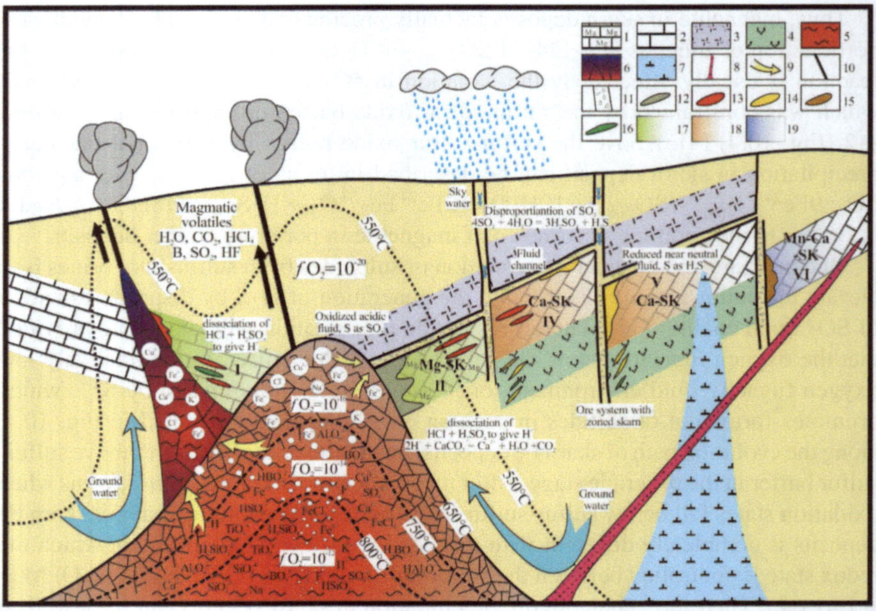

Fig. 10.5 The metallogenic model diagram of the Galinge deposit. 1. dolomitic marble; 2. marble; 3. silicious rocks; 4. Altered basaltic andesites; 5. Unconsolidated magma; 6. Consolidated magma; 7. Diorites; 8. Diorite porphyry dyke; 9. Hydrothermal fluid; 10. Fault; 11. Fractured zone; 12. magnetite orebody close to the contacted zone; 13. magnetite orebody out of contact zone; 14. Pyrrhotite-chalcopyrite-pyrite orebody; 15. Sphalerite-galena orebody; 16. Magnetite-chalcopyrite orebody; 17. Mg-SK zonation; 18. Ca-SK zonation; 19. Mn–Ca-SK zonation

References

1. Berman RG (1988) Internally-consistent thermodynamic data for minerals in the system Na_2O-K_2O-CaO-MgO-FeO-Fe_2O_3-Al_2O_3-SiO_2-TiO_2-H_2O-CO_2. J Petrol 29(2):445–522
2. Blundy J, Mavrogenes J, Tattitch B et al (2015) Generation of porphyry copper deposits by gas-brine reaction in volcanic arcs. Nat Geosci 8(3):235–240
3. Carmichael IS (1966) The iron-titanium oxides of salic volcanic rocks and their associated ferromagnesian silicates. Contrib Mineral Petrol 14(1):36–64
4. Carroll MR, Rutherford MJ (1987) The stability of igneous anhydrite: experimental results and implications for sulfur behavior in the 1982 El Chichon Trachyandesite and other evolved magmas. J Petrol 28(5):781–801
5. Cook NJ, Ciobanu CL (2001) Paragenesis of Cu-Fe ores from Ocna de Fier-Dognecea (Romania), typifying fluid plume mineralization in a proximal skarn setting. Mineral Mag 65(3):351–372
6. Fang J, Chen H, Zhang L et al (2015) Ore genesis of the Weibao lead-zinc district, Eastern Kunlun Orogen, China: constrains from ore geology, fluid inclusion and isotope geochemistry. Int J Earth Sci 1–25
7. Feng CY, Qu WJ, Zhang DQ et al (2009) Re–Os dating of pyrite from the Tuolugou stratabound Co (Au) deposit, eastern Kunlun Orogenic Belt, northwestern China. Ore Geol Rev 36(1):213–220

8. Feng CY, Li DS, Wu ZS et al (2010) Major types, time-space distribution and metallogenesis of polymetallic deposits in the Qimantage metallogenic belt, eastern Kunlun area. Northwestern Geol 43(4):10–17
9. Feng CY, Wang XP, Shu XF et al (2011) Isotopic chronology of the Hutouya skarn lead-zinc polymetallic ore district in Qimatage area of Qinghai Province and its geological significance. J Jilin Univ (Earth Sci Ed) 41(6):1806–1818
10. Feng CY, Wang S, Li GC et al (2012) Middle to Late Triassic granitoids in the Qimantage area, Qinghai Province, China: Chronology, geochemistry and metallogenic significances. Acta Petrol Sin 28(2):665–678
11. Frietsch R (1984) Formation of Mg-bearing magnetite and serpentine in skarn iron ores in northern Sweden. Gff 106(3):219–230
12. Gao YB, Li WY, Li K et al (2013) Fluid inclusions, isotopic geochemistry and genesis of the Hutouya Zn-Pb deposit in Qimantag, Qinghai Province. 1631–1642
13. He SY, Li DS, Li LL et al (2009) Re-Os age of molybdenite from the Yazigou copper (molybdenum) mineralized area in eastern Kunlun of Qinghai Province, and its geological significance. Geotecton Metallog 33(2):236–242
14. Helgeson HC (1978) Summary and critique of the thermodynamic properties of rock-forming minerals. Am J Sci 278:1–229
15. Hemley JJ, Cygan GL, Fein JB et al (1992) Hydrothermal ore-forming processes in the light of studies in rock-buffered systems; I, Iron-copper-zinc-lead sulfide solubility relations. Econ Geol 87(1):1–22
16. Kelley KA, Cottrell E (2009) Water and the oxidation state of subduction zone magmas. Science 325(5940):605–607
17. Landtwing MR, Pettke T, Halter WE et al (2005) Copper deposition during quartz dissolution by cooling magmatic-hydrothermal fluids: the Bingham porphyry. Earth Planet Sci Lett 235(1–2):229–243
18. Lehmann B (1990) Metallogeny of tin, vol 32. Springer
19. Liou JG (1974) Stability relations of andradite-quartz in the system calcium-iron-silicon-oxygen-hydrogen. Am Mineral 59
20. Liu P, Yao L, Liu G et al (2014) Fluid Inclusions Characteristics of the Hutouya Lead-Zinc Polymetallic Deposit in Qimantage Area, Qinghai Province, China. Acta Geol Sin (Engl Ed) 88(s2):1126–1127
21. Malaspina N, Poli S, Fumagalli P (2009) The oxidation state of metasomatized mantle wedge: insights from C–O–H-bearing garnet peridotite. J Petrol 50(8):1533–1552
22. Myers JT, Eugster HP (1983) The system Fe-Si-O: oxygen buffer calibrations to 1,500 K. Contrib Mineral Petrol 82(1):75–90
23. Qiu JT, Yu XQ, Santosh M et al (2013) Geochronology and magmatic oxygen fugacity of the Tongcun molybdenum deposit, northwest Zhejiang, SE China. Miner Depos 48(5):545–556
24. Richards JP (2014) Discussion of Sun et al (2013): the link between reduced porphyry copper deposits and oxidized magmas. Geochim Cosmochim Acta 126(2):643–645
25. Ridolfi F, Renzulli A, Puerini M (2010) Stability and chemical equilibrium of amphibole in calc-alkaline magmas: an overview, new thermobarometric formulations and application to subduction-related volcanoes. Contrib Mineral Petrol 160(1):45–66
26. Robie RA, Hemingway BS (1995) Thermodynamic properties of minerals and related substances at 298.15 K and 1 bar (10^5 Pascals) pressure and at higher temperatures, USGPO; For sale by US Geological Survey, Information Services
27. Rowe MC, Kent AJR, Nielsen RL (2009) Subduction influence on oxygen fugacity and trace and volatile elements in basalts across the Cascade volcanic arc. J Petrol 50(1):61–91
28. Sack RO, Carmichael I, Rivers ML et al (1980) Ferric-ferrous equilibria in natural silicate liquids at 1 bar. Contrib Mineral Petrol 75(4):369–376
29. Sun W, Arculus RJ, Kamenetsky VS et al (2004) Release of gold-bearing fluids in convergent margin magmas prompted by magnetite crystallization. Nature 431(7011):975–978
30. Sun WD, Liang HY, Ling MX et al (2013) The link between reduced porphyry copper deposits and oxidized magmas. Geochim Cosmochim Acta 103:263–275

31. Taylor BE, Liou JG (1978) The low-temperature stability of andradite in C-O-H fluids. Am Mineral 63:378–393
32. Tian CS, Feng CY, Li JH et al (2013) ^{40}Ar-^{39}Ar geochronology of Tawenchahan Fe-polymetallic deposit in Qimantag Mountain of Qinghai Province and its geological implications. Miner Depos 32(1):169–176
33. Timón SM, Moro MC, Cembranos ML et al (2007) Contact metamorphism in the Los Santos W skarn (NW Spain). Miner Petrol 90(1–2):109–140
34. Wang FC, Chen J, Xie ZY et al (2013) Geological features and Re-Os isotopic dating of the Lalingzaohuomolybdenum polymetallic deposit in East Kunlun. Geol China 40(4):1209–1217
35. Xiao Y, Feng CY, Liu JN et al (2013) LA-MC-ICP-MS zircon U-Pb dating and sulfur isotope characteristics of Kendekeke Fe-polymetallic deposit, Qinghai Province. Miner Depos 32(1):177–186
36. Yu M, Feng CY, Zhao YM et al (2014) Fluid inclusion geochemistry in the Kaerqueka copper polymetallic deposit, Qinghai Province and its genetic significances. Acta Geol Sin 88(5):903–917
37. Yu M, Feng CY, Liu HC et al (2015) ^{40}Ar-^{39}Ar geochronology of the Galinge large skarn iron deposit in Qinghai province and geological significance. Acta Geol Sin 89(3):510–521
38. Yu M, Feng CY, Zhu YF et al (2016) Multistage amphiboles from the Galinge iron skarn deposit in Qiman Tagh, western China: evidence of igneous rocks replacement. Miner Petrol 1–17
39. Yu M, Feng CY, Mao JW et al (2017) Multistage skarn-related tourmaline from the Galinge deposit, Qiman Tagh, Western China: a fluid evolution perspective. Can Mineral 55(1):3–19
40. Zhang AK, Liu GL, Feng CY et al (2013) Geochemical characteristics and ore-controlling factors of Hutouya polymetallic deposit, Qinghai Province. Miner Depos 32(1):94–108
41. Zhao CS, Yang FQ, Dai JZ (2006) Metallogenic age of the Kendekeke Co, Bi, Au deposit in East Kunlun Mountains, Qinghai Province, and its significance. Miner Depos S1
42. Zhao YM, Zhang YN, Bi CS et al (1998) The discovery of magnesioferrite from Au (Fe, Cu) magnesian skarn deposits and study of the magnesioferrite-magnesiomagnetite Series. Acta Geol Sin (Engl Ed) 72(4):382–391
43. Zhao SF, WU ZS, Zhang AK et al (2014) Geological Features, Deposit Genesis and Prospecting Potential of Changshan Molybdenum Deposit in Qimantage, Qinghai Province. Northwestern Geol 47(1):179–187
44. Zhou JH, Feng CY, Wang H et al (2014) Re-Os dating of molybdenite from the Yugouzi Fe-Cu (Mo) deposit in Qimantage, eastern Kunlun and its geological implications. Geol Explor 50(1):0001–0007

MIX
Papier aus verantwortungsvollen Quellen
Paper from responsible sources
FSC® C105338
FSC
www.fsc.org

If you have any concerns about our products,
you can contact us on
ProductSafety@springernature.com

In case Publisher is established outside the EU,
the EU authorized representative is:
Springer Nature Customer Service Center GmbH
Europaplatz 3, 69115 Heidelberg, Germany

Printed by Libri Plureos GmbH
in Hamburg, Germany